Climate Change Impacts on Toxins and Health Effects

Amal Saad-Hussein • Osama El-Tawil
El-Sayed A. El-Sheikh
Editors

Climate Change Impacts on Toxins and Health Effects

Springer

Editors

Amal Saad-Hussein
Department of Environmental and Occupational Medicine
Institute of Environment and Climate Change Research, National Research Centre
Giza, Egypt

Osama El-Tawil
Department of Toxicology
Faculty of Veterinary Medicine
Cairo University
Cairo, Egypt

El-Sayed A. El-Sheikh
Department of Plant Protection
Faculty of Agriculture, Zagazig University
Zagazig, Egypt

ISBN 978-981-96-3415-6 ISBN 978-981-96-3416-3 (eBook)
https://doi.org/10.1007/978-981-96-3416-3

This Springer imprint is published by the registered company Springer Nature Singapore Pte Ltd.
The registered company address is: 152 Beach Road, #21-01/04 Gateway East, Singapore 189721, Singapore

Preface

This book covers the aspects of global warming impacts on environmental toxins, occupational toxins, food toxins, marine toxins, and agricultural toxins. It discusses also the current knowledge on the environmental and health effects of these toxins, and how these toxins could be aggravated through global warming and the worsening environmental conditions. Step-by-step each chapter describes the impacts of climate change on a type of toxin and its health effects, accompanied by numerous helpful photographs and drawings. In addition, clear flow charts are included to aid in identifying the magnitude of the health problem among the target population. Essential information is provided on the physiology and pathology of these toxins on the human body. Further topics include the impacts of global warming on drugs and other different therapeutic medications.

The book provides an outlook on adaptive measures that could be taken to minimize the toxicity of these toxins, and how to minimize the health impacts. This book assists the medical persons and environmental scientists in negotiating the steep learning curve involved in gaining the skills needed to perform predictive and therapeutic strategies for proper adaptation with climate changes, which offers significant advantages in terms of avoidance of health impacts of climate change.

The 11 chapters of this book covers the latest evidence-based approach in diagnosis and management of global warming on toxic health impacts, provides easy-to-follow and key points in each chapter about the global warming impacts on different toxins, and includes special sections in each chapter on global warming health impacts due to aggravation of the toxins.

Chapter 1, "Climate Change and Toxins: A Comprehensive Overview," will describe the scientific basics overview about the undergoing profound and unprecedented transformation in climate due to human activities over the past century and inform changes in the pattern of temperatures and rainfall, which will put immense pressure on the vulnerable habitats and increase its consequences for human well-being and the future of life on Earth. Chapter 2, "Climate Change Impacts on Environmental Toxins, and Its Health Effects," will discuss the impacts of climate change on the environmental toxins, especially air and water chemicals and biological pollutants, and describe the association between climate change and the availability and the change in the constituents of these toxic pollutants. Also, the impact of these toxins on health will be targeted in this chapter, especially among vulnerable populations, with different influential factors. In addition, a general picture about infectious diseases related to climate change is provided.

Climate change faces an immediate threat to the expansion of the real gross domestic product (GDP) and to the working conditions and productivity of labor. Therefore, Chap. 3, "Climate Change Impacts on Occupational Exposed Toxins, and Its Health Effects," will address the impact of climate change on occupational toxins and the health of the workers in different exposures, with specific emphasis on the possible ways of adaptation and mitigation.

The challenges of climate change represent a major problem for food safety around the world. Chapter 4, "Climate Change Impacts on Food Toxins, and Its Health Effects," discusses the long-term changes in patterns of temperature, precipitation, humidity, and extreme weather event effects on agricultural activities, crop production, and nutritional quality of most food crops. In addition, it discusses the efforts made internationally to decrease the microorganisms and chemical and biological contaminants that affect the health of plants, animals, and humans. Moreover, Chap. 5, "Climate Change Impacts on Pillars of Food Security and Human Health," describes the threats related to climate change for the accessibility, availability, well-being, and food systems stability. Furthermore, it will explain the severities of socioeconomic impacts from climate change due to food insecurity in the vulnerable groups, especially in low- and middle-income countries, and how to prevent these health impacts.

The objective of Chap. 6, "Climate Change Impacts on Pesticides Use and Fate in the Environment," concentrates on investigating the impact of climate change on the pesticide use and their environmental fate, as well as its implications on environmental contamination, ecological effects, and human health. The chapter also aims to establish appropriate mitigation and adaptation strategies. It highlights the necessity of taking proactive steps to lessen the negative effects and explores the significance of implementing integrated pest management (IPM) practices, developing pesticide formulations that are climate-resilient, incorporating climate change considerations into pesticide policy and regulations, and promoting organic and precision agriculture approaches. Moreover, climate change may affect the toxicity of pesticides. Accordingly, Chap. 7, "Climate Change Impacts of Pesticide Toxicity," represents the area of growing concern as the delicate balance between environmental conditions, pesticide application practices, and changes in pesticide behavior as a result of climate change. The toxic effects of pesticides on ecosystems, wildlife, aquatic organisms, and humans are increasing as a result of climate change. Chapter 8, "Climate Change Impacts on Pharmaceutics, and Its Health Effects," discusses the role of pharmaceutical operations in altering climate through emissions and waste disposal, while climate change will affect the efficacy of drugs actions and increase drug resistance and residues in animal tissues. This chapter will also stress on the urgent need for collaborative action to tackle the interconnected challenges of climate change, pharmaceutical practices, and health outcomes. While, Chap. 9, "Climate Change Health Impacts on Drugs of Abuse," focuses on the illegal drug production and trafficking that cause serious environmental impacts in several ways. This chapter discusses its energy consumption results in carbon emissions, and related water usage, and the combined effects of climate change and abuse of drugs on human health.

Chapter 10, "Climate Change Impacts on Fresh Water Toxins, and Its Health Effects," discusses the potential risks associated with the effect of climate change on the proliferation of freshwater toxins in drinking water production and its impact on the transmission of waterborne diseases affecting human health and other aquatic organisms.

Chapter 11, "Role of Technology in Prediction of Climate Change Impacts on Health Toxicity," discusses traditional and new technologies that serve as a cornerstone in predicting and understanding the impacts of climate change on environmental toxins as well as on the health toxicity to enable data-driven analysis, modeling, surveillance, and decision-making processes.

Giza, Egypt — Amal Saad-Hussein
Cairo, Egypt — Osama El-Tawil
Zagazig, Egypt — El-Sayed A. El-Sheikh

Acknowledgments

This book was prepared by the members of the National Committee of Toxicology (NCT) at the Academy of Scientific Research and Technology, Ministry of Higher Education and Scientific Research, Egypt, with collaboration with Egyptian and international experts in the field of climate change and toxicology.

The NCT is established by a decision every three years from Academy of Scientific Research and Technology and includes approximately 15 experts in toxicology representing Egyptian universities and research centers. The NCT links the Egyptian Society of toxicology with the International Union of Toxicology, which provides scientific consultations in the field of environmental, agricultural, and medical toxins. The committee also holds conferences and workshops to raise awareness of the optimal handling of toxins. During one of these workshops under the theme of "climate change and toxins," the idea of cooperation came to produce and publish a manuscript on the impact of climate change on toxins and its health impacts for the benefit of workers in the environmental, agricultural, and health sectors.

Therefore, thanks are due to the Academy of Scientific Research and Technology for facilitating these collaborations and supporting this work to be ready for international publishing, and all the corresponding authors and coauthors for their efforts and time and collaboration with the editors to complete this work in time.

Contents

Editors and Contributors

About the Editors

Amal Saad-Hussein, M.D. She was graduated from Faculty of Medicine, Cairo University, Egypt. She obtained her M.D. in Public Health and Environmental Medicine from Faculty of Medicine, Cairo University, Egypt. She is Professor Emeritus of environmental and preventive Medicine, former Dean of Environment & Climate Change Research Institute (2016–2020), former Head of Environmental & Occupational Medicine Department (2011–2016). She has 94 peer-reviewed international publications, 22 peer-reviewed national publications, and 5 chapters in books published by Springer. Saad-Hussein was the corresponding editor of "Impact of Climate Change on Health in Africa: A Focus on Liver and Gastrointestinal Tract" published by Springer Nature. She is also an Associate Editor of the *Egyptian Journal of Public Health Association*. She was principal investigator of several national and international projects in environmental and occupational toxicology. In addition, she has delivered numerous oral presentations in numerous national and international meetings. She is a member of the Egyptian Environmental Research Council (since 2021), a member of the Egyptian Council for Climate Change, Egypt (since 2019), and a member of the National Committee of Toxicology (since 2011), Academy of Scientific Research and Technology (ASRT), Ministry of Higher Education and Scientific Research. She is also the vice-president of Egyptian Society of Environmental Mutagens, and a member of Egyptian Medical Women Association managerial board.

Osama El-Tawil, Ph.D. He is President of National Committee of Toxicology and Professor and Chairman of Toxicology and Forensic Medicine Department, Faculty of Veterinary Medicine, Cairo University. He is Adjunct Professor of Pharmacology and Toxicology, School of Medicine, University of Louisville, USA, WHO expert for updating the global air quality guidelines, and member and expert observer for Global European International Network to promote use of dust monitoring and forecasting products. He is Vice President for the Pan Africa Environmental Mutagen and Genomic Society. He is the initiator and main coordinator for the international cooperation and agreement between Cairo University and University of Louisville, and Maryland University, USA. He is a member of many scientific and research committees. He is reviewer and member of advisory board for governmental and nongovernmental authorities and institutions, and international reviewer for many scientific journals. He was awarded and honored with many national and international awards and prizes, and has more than 75 peer-reviewed international publications.

El-Sayed A. El-Sheikh, Ph.D. He graduated from the Faculty of Agriculture, Zagazig University, Egypt. He has conducted postdoctoral research in the University of California, Davis, and the University of Missouri. Currently he is a Professor of Pesticides Biotechnology and Environmental Toxicology at Faculty of Agriculture, Zagazig University. He is the coordinator of the National Committee of Toxicology at the Academy of Scientific Research and Technology (ASRT) and a member of Zagazig University Institutional Animal Care and Use Committee (ZU-IACUC). El-Sheikh has more than 50 national and international publications in the field of environmental toxicology and pesticide toxicology and residues. He is an editorial board member of several national and international journals and a peer reviewer for several international publishers. He is principal investigator of many national and international projects. El-Sheikh research is currently focusing on the detection of pesticide residues in environment using immunoassays and their relationship to climate change.

Contributors

Moustafa A. Abbassy Department of Plant Protection, Damanhour University, Damanhur, Egypt

Salwa M. Abdallah Center of Excellence for Toxicological Testing, Central Agricultural Pesticides Lab (CAPL), Agricultural Research Center (ARC), Giza, Egypt

Khaled Y. Abdel-Halim Mammalian & Aquatic Toxicology Department, Central Agricultural Pesticides Laboratory (CAPL), Agricultural Research Center (ARC), Giza, Egypt

Mosaad A. Abdel-Wahhab Department of Food Toxicology & Contaminants, National Research Centre, Dokki, Giza, Egypt

National Committee of Toxicology, Academy of Scientific Research and Technology (ASRT), Ministry of Higher Education and Scientific Research, Cairo, Egypt

Naeem Sheik Abdul Department of Biochemistry, Stellenbosch University, Stellenbosch, South Africa

Mohamed M. AlBatanony Occupational and Environmental Medicine Department, Faculty of Medicine, Cairo University, Cairo, Egypt

Stephane Alfaro Université de Paris-Est Créteil, (Laboratoire LISA), Créteil, France

Mahmoud M. Amr National Committee of Toxicology, Academy of Scientific Research and Technology (ASRT), Ministry of Higher Education and Scientific Research, Cairo, Egypt

Occupational and Environmental Medicine Department, Faculty of Medicine, Cairo University, Cairo, Egypt

Wagida Anwer National Committee of Toxicology, Academy of Scientific Research and Technology (ASRT), Ministry of Higher Education and Scientific Research, Cairo, Egypt

Department of Community, Environmental and Occupational Medicine, Faculty of Medicine, Ain Shams University, Cairo, Egypt

William Au School of Public and Population Health, University of Texas Medical Branch, Galveston, TX, USA

Fawzy Eissa Environment and Bio-Agriculture Department, Faculty of Agriculture, Al-Azhar University, Cairo, Egypt

Sameh Mohamed Mostafa El Nabtity Department of Pharmacology, Faculty of Veterrinary Medicine, Zagazig University, Zagazig, Egypt

Nesrine Salah El Dine El Sayed Department of Pharmacology and Toxicology, Faculty of Pharmacy, Cairo University, Cairo, Egypt

Hanan Salah Eldin El-Abhar Department of Pharmacology, Toxicology, and Biochemistry, Faculty of Pharmacy, Future University in Egypt (FUE), Cairo, Egypt

National Committee of Toxicology, Academy of Scientific Research and Technology (ASRT), Ministry of Higher Education and Scientific Research, Cairo, Egypt

Lamia S. Ellaithy Environmental and Occupational Medicine Department, Environment and Climate Change Research Institute, National Research Centre, Giza, Egypt

Aziza A. El-Nekeety Department of Food Toxicology & Contaminants, National Research Centre, Dokki, Giza, Egypt

National Committee of Toxicology, Academy of Scientific Research and Technology (ASRT), Ministry of Higher Education and Scientific Research, Cairo, Egypt

El-Sayed A. El-Sheikh National Committee of Toxicology, Academy of Scientific Research and Technology (ASRT), Ministry of Higher Education and Scientific Research, Cairo, Egypt

Department of Plant Protection, Faculty of Agriculture, Zagazig University, Zagazig, Egypt

Osama El-Tawil Department of Toxicology and Forensic Medicine, Faculty of Veterinary Medicine, Cairo University, Cairo, Egypt

National Committee of Toxicology, Academy of Scientific Research and Technology (ASRT), Ministry of Higher Education and Scientific Research, Cairo, Egypt

Marwa M. Fouad Occupational and Environmental Medicine Department, Faculty of Medicine, Cairo University, Cairo, Egypt

Omneya Osama Galal National Committee of Toxicology, Academy of Scientific Research and Technology (ASRT), Ministry of Higher Education and Scientific Research, Cairo, Egypt

Department of Pharmacology and Toxicology, Faculty of Pharmacy, Ahram Canadian University, Giza, Egypt

Doaa Y. Hammad Biological Anthropology Department, Institute of Medical Research and Clinical Studies, National Research Centre, Giza, Egypt

Nahed Mohamed Mohamed Ismail Department of Environmental Research, Theodor Bilharz Research Institute, Giza, Egypt

Wafaa A. Kandeel Biological Anthropology Department, Institute of Medical Research and Clinical Studies, National Research Centre, Giza, Egypt

Heba Mahdy-Abdallah Environmental & Occupational Medicine Department, Environment & Climate Change Research Institute, National Research Centre, Giza, Egypt

Jeanine L. Marnewick Applied Microbial and Health Biotechnology Institute, Cape Peninsula University of Technology, Cape Town, South Africa

Nayra Sh. Mehanna Dairy Science Department, Institute of Food Science and Nutrition, National Research Centre, Giza, Egypt

Atef M. F. Mohammed Air Pollution Research Department, Environment & Climate Change Research Institute, National Research Centre, Giza, Egypt

Abdel-Tawab H. Mossa Pesticide Chemistry Department, Chemical Industries Research Institute, National Research Centre, Giza, Egypt

Marco Neira Climate and Atmosphere Research Center (CARE-C), The Cyprus Institute, Nicosia, Cyprus

Amal Saad-Hussein National Committee of Toxicology, Academy of Scientific Research and Technology (ASRT), Ministry of Higher Education and Scientific Research, Cairo, Egypt

Department of Environmental & Occupational Medicine, Institute of Environment and Climate Change Research, National Research Centre, Giza, Egypt

Mohamed Salama National Committee of Toxicology, Academy of Scientific Research and Technology (ASRT), Ministry of Higher Education and Scientific Research, Cairo, Egypt

Institute of Global Health and Human Ecology, The American University in Cairo, Cairo, Egypt

Aisha M. Samir Occupational and Environmental Medicine Department, Faculty of Medicine, Cairo University, Cairo, Egypt

Sara Sayed Mahmoud Sayed Department of Environmental Research, Theodor Bilharz Research Institute, Giza, Egypt

Yasser S. Sayed Department of Veterinary Forensic Medicine and Toxicology, Faculty of Veterinary Medicine, Damanhour University, Damanhour, Egypt

Mona M. Taha Environmental & Occupational Medicine Department, Environment & Climate Change Research Institute, National Research Centre, Giza, Egypt

Marwa Tamim Ahmed Abdel-Wareth Department of Environmental Research, Theodor Bilharz Research Institute, Giza, Egypt

Nadia Gamal Eldin Zaki National Committee of Toxicology, Academy of Scientific Research and Technology (ASRT), Ministry of Higher Education and Scientific Research, Cairo, Egypt

Department of Narcotic Research, The National Center for Social and Criminological Research, Cairo, Egypt

Nermin H. Zawilla Occupational and Environmental Medicine Department, Faculty of Medicine, Cairo University, Cairo, Egypt

Hussein H. Zayet Occupational and Environmental Medicine Department, Faculty of Medicine, Cairo University, Cairo, Egypt

Nour El-Hoda A. Zidan Pesticides Chemistry and Toxicology Department, Faculty of Agriculture, Kafrelsheikh University, Kafr El-Sheikh, Egypt

National Committee of Toxicology, Academy of Scientific Research and Technology (ASRT), Ministry of Higher Education and Scientific Research, Cairo, Egypt

Moh Moh Thant Zin Division of Plant Science and Technology, University of Missouri, Columbia, MO, USA

Climate Change and Toxins: A Comprehensive Overview

1

Osama El-Tawil, Stephane Alfaro, Amal Saad-Hussein, and El-Sayed A. El-Sheikh

Abstract

The climate is undergoing a profound and unprecedented transformation due to human activities over the past century. Evidence suggests that humans are primarily responsible for the climate change crisis. The increasing burning of fossil fuels, the release of greenhouse gases (GHGs), and other industrial and agricultural practices have disrupted the delicate balance of the world's ecosystems. If current environmental practices continue, climate impacts are expected to worsen

O. El-Tawil (✉)
Department of Toxicology and Forensic Medicine, Faculty of Veterinary Medicine, Cairo University, Cairo, Egypt

National Committee of Toxicology, Academy of Scientific Research and Technology (ASRT), Ministry of Higher Education and Scientific Research, Cairo, Egypt
e-mail: osamaeltawil@cu.edu.eg

S. Alfaro
Université de Paris-Est Créteil, (Laboratoire LISA), Créteil, France
e-mail: stephane.alfaro@lisa.ipsl.fr

A. Saad-Hussein
National Committee of Toxicology, Academy of Scientific Research and Technology (ASRT), Ministry of Higher Education and Scientific Research, Cairo, Egypt

Department of Environmental & Occupational Medicine, Institute of Environment and Climate Change Research, National Research Centre, Giza, Egypt
e-mail: al.hussein@nrc.sci.eg

E.-S. A. El-Sheikh
National Committee of Toxicology, Academy of Scientific Research and Technology (ASRT), Ministry of Higher Education and Scientific Research, Cairo, Egypt

Department of Plant Protection, Faculty of Agriculture, Zagazig University, Zagazig, Sharkia, Egypt
e-mail: eaelsheikh@zu.edu.eg

A. Saad-Hussein et al. (eds.), *Climate Change Impacts on Toxins and Health Effects*,
https://doi.org/10.1007/978-981-96-3416-3_1

in the coming decades. Global temperatures are projected to rise several degrees Celsius by the end of this century, resulting in more frequent and intense heat waves, droughts, wildfires, and floods. High temperatures, ocean acidification, and changing rainfall patterns are also putting immense pressure on vulnerable habitats like coral reefs, rainforests, and the Arctic tundra, leading to their extinction. The loss of biodiversity and the degradation of natural systems will have dire consequences for human well-being and the future of life on Earth. Addressing these environmental challenges requires a comprehensive global effort to shift away from fossil fuels, protect and restore natural ecosystems, and reduce pollutant emissions. Bold and coordinated action is necessary to avoid worst-case scenarios and ensure a livable future for future generations. This chapter is discussing all that in more detail.

Keywords

Climate change · Greenhouse gases · Temperature · Precipitation · Human well-being · Ecosystems

1.1 Introduction

This book will cover the aspects of global warming impacts on environmental toxins, occupational toxins, food toxins, marine toxins, and agricultural toxins. It will also discuss the current knowledge on the environmental and health effects of these toxins and how these toxins could be aggravated through global warming and the worsening environmental conditions. Step by step, each chapter will describe the impacts of climate change on a type of the toxins and its health effects, accompanied by numerous helpful photographs and drawings. In addition, clear flow charts are included to aid in identifying the magnitude of the health problem among the target population. Essential information is provided on the physiology and pathology of these toxins on the human body. Further topics include the impacts of global warming on drugs and other different therapeutic medications.

This book will also provide an outlook on adaptive measures that could be taken to minimize the toxicity of these toxins and how to minimize the health impacts. This book will assist the medical persons and environmental scientists in negotiating the steep learning curve involved in gaining the skills needed to perform predictive and therapeutic strategies for proper adaptation with climate changes, which offers significant advantages in terms of avoidance of health impacts of climate change. .

So, this chapter will give an overview of what happened to the climate of our Earth in detail and the future expectations of the global climate with the implementation of the global strategies.

1.2 Observations: A Changing Climate

In the eighteenth century, attempts to measure weather-related physical quantities with instruments were made, but they were disorganized. It was only in the nineteenth century that the first national meteorological societies were created. These societies not only promoted the establishment of networks of perennial meteorological stations around the world, but they also advocated international cooperation. By doing so, they paved the way for the organized observation of the state of the Earth's climate. These efforts culminated in the 1970s when satellites provided the first consistent, direct, and uninterrupted global-scale observations of our planet.

The analysis of the data collected since the beginning of the instrumental era (circa 1850) leaves no place for doubt: our climate is changing rapidly, the change is widespread, and it affects the atmosphere, the ocean, and the land (IPCC 2021).

1.2.1 Temperature

As compared to the average of 1850–1900, usually chosen as the reference, the global temperature was 1.09 °C (CI: 0.95–1.20 °C) warmer in 2011–2020 (Fig. 1.1). Over this period of more than 1.5 centuries, the warming has not been regular but accelerating. As of 1980, every decade is warmer than any other preceding it since

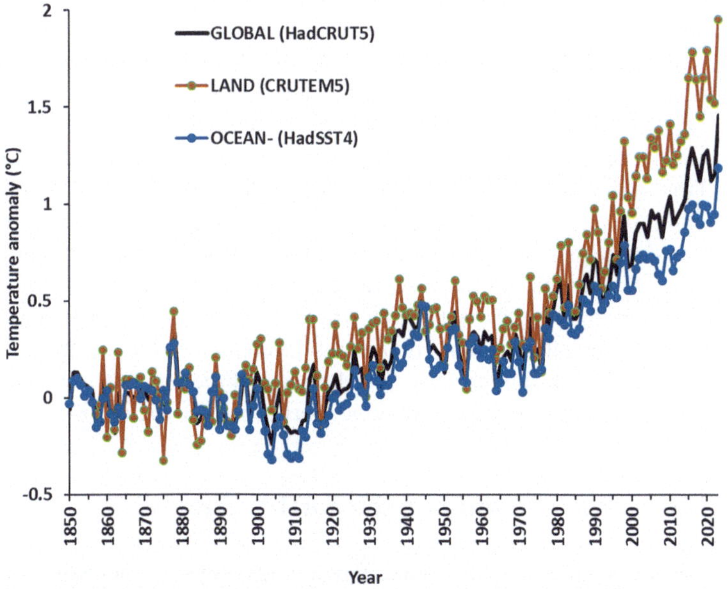

Fig. 1.1 Evolution between 1850 and 2023 of the global temperature anomalies (relative to 1850–1899). The cases of the continental (brown curve) and marine (blue curves) surfaces are distinguished from the global Earth (red curve). The observation data used for this figure are those of the Met Office Hadley Centre (https://www.metoffice.gov.uk/hadobs/)

1850. The rate of temperature increase of the last decades (0.20 °C/decade between 1980 and 2023) is unprecedented in more than 2000 years, and the current global temperature is above that of the warmest multicentury period in the last 100.000 years (IPCC 2021).

At the surface of the planet, the warming is not spatially homogenous. It is faster globally across continents and comparatively slower in the oceans than the global average (Fig. 1.1). For instance, in 2011–2020, the average surface temperature over land was 1.60 °C above the 1850–1900 reference. In the same period, the difference was only 0.86 °C for the ocean.

The polar regions, particularly the Arctic, are the most rapidly affected by the warming. This effect has been called the polar amplification of climate change. In the Arctic, it is estimated that the rate of temperature increase since 1980 is about 0.75 °C/decade (Rantanen et al. 2022), which is three to four times larger than at a global level. Although to a lesser extent, the warming observed over the Tibetan Plateau is also faster than the average (0.34 °C/decade, in the same period). The existence of this amplification has led some authors to use the nickname "Third Pole" for the Tibetan region (You et al. 2021).

1.2.2 Precipitations

Global warming influences precipitations directly. On the continents, more evaporation leads to drier surfaces and increases the risk of drought. Conversely, evaporation increases the precipitable water content in the atmosphere, and all the more so as the water-holding capacity of air increases with the temperature. At the global scale, this increased atmospheric moisture leads to more intense precipitations, included in the form of extreme events (storms, cyclones).

The analysis of the twentieth-century precipitation record reveals the existence of some large-scale patterns of systematic change (Trenberth et al. 2011): in general, precipitations have tended to decrease in the subtropics and tropics, but they have increased over land at higher latitudes. In this global pattern, some regions are more affected than others: for instance, the eastern basin of the Mediterranean Sea and the neighboring countries is a region where the number of yearly rain days as well as the total precipitations has significantly decreased in the last three decades (Philandras et al. 2011). Interestingly, this decrease of the precipitation is not contradictory to the increase of the atmospheric moisture in the region (Fig. 1.2).

1.3 Humans Are Modifying the Earth's Climate

Thanks to the efforts of scientists from different fields (physics, chemistry, and other natural sciences), the main forces that govern the global climate system and the numerous feedbacks involved in its functioning are now well understood. Empirical measurements documenting in real time the state of the climate system are also continuously acquired around the globe. In combination with the rapid increase of

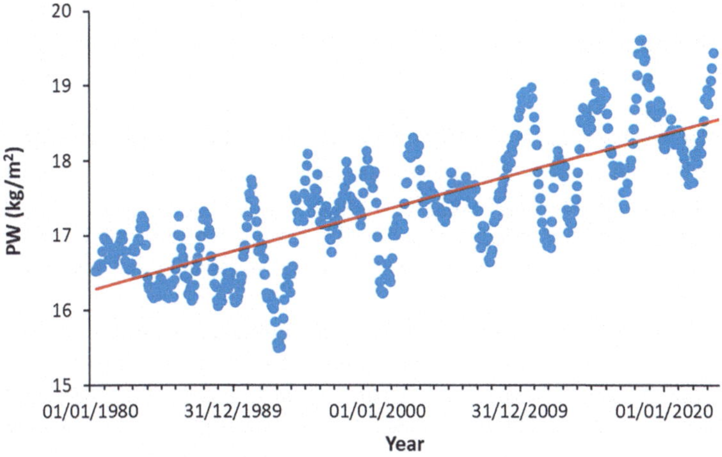

Fig. 1.2 Increase of the atmospheric columnar content in precipitable water (PW, in kg/m^2) over the eastern basin of the Mediterranean Sea between 1980 and 2023. The data are those of the MERRA-2 reanalysis (data downloaded from the NASA Giovanni website: https://giovanni.gsfc.nasa.gov/)

the computer power, this understanding has allowed the development of numerical models simulating the complexity of the climate system. These climate models are also referred to as "general circulation models" (GCM), and the assessment of their performance is made by comparison of their predictions with current or past observations.

Since direct experiments on the climate system are not possible, simulations performed with the GCMs are the only possibility to explore and quantify the effect of the modification of individual variables such as solar irradiance, content of the atmosphere in greenhouse gases (GHGs), aerosols, or combinations thereof. This is fundamental for attribution studies that aim to isolate and quantify the share of the climate change that can be explained by variations of natural (solar activity, volcanism) and/or anthropogenic (GHG, aerosols, modification of surface albedo) forcing.

Because all the existing individual climate models differ (by their grid type, parameterizations, etc.), the international climate modeling community is performing more and more sophisticated experiments in the frame of successive coupled model intercomparison projects (CMIPs). Using an ensemble of models rather than single ones leads to more robust simulations and also allows us to estimate the uncertainty on the products of these simulations. Indeed, the results of these CMIP experiments performed with dozens of models show that the multimodel ensembles are able to simulate successfully the global surface temperature trends between 1850 and 2020 when both the anthropogenic and natural forcing are taken into account. This is not the case when only natural forcing is considered. In particular, in the latter case, the multimodel ensembles fail to reproduce the sharp increase of the temperature observed after 1970 (Fig. 1.1). This provides "unequivocal" evidence (IPCC 2021) that human influence is responsible for this rapid warming of the past five decades.

In its last Assessment Report (AR6, issued in 2021), the IPCC synthesized the outcomes of the most recent attribution studies: the 1.09 °C warming of the 2010–2019 period relative to 1850–1900 is primarily due to the man-induced increase of the atmospheric concentration of greenhouse gases (carbon dioxide, methane, nitrous oxide, volatile organic compounds, carbon monoxide, halogenated gases, etc.). Considered alone, GHGs would cause a warming of 1.5 °C, but the cooling due to other human drivers (aerosols, land use changes, etc.) partially masks this effect.

Attribution studies also allow quantifying the influence of climate change on the frequency of extreme events. There is evidence that tropical cyclones, floods, heat waves, wildfires, and droughts have become more frequent since 1950, and this is due to the human-induced climate change (Seneviratne et al. 2021). The probability of observing compound events (e.g., heat waves, drought, and wildfires, or intense precipitations and floods, occurring at the same time) has also increased.

1.4 Toward a Bleak Future?

Once its reliability has been validated by comparison with historical observations (1850–present), ensemble modeling can be used to predict the evolution of the Earth's climate in the coming decades. However, these future projections need to be based on scenarios that take into account not only the socioeconomic development of the world population (this includes such diverse factors as population size, level of education, urbanization, energy type and production, land use, extent of regional collaboration, etc.) but also the radiative forcing levels achieved at the end of this century. Therefore, the AR6 scenarios are based on a matrix of five shared socioeconomic pathways (from SSP1, corresponding to the "'green road,"'to SSP5, where there is no real adaptation strategy and the economy is mainly dependent on fossil fuels. For more details, see Riahi et al. 2017) and five representative concentration pathways (RCPs) corresponding to global forcing levels in 2100 increasing from 1.9 to 8.5 $W.m^{-2}$. Because some combinations of SSP and RCP are implausible—such as that of the largest radiative forcing (RCP8.5) with the most environmentally friendly socioeconomic pathway (SSP1)—the IPCC retained only five illustrative cases in its sixth assessment report: SSP1–1.9, SSP1–2.6, SSP2–4.5, SSP3–7.0, and SSP5–8.5 (Lee et al. 2021). Considered together, these five cases span the range of possibilities, from the most optimistic to the most depressing.

In all scenarios, the concentration of GHGs, particularly CO_2, plays a key role. CO_2 is the main contributor to radiative forcing, and its residence time in the atmosphere is particularly long (it is estimated to be between 300 and 1000 years). Because of these two combined factors, it was already shown in the IPCC fifth assessment report (IPCC 2013) that the average global increase of surface temperature is approximately proportional to the CO_2 emissions accumulated since the beginning of the industrial era, and that, even if CO_2 emissions were stopped immediately, the Earth's climate would continue to change for centuries. This is confirmed by the results of AR6 (Table 1.1), which indicate that, under all scenarios, the 1.5 °C warming limit (relative to 1850–1900) agreed to by the 196 Parties of the 2015 Paris Agreement will be exceeded between 2041 and 2060. Afterwards, only the two most favorable scenarios (SSP1–1.9 and SSP1–2.6) would lead to near stabilization of the temperature in 2081–2100, but they need to achieve net-zero CO_2 emissions by 2050 or 2070 and to divide by more than 2 the methane emissions of 2015 (400 MtCH4/year). Under SSP3–7.0 (SSP5–8.5), no real action is taken to mitigate the release of GHGs into the atmosphere, the net CO_2 emissions continue to grow, and they double (treble) between 2015 and 2100. In this case, by the end of the century the global temperature will be 3.6–4.4 °C above the 1850–1900 value. By comparison, the rise of temperature separating the last glacial maximum (ca. 20,000 years ago) from the Holocene is estimated at 4.5–5.5 °C, but this rise took several thousand years to occur, not 150 years. This comparison emphasizes how dramatic global warming already is or could become if no rapid action is taken to reduce it.

As already observed in the current climate, ensemble simulations show that the future warming will not be spatially homogenous. Whatever the level of global temperature increase in the future compared to the period 1850–1900, all regions of the globe will be affected, but the land will warm faster than the oceans, and the polar regions will face the sharpest rise of temperature (Fig. 1.3).

Regarding precipitations, the simulations show that—as expected—the amplitude of the changes will depend on the degree of warming, and there will be a further amplification of what is already currently observed. More precisely, total precipitations will increase almost everywhere except in some parts of the subtropics (Fig. 1.4, top row). Over land in the latter regions, the reduction of the precipitations combined with enhanced evaporation will lead to a dramatic decrease of the soil moisture (Fig. 1.4, bottom row). This is the case for most of South America, Central America, the southwestern part of the USA, South Africa, eastern and southeastern Asia, Australia, and all the Mediterranean region.

Finally, climate extremes (high temperatures, heavy precipitations, droughts in drying regions) will become more frequent. For instance, Table 1.2 displays—for different levels of global warming—the increasing frequency and intensity of decadal events (i.e., events that occurred only once in 10 years in the reference period 1850–1900). Even in the most optimistic case (warming limited to 1.5 °C),

Table 1.1 Rise of the global temperature (and associated uncertainty, both in °C) above the 1850–1900 values for the five illustrative scenarios selected by the IPCC in its sixth Assessment Report. The projections are given for the short- (2021–2040), middle- (2041–2060), and long-term (2081–2100) 20-year periods of this century

Scenario	SSP1–1.9	SSP1–2.6	SSP2–4.5	SSP3–7.0	SSP5–8.5
2021–2040	1.5 (±0.3)	1.5 (±0.3)	1.5 (±0.3)	1.5 (±0.3)	1.6 (±0.3)
2041–2060	1.6 (±0.4)	1.7 (±0.4)	2 (±0.4)	2.1 (±0.4)	2.4 (±0.5)
2081–2100	1.4 (±0.4)	1.8 (±0.5)	2.7 (±0.6)	3.6 (±0.9)	4.4 (±1.2)

Adapted from IPCC (2021)

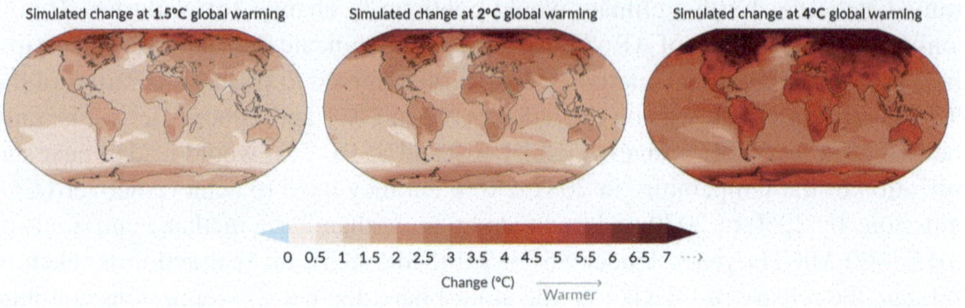

Fig. 1.3 Change of annual mean temperature (°C, relative to 1850–1900) for three different levels of global warming (1.5, 2, and 4 °C). This figure is extracted from Figure SPM.5 in IPCC 2021: Summary for Policymakers. In: *Climate Change 2021: The Physical Science Basis. Contribution of Working Group I to the Sixth Assessment Report of the Intergovernmental Panel on Climate Change* [Masson-Delmotte, V., P. Zhai, A. Pirani, S.L. Connors, C. Péan, S. Berger, N. Caud, Y. Chen, L. Goldfarb, M.I. Gomis, M. Huang, K. Leitzell, E. Lonnoy, J.B.R. Matthews, T.K. Maycock, T. Waterfield, O. Yelekçi, R. Yu, and B. Zhou (eds.)]. (Cambridge University Press, Cambridge, UK and New York, NY, USA, pp. 3–32, doi: https://doi.org/10.1017/9781009157896.001)

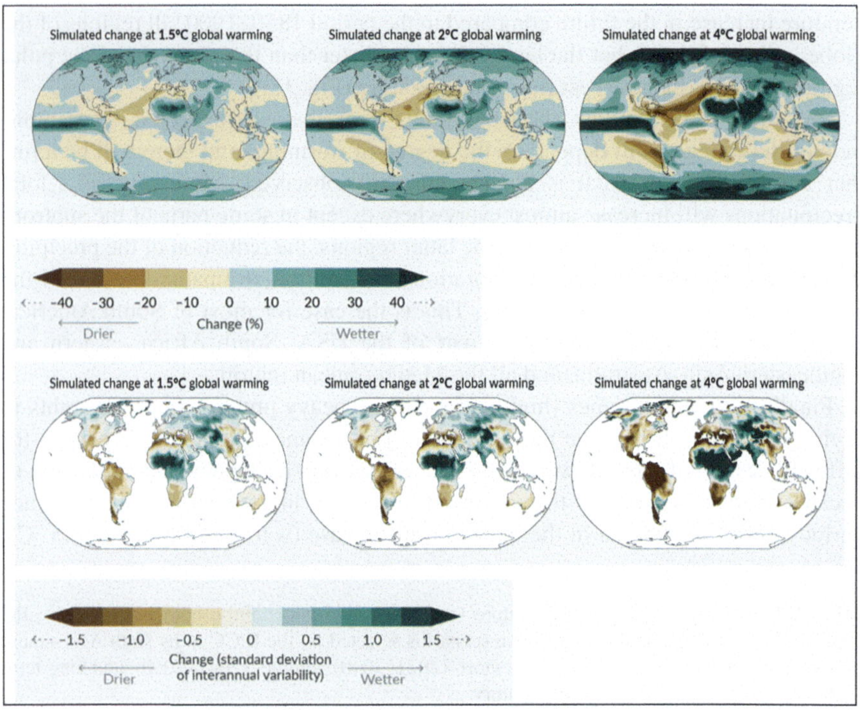

Fig. 1.4 Annual mean precipitation (in %, top row) and soil moisture change (standard deviation, bottom row) for the three levels of global warming considered in Fig. 1.3 (these figures were extracted from the same IPCC report as Fig. 1.3)

Table 1.2 Predicted increase of the frequency and intensity of extreme temperature, heavy precipitations, and severe droughts in drying regions for the current situation (1 °C above 1850–1900) and three additional increments of global warming (1.5, 2, and 4 °C). For heat extremes, the intensity is expressed in °C, for precipitations in % wetter, and for droughts in fractions of the standard deviation of the annual soil moisture

Extremes	Level of global warming	1 °C	1.5 °C	2 °C	4 °C
Hot	Frequency	2.8	4.1	5.6	9.4
	Intensity (°C hotter)	1.2	1.9	2.6	5.6
Precipitations	Frequency	1.3	1.5	1.7	2.7
	Intensity (% wetter)	6.7	10.5	14	30.2
Droughts	Frequency	1.7	2.0	2.4	4.1
	Intensity (sd)	0.3	0.5	0.6	1.0

Source of data: IPCC (2021)

extreme temperature will be 4.1 times more frequent and 1.9 °C hotter than in 1850–1900, and heavy precipitations will be 1.5 times more frequent and 10.5% wetter. In drying regions, the droughts will double in frequency and in severity (0.5 standard deviation of annual soil moisture dryer).

As mentioned previously, the impact of climate change is not spatially homogenous, and some regions of the world will be particularly affected by climate change in the coming decades. Identifying these regions, also referred to as "hotspots" of climate change (Giorgi 2006), is crucial to prepare mitigation measures. Hotspots do not coincide with regions where the evolution of mean temperature and precipitation is expected to be important, but to locations with the highest probability of observing single and complex climate extremes. Indeed, these are the concurrent extreme events that cause the strongest damage to human societies and natural ecosystems. In their recent analysis of the CMIP6 simulations, Vogel et al. (2020) showed that, independently of the degree of global warming (from 1.5 to 3 °C), the Amazon rainforest, the Mediterranean region, and Southern Africa appeared as hotspots of combined hot and dry extremes. Conversely, tropical Africa, the Sahel, Indonesia, the Andes, and the Himalayas have been identified as hotspot regions of simultaneous wet and hot extremes.

1.5 Climate Change and Ecosystems

By modifying their habitat or overharvesting, humans have already put the natural ecosystems and biodiversity directly under great stress. Through the effects of climate change, they also exert indirectly on them an additional pressure that is already visible (shift of climate zones, bleaching of corals as a result of the acidification of the oceans, giant wildfires, etc.) and can only increase in the next decades. However, because the complex interactions between individual species are often poorly understood and highly dynamic, it is hard to predict exactly how ecosystems will respond to climate change and recover from extreme events (Malhi et al. 2020). Because ecosystems control at least in part the carbon, water, and biogeochemical cycles, this incomplete understanding introduces uncertainty in the prediction of the future evolution of the Earth's climate.

1.6 Conclusion

It has become certain that human activities are radically changing the Earth's climate, with potentially devastating consequences in the future. The continuous release of GHGs such as carbon dioxide and methane into the atmosphere over the past century has led to a rapid and unprecedented warming of the global climate. This climate change poses an existential threat to life on planet Earth.

As temperatures continue to rise, the world is rapidly heading toward a bleak future. Sea levels are rising, environmentally destructive weather events are becoming more frequent and intense, and ecosystems around the world are being disturbed and degraded. Rising temperatures, ocean acidification, changing precipitation patterns, and other climate-induced changes are putting enormous pressures on the natural world. Coral reefs, rainforests, and Arctic tundra are already undergoing major transformations, with cascading impacts on the web of life they support. Many scientists warn that without drastic action to mitigate these trends, the results could be disastrous. Entire areas may become uninhabitable, food and water supplies may be severely compromised, and mass extinction of plant and animal species may occur.

Addressing the climate change crisis will require a comprehensive global effort to transition from fossil fuels to clean, renewable energy sources. At the same time, we must work to protect and restore the Earth's ecosystems, which play a vital role in regulating the climate and sustaining all life. The future of our planet and our survival as a species is at stake, and decisive action is needed now to avoid the looming environmental catastrophe.

References

Giorgi F (2006) Climate change hot-spots. Geophys Res Lett 33:L08707. https://doi.org/10.1029/2006GL025734

IPCC (2013) In: Stocker TF, Qin D, Plattner G-K, Tignor M, Allen SK, Boschung J, Nauels A, Xia Y, Bex V, Midgley PM (eds) Climate change 2013: the physical science basis. Contribution of Working Group I to the Fifth Assessment Report of the Intergovernmental Panel on Climate Change. Cambridge University Press, Cambridge, p 1535

IPCC (2021) Summary for policymakers. In: Masson-Delmotte V, Zhai P, Pirani A, Connors SL, Péan C, Berger S, Caud N, Chen Y, Goldfarb L, Gomis MI, Huang M, Leitzell K, Lonnoy E, Matthews JBR, Maycock TK, Waterfield T, Yelekçi O, Yu R, Zhou B (eds) Climate Change 2021: The physical science basis. Contribution of Working Group I to the Sixth Assessment Report of the Intergovernmental Panel on Climate Change. Cambridge University Press, Cambridge, pp 3–32. https://doi.org/10.1017/9781009157896.001

Lee J-Y, Marotzke J, Bala G, Cao L, Corti S, Dunne JP, Engelbrecht F, Fischer E, Fyfe JC, Jones C, Maycock A, Mutemi J, Ndiaye O, Panickal S, Zhou T (2021) Future global climate: Scenario-based projections and NearTerm information. In: Masson-Delmotte V, Zhai P, Pirani

A, Connors SL, Péan C, Berger S, Caud N, Chen Y, Goldfarb L, Gomis MI, Huang M, Leitzell K, Lonnoy E, Matthews JBR, Maycock TK, Waterfield T, Yelekçi O, Yu R, Zhou B (eds) Climate change 2021: The physical science basis. Contribution of Working Group I to the Sixth Assessment Report of the Intergovernmental Panel on Climate Change. Cambridge University Press, Cambridge, pp 553–672. https://doi.org/10.1017/9781009157896.006

Malhi Y, Franklin J, Seddon N, Solan M, Turner MG, Field CB, Knowlton N (2020) Climate change and ecosystems: threats, opportunities and solutions. Phil Trans R Soc B 375:20190104. https://doi.org/10.1098/rstb.2019.0104

Philandras CM, Nastos PT, Kapsomenakis J, Douvis KC, Tselioudis G, Zerefos CS (2011) Long term precipitation trends and variability within the Mediterranean region. Nat Hazards Earth Syst Sci 11:3235–3250. https://doi.org/10.5194/nhess-11-3235-2011

Rantanen M, Karpechko AY, Lipponen A et al (2022) The Arctic has warmed nearly four times faster than the globe since 1979. Commun Earth Environ 3:168. https://doi.org/10.1038/s43247-022-00498-3

Riahi K, van Vuuren DP, Kriegler E, Edmonds J, O'Neill BC, Fujimori S, Bauer N, Calvin K, Dellink R, Fricko O, Lutz W, Popp A, Crespo Cuaresma J, KC S, Leimbach M, Jiang L, Kram T, Rao S, Emmerling J, Ebi K, Hasegawa T, Havlik P, Humpenöder F, Aleluia Da Silva L, Smith S, Stehfest E, Bosetti V, Eom J, Gernaat D, Masui T, Rogelj J, Strefler J, Drouet L, Krey V, Luderer G, Harmsen M, Takahashi K, Baumstark L, Doelman J, Kainuma M, Klimont Z, Marangoni G, Lotze-Campen H, Obersteiner M, Tabeau A, Tavoni M (2017) The shared socio-economic pathways and their energy, land use, and greenhouse gas emissions implications: an overview. Glob Environ Chang 42:153–168. https://doi.org/10.1016/j.gloenvcha.2016.05.009

Seneviratne SI, Zhang X, Adnan M, Badi W, Dereczynski C, Di Luca A, Ghosh S, Iskandar I, Kossin J, Lewis S, Otto F, Pinto I, Satoh M, Vicente-Serrano SM, Wehner M, Zhou B (2021) Weather and Climate Extreme Events in a Changing Climate. In: Masson-Delmotte V, Zhai P, Pirani A, Connors SL, Péan C, Berger S, Caud N, Chen Y, Goldfarb L, Gomis MI, Huang M, Leitzell K, Lonnoy E, Matthews JBR, Maycock TK, Waterfield T, Yelekçi O, Yu R, Zhou B (eds) Climate Change 2021: The Physical Science Basis. Contribution of Working Group I to the Sixth Assessment Report of the Intergovernmental Panel on Climate Change. Cambridge University Press, Cambridge, pp 1513–1766. https://doi.org/10.1017/9781009157896.013

Trenberth KE (2011) Changes in precipitation with climate change. Clim Res 47(1–2):123–138

Vogel MM, Hauser M, Seneviratne SI (2020) Projected changes in hot, dry and wet extreme events' clusters in CMIP6 multi-model ensemble. Environ Res Lett 15. https://doi.org/10.1088/1748-9326/ab90a7

You Q, Cai Z, Pepin N, Chen D, Ahrens B, Jiang Z, Wu F, Kang S, Zhang R, Wu T, Wang P, Li M, Zuo Z, Gao Y, Zhai P, Zhang Y (2021) Warming amplification over the Arctic Pole and Third Pole: trends, mechanisms and consequences. Earth Sci Rev 217:103625. https://doi.org/10.1016/j.earscirev.2021.103625

Climate Change Impacts on Environmental Toxins, and Its Health Effects

2

Amal Saad-Hussein, Wagida Anwer, William Au, Marco Neira, Jeanine L. Marnewick, and Naeem Sheik Abdul

Abstract

Since the beginning of the industrial era, greenhouse gas emissions with the extensive anthropogenic changes to the earth globally have led to crucial changes in environmental and meteorological parameters that result in climate changes.

A. Saad-Hussein (✉)
Department of Environmental & Occupational Medicine, Institute of Environment and Climate Change Research, National Research Centre, Giza, Egypt

National Committee of Toxicology, Academy of Scientific Research and Technology (ASRT), Ministry of Higher Education and Scientific Research, Cairo, Egypt
e-mail: al.hussein@nrc.sci.eg

W. Anwer
National Committee of Toxicology, Academy of Scientific Research and Technology (ASRT), Ministry of Higher Education and Scientific Research, Cairo, Egypt

Department of Community, Environmental and Occupational Medicine, Faculty of Medicine, Ain Shams University, Cairo, Egypt

W. Au
School of Public and Population Health, University of Texas Medical Branch, Galveston, TX, USA

M. Neira
Climate and Atmosphere Research Center (CARE-C), The Cyprus Institute, Nicosia, Cyprus
e-mail: meneira@cyi.ac.cy

J. L. Marnewick
Applied Microbial and Health Biotechnology Institute, Cape Peninsula University of Technology, Cape Town, South Africa
e-mail: marnewickj@cput.ac.za

N. S. Abdul
Department of Biochemistry, Stellenbosch University, Stellenbosch, South Africa
e-mail: sheikn@sun.ac.za

A. Saad-Hussein et al. (eds.), *Climate Change Impacts on Toxins and Health Effects*,
https://doi.org/10.1007/978-981-96-3416-3_2

This chapter will discuss the general picture of the impacts of climate change on human health in the form of extreme warming days, water scarcity, and water and air pollutions. Therefore, it will give a picture of infectious diseases related to climate change, especially vector- and water-borne diseases. Mental health problems related to climate change cannot be neglected. After that, this chapter will discuss in detail the impacts of climate change on the environmental toxins, especially air and water chemical and biological pollutants, and describe the association between climate change and the availability and the change in the constituent of these toxic pollutants. Also, the impact of these toxins on health will be targeted in this chapter, especially among vulnerable populations with different influential factors. The health effects from exposure to the chemical toxin affected by climate change will be mentioned in some detail with examples, polycyclic aromatic hydrocarbons, particulate matters, and volatile organic compounds. The impact of climate change on the biological toxins will also be discussed in detail, with the focus on health impacts of the inhaled mycotoxins and the effect of climate change on its molecular toxicity.

Keywords

Climate change · Air pollutants · Water pollutants · Environmental chemical toxins · Environmental biological toxins · Health impacts

Abbreviations

AFB1	Aflatoxin B_1
AFB2	Aflatoxin B_2
AFG1	Aflatoxin G_1
AFG2	Aflatoxin G_2
AFs	Aflatoxins
As	Arsenic
BALF	Bronchoalveolar lavage fluid
BaP	Benzo[a]pyrene
CH_4	Methane
CNS	Central nervous system
CO_2	Carbon dioxide
COPD	Chronic obstructive pulmonary disease
Cr	Chromium
CVDs	Cardiovascular diseases
CYP450	Cytochrome P450
DEPs	Diesel exhaust particles
DON	Trichothecenes T-2 toxin and deoxynivalenol
ECRS	Eosinophilic chronic rhinosinusitis
FBs	Fumonisins
GDP	Gross Domestic Product
GHGs	Greenhouse gases

GSH	Glutathione
hAECB	Human airway epithelial cells of bronchial origin
hAECN	Human airway epithelial cells of nasal origin
HCFCs	Hydrochlorofluorocarbons
Hg	Mercury
HVAC	Heating, ventilation, and air conditioning systems
MAPK	Mitogen-activated protein kinases
MDA	Malondialdehyde
NAAQS	National Ambient Air Quality Standards
NO	Nitric oxide
NO_2	Nitrogen dioxide
O_3	Ozone
OTA	Ochratoxin A
OTB	Ochratoxin B
OTC	Ochratoxin C
OTs	Ochratoxins
PAHs	Polycyclic aromatic hydrocarbons
PCDD	Polychlorinated dibenzo-p-dioxins
PCDF	Dibenzofurans
PHC	Primary health care
PM	Particulate matter
PM_{10}	Particulate matter with a diameter of up to 10μm
$PM_{2.5}$	Particulate matter with an aerodynamic diameter $\leq$2.5μm
POPs	Persistent organic pollutants
ROS	Reactive oxygen species
RR	Relative risk
TCT	Trichothecenes
UHC	Urban health care
VBDs	Vector-borne diseases
VOCs	Volatile organic compounds
WBDs	Water-borne diseases
WHO	World Health Organization
WTC	World Trade Centre
ZEA	Zearalenone

2.1 Introduction

Since the beginning of the industrial era, the increasing emissions of greenhouse gases (GHGs), together with extensive anthropogenic changes to the landscape, such as mass deforestation and urban growth, have caused profound changes to crucial environmental and meteorological parameters, such as the earth's surface temperature and rainfall patterns (Zittis et al. 2022). At present, it is estimated that the mean surface temperature is approximately 1.1 °C over preindustrial values (Romanello et al. 2022). As a consequence of these changes, both the magnitude and frequency of extreme weather events, in the form of extreme heatwaves, unexpected rains, or droughts, have also increased in many regions, creating conditions that threaten the life and well-being of populations around the world.

The interactions between climatic factors and human life are diverse and complex. Climate change can influence health in several ways. Direct effects include the exposure to extreme temperatures, water scarcity, and flooding, among others. Indirect effects can also be exerted through changes in factors such as the ecological landscape of infectious diseases, food availability, air quality. Importantly, it is necessary to keep in mind that these interactions are also modulated by socio-economic, biological, and ecological factors such as economic status, personal and occupational criteria, and geographic location. Furthermore, climatic factors can also impact human health by fostering the displacement of entire populations and by influencing the course of preexisting armed conflicts (Al-Delaimy 2020; Watts et al. 2021; Neira et al. 2023).

Below, we summarize the health effects of some of the prominent consequences of climate change, including exposure to extreme temperatures, water scarcity, air pollution, epidemiological factors for infectious diseases, and mental problems associated with climatic factors. In addition, climate change impacts on environmental toxins were discussed, focusing on the main health effects.

2.2 General Picture of Climate Change Impacts on Human Health

2.2.1 Extreme Temperatures

Human body must keep its internal temperature within a relatively narrow range—roughly between 36.5 °C and 40.5 °C for normal physiological functions (Byrne et al. 2006; Lim et al. 2008; Gauer and Meyers 2019; Racinais et al. 2019). Therefore, when exposed to increased environmental temperatures, humans display an array of physiological and behavioural adaptations aimed at facilitating efficient heat dissipation. In humans, the primary physiological mechanism for heat dissipation is evaporative heat loss, through the evaporation of sweat. Because the efficiency of sweat evaporation is inversely related to the amount of moisture in the air (i.e. humidity), evaporative heat loss is progressively impeded with increasing environmental humidity (Lim 2020). As a result, the harmful effects of extreme temperatures are amplified under high-humidity conditions.

It has been proposed that the human body reaches its limit for thermal regulation when environmental ‘wet bulb’ temperature (a scale that combines the effect of both temperature and humidity) reaches 35 °C and that unmitigated exposure to conditions above this, even for a few hours, would result lethal for most humans (Sherwood and Huber 2010) (Fig. 2.1). Unfortunately, predictive models suggest that this limit is likely to be surpassed in several large cities before the end of the century, posing the risk of mass mortality due to hyperthermia (Pal and Eltahir 2016).

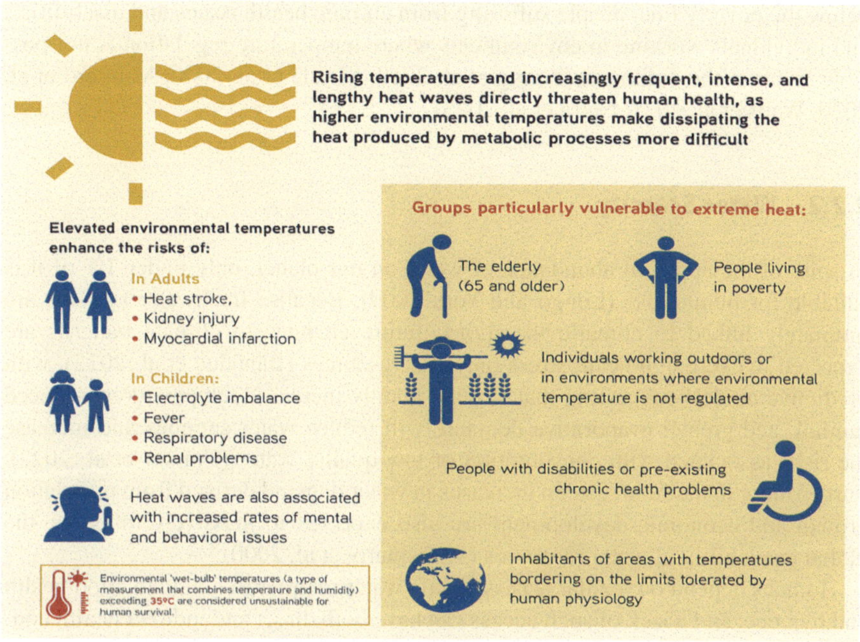

Fig. 2.1 Health effects of exposure to extreme temperatures. (Source: Neira et al. 2023)

Various studies detected the association between heatwaves and mortality increases (Basu 2009; Baccini et al. 2011; Son et al. 2012; Wang et al. 2015; Guo et al. 2017), with densely populated urban communities being particularly affected (Medina-Ramón and Schwartz 2007). This effect is due, at least in part, to the "urban heat island" effects, whereby environmental temperatures in urban regions are always higher than temperatures in the surrounding rural or planted areas (Oke 1982; Guo et al. 2017). The death toll of heatwaves can be staggering. In Europe alone, heat-related mortality exceeded 70,000 fatalities in the summer of 2003, and 60,000 in 2022 (Robine et al. 2008; Ballester et al. 2023).

However, the negative effects of extreme heat are not limited to mortality. Exposure to temperatures below these physiological limits can also result in severe health problems, especially if sustained over long time. Health problems associated with exposure to extreme heat in adults include heat stroke, renal injury, and myocardial infarction (MI), while in children, it can trigger electrolyte imbalance, fever, respiratory disease, and kidney problems, among others (Parker et al. 2019; Watts et al. 2019) (Fig. 2.1). Furthermore, mental health problems (including sleep disturbances, cognitive deficits, violent behaviours, and suicide) are also known to be triggered and/or exacerbated by exposure to extreme heat over sustained periods of time (Levy et al. 2017; Watts et al. 2019; Palinkas and Wong 2020).

Therefore, extreme temperature either heating or cooling to unexpected levels will affect human health not only directly, but also indirectly. Particularly vulnerable groups include elders (aged 65 years and older), pregnant women, people living below the poverty line, people suffering from chronic health issues and disabilities, and individuals working in environments where temperature regulation is not possible (such as the outdoors) (Zhang et al. 2019; Meade et al. 2020; Millyard et al. 2020; Watts et al. 2021; Neira et al. 2023).

2.2.2 Water Scarcity

In spite of the apparent abundance of water on our planet, only under 1% of it is suitable for human use (Longo and York 2009). Because freshwater systems are intimately linked to climatic variations, future changes to climate patterns are expected to exert significant stress on water resources (Bănăduc et al. 2022), with predictive models suggesting that factors, mainly increased temperatures, reduced rainfall, and greater evaporative demand, will reduce water supplies and increase the risks associated with consumption of low-quality water (Comair et al. 2022). Furthermore in the future, steep increases in water demand derived from population growth and economic development are also expected to massively influence the global availability of water resources (Vörösmarty et al. 2000).

Humans depend on regular access to safe water in order to maintain good health and hygiene, and a lack of such access can have both direct and indirect health consequences (Box 2.1) (Hunter et al. 2010; Neira et al. 2023). Maintaining an adequate water balance in the body is crucial for the appropriate functioning of most basic biological processes, including metabolism, digestion, circulation, transmembrane transport, and osmotic balance (Armstrong et al. 2024). As a consequence, severe dehydration can lead to speech incoherence, extreme weakness, hypotonia, hypotension, tachycardia, fever, unconsciousness, and death (Jéquier and Constant 2010). Additionally, the inability to meet appropriate hydration levels over long periods of time is associated with a plethora of health problems, including increased markers of inflammation and coagulation, dementia, heart failure, chronic lung disease, high blood pressure, diabetes, and kidney problems (Allen et al. 2019; Armstrong et al. 2024).

Because regular water intake is crucial for daily subsistence, when a supply of safe potable water is not readily available and accessible, people are often forced to consume water of substandard quality. Therefore, one of the most important health effects associated with water scarcity is the exposure to water contaminated with either microbial pathogens or toxic chemicals (Box 2.1). Diarrhoeal disease caused by the consumption of water contaminated with pathogenic viruses or bacteria is contributed to one of the top worldwide disease burdens, and one of the top three causes of child mortality globally (Abbafati et al. 2020). In fact, restricted access to safe water services is a key risk factor for infant mortality, even when controlling for economic inequalities (Hunter et al. 2010). Other pathogens which can be transmitted through contact with unsafe water include the parasites responsible for dracunculiasis, amoebiasis, giardiasis, and schistosomiasis (Tarrass and Benjelloun 2012; Colley et al. 2014; Elmorshedy et al. 2020; Ngowi 2020).

Box 2.1: Effects of water scarcity on health

Direct impacts:

- Dehydration
- Symptoms ranging from mild to severe and life-threatening
- Consumption of unsafe water
  - Diarrhoeal disease
  - Parasitic disease
  - Chemical poisoning
- Water storage under unsafe conditions
  - Increases in vector-borne disease incidence

Indirect impacts:

- Reduced agricultural productivity
- Reduced income for agricultural workers
- Increase in food prices (worsening of nutritional status)
- Reduced financial resources for public health services

Besides pathogens, unsafe water can also be contaminated with toxic chemicals. These can be either inorganic (e.g. arsenic, copper, fluoride, lead, nitrate) or organic (e.g. pesticides, phenols, and endocrine-disrupting chemicals). The effects of such contaminants range from relatively mild, such as nausea or skin rashes, to severe; in the form of foetal abnormalities, and even life threatening from renal failure or cancers (Hunter 1997; Saha 2003; Hunter et al. 2010). Additionally, conditions of prolonged water scarcity often force people to store large quantities of water for later consumption. If not properly secured in safe covered containers, stored water can easily provide breeding sites for disease vectors, such as the *Aedes* mosquitoes that transmit dengue and yellow fever, among other maladies (Neira et al. 2023).

Water scarcity can also affect human health indirectly in several ways. Most importantly, reductions in water availability are usually linked to reduced agricultural productivity, which can in turn have a multifaceted negative effect: at the level of consumers, this reduced productivity not only increases the prices of food (making it harder for people to obtain a balanced and nutritious diet), but also negatively affects the overall economy of the population, particularly of families working on the agricultural sector. Furthermore, at the country level, reduced agricultural productivity can cause severe reductions in Gross Domestic Product (GDP) and export revenues, thereby reducing the financial resources available for funding public health services. As an example, water scarcity in the Middle East is expected to cause reductions of up to 60% in the agricultural productivity of some countries in the region by the year 2050, shrinking national GDPs by up to 14% and affecting the livelihood of almost 30% of the region's population (Borgomeo et al. 2018; North Atlantic Treaty Organization (NATO) 2019; Neira et al. 2023).

2.2.3 Air Pollution

The wide majority of pollutants found in the air are by-products of large-scale human activities and infrastructure, such as fossil fuel combustion, agricultural development, power stations, refineries, and chemical industries. Among these, emissions linked to human mobility (automobiles, trains, airplanes, etc.) are estimated to be the main sources of pollution (Moller et al. 1994; Kinney 2018a, b; Manisalidis et al. 2020). In addition, pollutants emitted by natural phenomena such as smoke from wildfires, airborne dust, and pollen can also be influenced by climatic patterns.

Airborne particles of very small diameter (known as "particulate matter" [PM]) are among the most important pollutants from a health perspective. PM with an aerodynamic diameter ≤2.5μm ($PM_{2.5}$) is particularly dangerous to our health: due to their minimal size, these particles can remain airborne for days after being generated, increasing their environmental persistence and spread. Once inhaled, they are able to pass through the defensive barriers of our respiratory system, enter the lungs, and, eventually, access the bloodstream, whereby they can be spread throughout the whole body, affecting critical tissues and organs (Kinney, 2018a, b).

Exposure to increased concentrations of $PM_{2.5}$ in the air is known to cause health issues such as respiratory and cardiovascular diseases (CVDs), cancers, neurological disorders, and deaths (Burnett et al. 2014; Lelieveld et al. 2015; Yang et al. 2022). It is estimated that on a global scale, outdoor air pollution with $PM_{2.5}$ generates up to nine million fatalities per year (Cohen et al. 2017; Burnett et al. 2018). Furthermore, air pollutants with diameters above 2.5μm are also associated with health problems: exposure to high concentrations of PM with a diameter of up to 10μm (PM_{10}) pollution (i.e. airborne PM_{10}) has been linked to reduced lung function, increased respiratory disease, increased hospital admissions, and increased mortality (Archer 1990; Pope and Dockery 1992). Interestingly, a large, multicountry study about the impact of exposure to PM on mortality found a concentration-response function displaying a direct correlation between PM concentration and daily mortality, without a discernible threshold (Liu et al. 2019). The same study reported that an increase of 10μg/m^3 PM_{10}, over a two-day period, was associated with a daily increase in mortality rates (0.44% of all mortality causes per day). A similar increment in $PM_{2.5}$ resulted in an increment of 0.68% in daily all-cause mortality.

Another source of air pollution linked to climate change is dust. Predicted water shortages, increasing temperatures, erosion, deforestation, and desertification are likely to expand the surface of dry, deserted lands that serve as sources of dust, thereby increasing the intensity and frequency of dust storms in some world regions (Keyantash and Dracup 2002; Keramat et al. 2011; Zittis et al. 2016; Gat et al. 2017; Ahmady-Birgani et al. 2018a, b; Mousavi et al. 2020). Additionally, changes in wind patterns triggered by the expected changes in surface temperature can result in increased dust dispersion (Hamidianpour et al. 2021; Neira et al. 2023).

The reported health outcomes of dust exposure include increased morbidity and mortality due to respiratory diseases, including bronchial asthma and chronic obstructive pulmonary disease (COPD), and circulatory diseases, including acute myocardial infarction (Samoli et al. 2011; Lorentzou et al. 2019; Bonyadi et al. 2020; Linares et al. 2020). Furthermore, the dispersion of dust can facilitate the transport of microbial agents that facilitate spread and frequency of infectious diseases, as well as toxic chemicals such as herbicides, pesticides, fertilizers, and heavy metals (Falkovich and Rudich 2001; Thomson et al. 2009; Mazar et al. 2016; Gat et al. 2017; Pardo et al. 2017; Aminharati et al. 2018; Schweitzer et al. 2018; Lang-Yona et al. 2020; Mousavi et al. 2020).

2.2.4 Infectious Diseases

The impacts of climate change on infectious disease epidemiology are significantly multifactorially increasing. Climate change is strongly associated with alterations in the transmission of infectious diseases through impacts on pathogens, vectors, and animal hosts/reservoirs. These alterations respond to factors such as changes in the density and distribution of disease vector species, changes in the climatological parameters that define the geographic distribution of pathogens, and changes in the migration patterns of vertebrate hosts and reservoirs (Altizer et al. 2011; Caminade et al. 2019).

Weather phenomena, such as fires, drought, or flooding, can increase the frequency of zoonotic disease transmission that was facilitated through the increased human-to-animal contacts (Watson et al. 2007; Edelson et al. 2023). Alternatively, these phenomena can cause the displacement of entire populations, driving people into areas where they are exposed to pathogens which they have never encountered before, and against which they are defenceless both immunologically and behaviourally (Neira et al. 2023). Extreme weather events can directly impact infectious disease transmission by mechanisms such as increased pathogen dispersal (e.g. the effects of flooding on water-borne bacteria) or through the alteration of environments suitable for pathogen transmission (such as the increase in vector breeding sites following periods of extreme rainfall) (Chowell et al. 2019; Edelson et al. 2023). Mid- and long-term changes in climatologic patterns can affect the epidemiological landscape of infectious diseases either geographically (by altering the size, number, and distribution of areas suitable for transmission) or temporally (by changing the number of yearly days suitable for transmission) (Ciota and Keyel 2019; Semenza and Paz 2021; Neira et al. 2023; Sips et al. 2023).

Among infectious diseases, two categories are particularly susceptible to rising temperatures and altered precipitation patterns: vector-borne diseases (VBDs) and water-borne diseases (WBDs). Below, we address each one of these categories separately.

2.2.4.1 Vector-Borne Diseases (VBDs)

Arthropod vectors of human disease, such as mosquitoes, ticks, and sand flies, are ectotherms.[1] For this reason, the biology of disease vectors is profoundly dependent on environmental conditions, and any changes in temperature can affect their physiology, behaviour, and disease-transmission capacity. Parameters of vector biology affected by temperature include developmental rates, longevity, and survival, among others (de Souza and Weaver 2024).

It has been shown that warmer temperatures can accelerate the gonadotrophic cycle[2] of mosquitoes. Because the biting rate of these insects is directly related to the duration of their gonadotrophic cycle, increasing temperatures can result in more frequent biting, which increases the likelihood of transmission of the infectious diseases in endemic areas (Siraj et al. 2017). Likewise, increased metabolic rates associated with warmer temperatures could reduce the time required for completion of the extrinsic incubation period[3] of several arthropod-borne viral diseases and increase their transmission risk, e.g. dengue and chikungunya (Ernst et al. 2023).

Besides impacting vectors, changes in ambient temperature can also affect the biological characteristics of pathogens. As previously stated, increased temperature globally could lead to an increase in the number of new arboviral infections; because of the high genetic mutation rates characteristic of viruses, each new infection has the potential to increasing the genetic diversity of these pathogens, thereby increasing the possibility of appearance of novel virus variants, with diverse characteristics regarding their virulence and transmissibility (Tozan et al. 2020).

Environmental temperature was associated with VBD transmission in a complex pathway that can follow nonlinear paths and depend on both the vector and pathogen species involved. While it is generally accepted that in some cases, warming environmental temperatures can lead to increased disease incidence (by allowing vectors to thrive in areas and/or periods that were previously too cold), it is also possible that in specific regions or periods, temperatures could rise above the thresholds critical for vector or pathogen survival, resulting in reduced disease transmission (Murdock et al. 2016; António et al. 2018; Mordecai et al. 2020). Additionally, other environmental factors, including relative humidity, wind speed, and photoperiod, can influence the epidemiology of VBDs (Reiter 2001; Negev et al. 2015). Socio-economic factors, mainly access to basic healthcare services, population density, housing characteristics, and public knowledge, attitudes, and practices, can also have impacts on the VBD epidemiology (Padmanabha et al. 2010; Schmidt

[1] Ectotherms are defined as animals that do not regulate their body temperature by physiological processes, and are therefore thermally dependent on their external environment (Akin 2011).

[2] The gonadotrophic cycle is the process required for egg maturation, which in mosquitoes is usually triggered by the ingestion of a blood meal.

[3] For an arbovirus to be transmitted by bite, a vector must first ingest infective viral particles with a blood meal, and the virus must then replicate and spread within the vector, infect the salivary glands, and subsequently be injected into a host during the next episode of blood feeding (Oviedo et al. 2011). This process is known as the 'extrinsic incubation period'.

et al. 2011; Heydari et al. 2017; Kenneson et al. 2017; Horstick and Runge-Ranzinger 2019).

2.2.4.2 Water-Borne Diseases (WBDs)

As previously mentioned, the transmission of WBDs can also be influenced by climate change. Most notably, flooding events can cause the contamination of drinking water, leading to the transmission of enteric pathogens such as *Vibrio cholera, Escherichia coli, Salmonella enterica, Cryptosporidium parvum,* norovirus, and hepatitis viruses (Watson et al. 2007). Therefore, outbreaks of diarrhoeal disease are one of the main health consequences of extreme weather events involving flooding, such as torrential rains, tsunamis, and hurricanes. Case examples include the large cholera outbreaks reported following flooding in West Bengal in 1998 (Sur et al. 2000) and Bangladesh in 2004 (Qadri et al. 2005).

Additionally, flooding can facilitate the spread of bacteria of the genus *Leptospira*, a zoonotic pathogen normally spread through contact of the skin or mucous membranes with water, soil, or vegetation contaminated with the urine from infected mammals (mainly rodents) (Watson et al. 2007). Symptoms of leptospirosis range from undifferentiated febrile illness to renal failure and lethal pulmonary haemorrhage (Bharti et al. 2003). Outbreaks of leptospirosis following flooding events have been recorded in Brazil in 1996 (Barcellos and Sabroza 2001), India in 2000 (Karande et al. 2003), and Taiwan in 2001 (Yang et al. 2005), among others.

2.2.5 Mental Health

The effects of climate change on mental health can be exerted by exposure to (a) extreme weather events or natural disasters, (b) moderate but long-lasting weather events (including droughts and extended heatwaves), and (c) persistent/permanent changes, such as the progressive rise in global average temperatures, sea-level rise, and permanent environmental degradation (Palinkas and Wong 2020).

Extreme weather events, such as droughts, floods, and hurricanes, can trigger stress-related anxiety, depression, post-traumatic stress disorder, substance abuse, and suicidal tendencies (Padhy et al. 2015; Palinkas and Wong 2020; Lawrance et al. 2022; Walinski et al. 2023). These effects occur in people of all ages and can be particularly pronounced in those who experience life-threatening situations, making them also more prone to develop acute stress reaction and adjustment disorder. Staff working on disaster-relief efforts are also at risk for these conditions (Padhy et al. 2015; Lawrance et al. 2022).

When extreme events involve the loss of a home, social structures, or loved ones, the psychological consequences can involve bereavement or depression—an effect that is likely to be more pronounced for people living in small rural communities than for those living in cities (Ginexi et al. 2000; Padhy et al. 2015). It has been estimated that following a natural disaster, the number of people suffering mental health problems can vastly surpass those suffering physical injuries, with an estimated 25% to 50% of people exposed to an extreme weather event suffering negative psychological effects (Hayes et al. 2018; Palinkas and Wong 2020).

Heatwaves have been reported to increase rates of mental health disturbances such as sleep disorders, violence, assault, homicide, and suicide (Levy et al. 2017; Watts et al. 2019; Palinkas and Wong 2020). Generally, the severity of mental health effects seems to be related to the length of exposure. Examining the correlation between climatic conditions and mental health on a random sample of US residents, et al. Obradovich et al. (2018) found that within a 5-year period, a 1 °C increase in the average maximum temperature resulted in approximately a 2% increase in mental health problems' prevalence. People with preexisting psychological conditions, such as dementia, schizophrenia, and a history of substance abuse can be vulnerable to heat stress, due to heat-related cognitive impairment and/or physiological secondary effects of psychotropic medication. Therefore, heatwaves can trigger increases in morbidity, mortality, and hospitalization rates in these groups (Schmeltz and Gamble 2017; Palinkas and Wong 2020).

Prolonged periods of drought are also linked to deleterious effects on mental health. The emotional burden of watching crops fail, livestock perish, and water sources dry up without a foreseeable end can lead to deep psychological distress, causing stress, anxiety, depression, and even suicidal tendencies, particularly among those whose livelihoods are profoundly tied to the land (such as farmers) and the elderly (Padhy et al. 2015; Palinkas and Wong 2020). Furthermore, extended periods of water scarcity often led to population displacement, as people whose subsistence is mainly dependent on agricultural activities are forced to leave their land in search of better conditions. The mental and emotional effects of forced migration include increases in stress and anxiety, loss of identity, and loss of supportive social connectivity (Lawrance et al. 2022).

Psychological trauma is different from physical injury in that it has very long-lasting effects, which can even be transmitted transgenerationally (East et al. 2018). Interestingly, it has been proposed that immigrants and their children display double the risk of being affected by psychotic disorders than nonimmigrants (Bourque et al. 2011). For these reasons, mental health problems are considered among the most important effects of climate change (Al-Delaimy 2020; Neira et al. 2023).

2.3 Impacts of Climate Change on Environmental Toxins

Many sectors (particularly agriculture, industrial, water, air, and health) are very much interconnected. Besides the impacts of climate change on infectious diseases, further risk arises from the impacts of climate change on environmental chemical toxins, such as persistent organic and heavy metal pollutants, circulating within aquatic and terrestrial ecosystems, which are the major concern for reproductive hazards for human health (Dudley et al. 2015). Extreme weather events due to climate change, in the form of heatwaves, wildfires, droughts, and floods, can have direct and indirect impacts on the various environmental pollutants (Bolan et al. 2024).

2.3.1 Impacts of Climate Change on Air Chemical Pollutants

Climate change directly affects the concentrations and distributions of outdoor air pollutants emitted from intense exhaust and energy consumption from heavy road traffic in urban areas; moreover, climate change affects seasonal variation of allergenic pollens, and may prolong the release periods. These chemical pollutants and aeroallergens were found to cause an increase in the health problems of the respiratory system (Fig. 2.2), in the form of increasing airway inflammation and the prevalence of allergic respiratory diseases in polluted urban areas (D'Amato et al. 2010), as well as cardiopulmonary diseases among individuals with preexisting cardiac problems (D'Amato et al. 2014). Moreover, exposure to chemical air pollutants leads to impairment in the mucociliary clearance system and may lead to access inhalation of aeroallergens to the cells of the immune system (D'Amato 2008).

Therefore, due to climate change and persistence to emissions of pollutants, from the anthropogenic air pollutants from burning of biomass, combustion of fossil fuels, and agricultural activities, air pollution will be directly influenced, and due to stratospheric ozone (O_3), cosmic radiation will be an essential factor for the formation of photochemical smog, which includes tropospheric O_3 and aerosols, in addition to the production of hydroxyl radicals, which control the amount of many climate- and O_3-relevant gases, such as hydrochlorofluorocarbons (HCFCs) and methane (CH_4) in the atmosphere. Moreover, the predicted future changes in cosmic radiation and climate, detected through numerical models, will modify the trends and geographic distribution of hydroxyl radicals that will affect the formation of photochemical smog in many urban and regional areas (Tang et al. 2011).

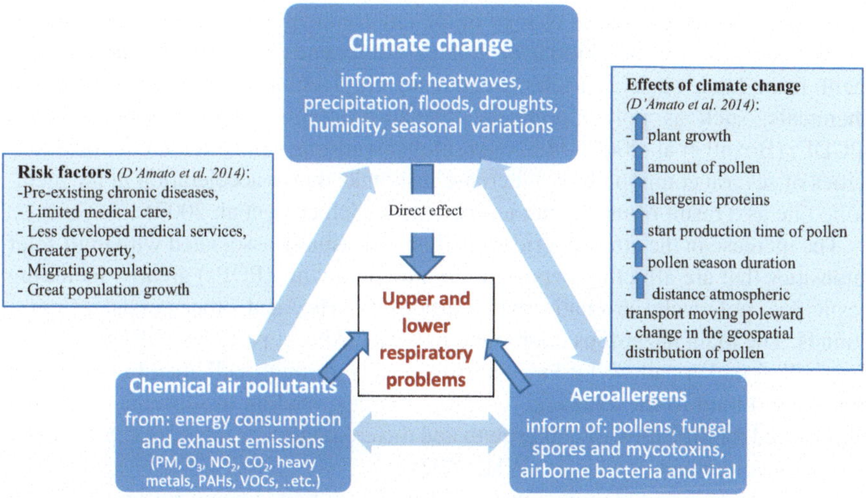

Fig. 2.2 Summary of the impacts of climate change on chemical and aeroallergen air pollutants and their role in the respiratory system. (Source: self-created)

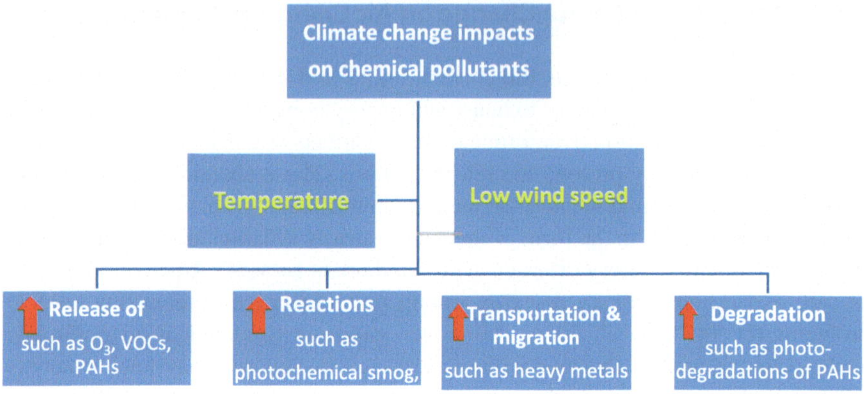

Fig. 2.3 Main effects of temperature and wind speed on chemical air pollutants. (Source: self-created)

Climate change can have hazardous effects on air pollutants by different mechanisms (Fig. 2.3). The exhaust of heavy traffic is the most significant source of outdoor nitrogen dioxide (NO_2), which is one of the greenhouse gases (GHGs, and is a precursor of photochemical smog found in outdoor air in urban and industrial regions, and may conjunct with the sunlight and hydrocarbons resulting in the production of O_3 (D'Amato et al. 2016). In addition, increased atmospheric temperature is associated with elevation of the air O_3 levels (D'Amato et al. 2010), and low wind speed increases air pollutant concentrations at the ground level (D'Amato et al. 2014). Acute exposure to high O_3 levels is always linked to asthma and asthma-like symptoms, and chronic exposure will lead to impairment in the respiratory functions (D'Amato et al. 2005).

Climate change can cause change in the environmental temperature and humidity that can increase the amount of volatile compounds' release into the environment from land, such as trichloroethylene, vinyl chloride, and alter persistent chemicals, such as polychlorinated dibenzo-p-dioxins (PCDD), dibenzofurans (PCDF) (Boxall et al. 2009). Moreover, global warming can increase chemical toxicities of several compounds and increase its uptake due to acceleration in the metabolic rate as a result of increasing temperatures (Schiedek et al. 2007).

The increase in the atmospheric temperature is usually associated with high solar intensities that are affecting persistent organic pollutants (POPs) greatly, including pesticides, polycyclic aromatic hydrocarbons (PAHs), and other persistent compounds. The main hazardous compound in diesel exhaust particles (DEPs) is PAHs that will found to allow DEPs to diffuse easily through cell membranes when inhaled and bind to the cytosolic receptor complex, leading to subsequent nuclear activity and modification of cell growth and differentiation programmes that predispose to development of cancers (WHO 2021).

Moreover, increased temperature could alter the organic composition of the soil leading to organic matter destruction at temperatures between 600 °C and 700 °C, leading to frequent erosions that enhance transportation and remobilization of heavy

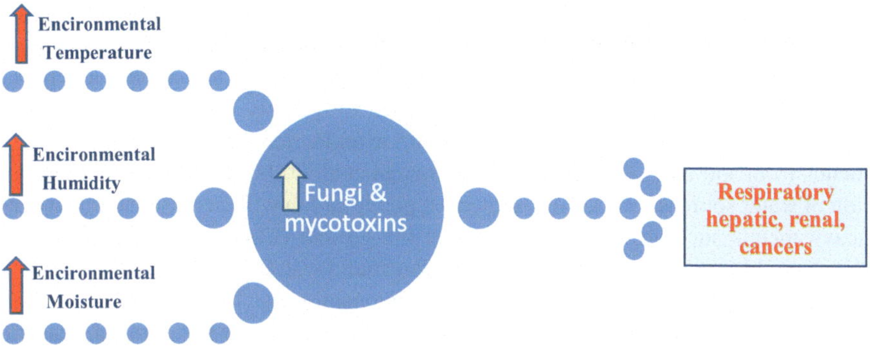

Fig. 2.4 Climate change impacts on fungi and mycotoxins and development of cancers. (Source: self-created)

metals through soil loss and landslides, e.g. the change of mercury to methyl mercury (Frogner-Kockum et al. 2020).

The concentration of PAHs in the environment is also projected to be increased by the climatic events of wildfires of the forests, when the temperature and light intensity increase to unusual levels between 350 °C and more than 1200 °C, and PAHs with low molecular weight volatilize easily (Bolan et al. 2024). Under these high solar intensities, the photodegradation of PAHs, including naphthalene, anthracene, phenanthrene, and pyrene, can occur and can develop intermediate products, which could be more toxic compared to the original molecules (Kumar et al. 2021).

Climate change has an impact on the growth of air biological microorganisms, such as fungi with the increase in mycotoxin production (Saad-Hussein and Ramadan 2023). Fungal growth depends mainly on the environmental temperature and humidity (Talley et al. 2002). Therefore, climate warming will increase the growth of fungi, especially if there is an increase in the environmental humidity (Saad-Hussein et al. 2011). It was found that elevation of mycotoxin levels in biological fluids of workers was associated with occupational exposure to high concentrations of fungi in their working environment (Saad-Hussein et al. 2012, 2016, 2022). So, climate change can increase the health hazards of environmental exposure to high concentrations of fungi and its mycotoxins, and increase the risk of development of cancers among the exposed populations (Fig. 2.4).

2.3.2 Impacts of Climate Change on Water Chemical Pollutants

Climate change can result in change in the temperature and precipitation patterns, causing both droughts and floods, that may result in migration, dilution, or accumulation of pollutants (Bolan et al. 2024). The main impact of climate change on environmental contamination is the mobility of different pollutants, such as mobility of heavy metals resulting from redox reactions due to floods. Floods could impact the concentrations of potentially toxic elements in the groundwater, e.g. As and chromium (Cr). The excess floods' water reduces the less mobile As(V) to become more mobile As(III), and the more mobile Cr(VI) to the less mobile Cr(III), which will affect their toxicity, bioavailability, and transport in soil and aquatic environments (Ayotte et al. 2011). Moreover, increased CO_2 levels in the atmosphere will lead to increased chemical reactions resulting in an increase in the release of heavy metals from soil and earth crust (Prathumratana et al. 2008), leading to contamination of water sources. In addition, increased precipitation will increase the release of heavy metals such as mercury (Hg) from the ground (Chen and Driscoll 2018), which could result in contamination of underground water.

Another example is alteration of rainfall intensities was found to affect the net carbon release from the soil. Droughts as an example, due to low rainfall, lead to release of CH_4 and CO from acidic soil, and to excess release of N_2O from natural soil and inorganic carbon, in the form of carbonates and bicarbonates (IPCC 2019). Inorganic nitrogen pollution of ground and surface waters can also induce adverse effects on human health and economy and may develop acid rains (Camargo and Alonso 2006).

Acid precipitation will damage freshwater ecosystems, reducing fish population and thus relieving mosquito larvae; in marine systems, it leads to eutrophication and formation of toxic algal bloom. According to Camargo and Alonso (2006), three major environmental problems will be developed: (1) an increase in hydrogen ions' concentration in freshwater ecosystems, resulting in its acidification; (2) enhanced development of eutrophication of aquatic ecosystems; and (3) reaching toxic levels that impair the ability of aquatic animals to survive, grow, and reproduce.

Moreover, 1 °C temperature rise is projected to increase mobility of POPs from its sources by 10% to 15% increase in its volatility (Lamon et al. 2009), and reduced rainfall leads to persistence of these POPs in the soil (Casas et al. 2021), while increased rainfall facilitates the movement of these contaminants, in addition to metals, hydrocarbons, and dioxins, from polluted areas to nonpolluted areas (Lake et al. 2005). Another example is the impact of the increased temperature and precipitation patterns on potential effects of pesticides is projected to be increased by 1.1 to 2.5 times by the year 2040, and by 2.4 to 9.1 times by the year 2070 (Zhang 2018).

Climate change also has significant effects on the occurrence of aquatic pollutants, such as microplastics from different plastic sources, and on the fate of aquatic ecosystems (Abdelhak et al. 2023). They expected that the distribution pattern of plastics will be changed in the coming decades, due to alterations in air and water movement, volume, ocean salinity due to climate change. Factors including increased floods, increased runoffs, changes in water density, and strong winds can significantly impact the fate of aqueous plastics (Abdelhak et al. 2023). Climate change indirectly impacts on the micropollutants such as microplastic concentrations due to the importance of using recycled wastewater in the areas affected by increased dryness (Lenka et al. 2021).

2.4 Impacts of Environmental Toxins on Health

2.4.1 Chemical Toxins

Concern over the extraordinary rate of climate change has spread over the world, and many countries and organizations have issued recommendations for addressing the numerous issues related to climate change (WHO 2023). Among global problems are extreme weather changes, erosions and floods, disruption of agriculture and food delivery, release of toxic pollutants, and impact on human health and death. However, little attention has been focused on climate change impacts on human exposures to environmental toxic chemicals, and on such exposure towards health consequences.

It has been generally agreed that climate change negatively impacts human health and death. Indeed, climate change can affect human health in many ways. For example, climate change can cause unprecedented, more severe, longer duration of, and higher frequency of health problems than anticipated, as well as in new places and populations (National Institute of Environmental Health Sciences 2022). In one report, extreme temperature changes were associated with increased pollution and with increased incidences of stroke, migrant headaches, and hospitalization of severely ill patients (Louis et al. 2023). However, it has been very difficult to determine the extent of health problems and deaths that are caused by the impact from climate change. Nevertheless, the World Health Organization (WHO) has conservatively projected that 250,000 additional yearly deaths were due to climate change impacts on diseases like malaria and coastal flooding alone by the 2030s (WHO 2023).

2.4.1.1 Association Between Climate Change and Availability of Toxic Pollutants

Very few reports have been published to evaluate associations between climate change and levels of toxic pollutants in the environment. In one report, investigators analysed three publicly available datasets: Notre Dame Global Adaptation Index (ND-GAIN), Yale Environmental Performance Index (EPI), and Global Alliance on Health and Pollution (GAHP) to test such associations (Marcantonio et al. 2021). These datasets contained information from 176 countries since 2018.

In their study, a significant association was found between the spatial distribution of global climate risk and the presence of toxic pollutants ($r_s = -0.798$; 95% CI -0.852, -0.727; $p < 0.0001$). Their geographical analysis showed that around the world, countries which were at risk for impacts of climate change were often at risk for high toxic pollutions.

The same three datasets were also used to measure other variables, e.g. correlations between spatial distribution of toxic environments due to climate change and total mortality due to pollution. They found a good association between the two factors; however, the collected data were insufficient to generate statistical significance.

Other well-established studies also clearly indicate additional relationships between climate change and changes in environmental pollution. For example, global warming increases ice melt which releases previously accumulated pollutants. Warmer water temperatures can also increase the release of pollutants from sediments. These pollutants include broad categories of common and persistent environmental chemicals such as PAHs, airborne PMs, and volatile organic compounds (VOCs). Eventually, these pollutants are released into the environment which can be taken up by vegetation, animals, and fish. The pollutants can therefore be passed onto humans via the food chain and cause an increase in health problems. These studies strongly suggest relationships between climate changes and enhancement of toxic pollutants which can cause increased health hazards and mortality in humans.

2.4.1.2 Vulnerable Populations

Problems associated with the combination of climate change, release of toxic substances, and health effects are not equitably shared by all populations (https://www.cnn.com/2023/12/01/us/black-communities-greater-climate-risk-reaj/index.html). Residents in poor communities are likely to take up a larger burden than those in more affluent communities. There are many factors which contribute to the increased burden. For example, poor communities are likely to be in less desirable neighbourhoods such as municipal waste dumps, hazardous waste sites, and flood-prone areas. In addition, these residents have fewer personal resources for prevention and intervention of the problems.

Other groups of susceptible populations include the elderly and the very young, including those in utero. The former group tends to have underlying health problems which make them more sensitive to the induction of health effects. They also have less resources and physical ability to avoid exposures and to intervene in health impacts. Young children and foetuses are susceptible because they have higher physical and metabolic activities which may cause them to be more exposed to toxic substances but have less developed detoxification mechanisms which may compromise their protective barriers.

Around the world, citizens from certain countries are more susceptible to the impact on health from climate change than those in other countries. The former countries typically include those with poor infrastructures such as transportation and medical facilities.

2.4.1.3 Climate Change and Changes in Chemical Toxicity

It is well established that warmer water temperature can affect pH, dissolved oxygen, and salinity which can therefore alter toxicity of pollutants (https://coastalscience.noaa.gov/project/effects-climate-change-chemical-contaminant-toxicity/). For example, some chemicals are more toxic at warmer temperatures, and others are reversed. Although little is known about interactive effects among chemical toxins under climate change conditions, it is prudent to assume that the impact on health is elevated due to the contributions from many stressors that are associated with climate change.

2.4.1.4 Other Influential Factors

There are a variety of other factors which affect climate change-related toxicity and human health effects. These include poverty, malnutrition, exposure to other chemical toxins at work, dysfunctional social orders, and poor public health infrastructure. In addition, changes in climate may alter people's lifestyle to enhance their exposure to chemical toxins.

2.4.2 Health Effects from Exposure to Chemical Toxins

2.4.2.1 Health Effects from Exposure to polyaromatic hydrocarbons

Aromatic substances with two or more fused benzene rings in their structural configurations are known as polycyclic aromatic hydrocarbons, or PAHs. The following are some examples of PAHs: benzo[a]pyrene (BaP), chrysene, benzo[b]fluoranthene, naphthalene, anthracene, phenanthrene, acenaphthylene, acenaphthene, fluorene, fluoranthene, pyrene, benz[a]anthracene. Having the formula C20H12, BaP is a typical PAH that is found in coal tar. It is created when incomplete combustion of organic molecules combines benzene and pyrene, and it is then discharged into the atmosphere. Figure 2.5 depicts the compound's structure.

In fact, a variety of environmental sources, including the air, water, soil, and food, contain PAHs. Numerous health impacts have been linked to PAH exposure; the degree of these effects varies depending on individual sensitivity, exposure length and intensity, and other factors. As a result, to regulate, anticipate, and prevent the harmful consequences of PAH pollution, the release of PAHs must be controlled through specialized air quality management, such as monitoring release and health issues (Sahoo et al. 2020).

According to toxicological studies, by-products of coal combustion include endocrine disruptors, carcinogens, and poisons that affect the cardiorespiratory system. A thorough search of the literature turned up 2152 publications over the previous several years. These studies unequivocally show that excessive exposure to PAHs has a negative impact on mortality and morbidity rates in human populations. Future studies should concentrate on enhancing exposure assessment models, particularly those that use geographic information system techniques and source apportionment to simulate emissions unique to power plants (Amster 2021).

The following are some possible health consequences linked to PAH exposure:

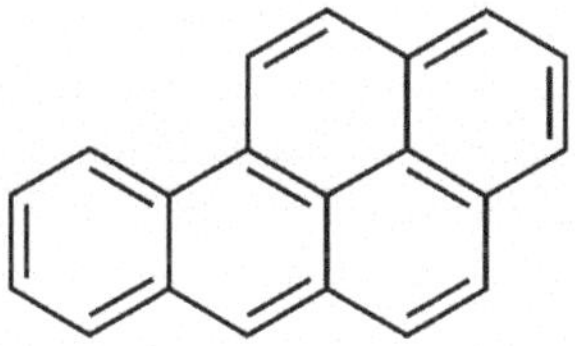

Fig. 2.5 By Calvero. Self-made with ChemDraw, Public Domain, https://commons.wikimedia.org/w/index.php?curid=1696698 (Wikimedia Commons contributors 2025)

Carcinogenicity

According to the Illinois Department of Public Health http://www.idph.state.il.us/cancer/factsheets/polycyclicaromatichydrocarbons.htm), few PAHs are known or probable human carcinogens. Chronic high-level exposure to several PAHs has been associated with a higher risk of cancer, especially gastrointestinal, bladder, skin, and lung cancers. When it comes to the potential for cancer, BaP is one of the PAHs that is most researched.

Effects on the Respiratory System

Breathing in PAHs, particularly when exposed to air pollution, has been linked to respiratory issues such as coughing, wheezing, and reduced lung function. People who have asthma may be more vulnerable to the respiratory consequences of exposure to PAHs. Furthermore, PAHs may exacerbate asthmatic problems by interacting with other environmental toxicants (Mattila et al. 2021).

Effects on Development and Reproduction

Research indicates that exposure to specific PAHs may have negative impacts on health of the reproductive system and foetal development. High amounts of PAH exposure during pregnancy may raise the chance of unfavourable pregnancy outcomes, such as low birth weight and aberrant development. PCOS, POF, and reproductive hormone imbalance are considered unfavourable reproductive outcomes (publicmed.ncbi.nlm.nih.gov).

Immunotoxicity

Exposure to PAHs has been associated with immunosuppression, which means that the immune system's capacity to fight off infections and illnesses may be weakened (Yu et al. 2022).

Dermal Effects

Skin irritation and rashes can result from encountering materials that contain polycyclic aromatic hydrocarbons (PAHs), such as some contaminated soil or water. Extended skin contact with high PAH concentrations may raise the chance of developing skin cancer (Diggs et al. 2011).

Cardiovascular Effects

A new study indicates that atherosclerosis and heart attacks are among the cardiovascular conditions that may be linked to exposure to specific PAHs. Human health problems, such as impaired cardiovascular function, have been linked to exposure to PAHs (Marris et al. 2020). There is much more to learn about the underlying mechanisms. Examining the data that currently links pollution to cardiovascular disease, with a particular emphasis on PM and the low molecular weight PAHs that are associated with it as important mediators of cardiotoxicity, the authors talked about the present understanding of the molecular relationships between exposure to PM and PAHs and the pathophysiology of the mammalian cardiovascular system, including atherosclerosis, cardiac hypertrophy, arrhythmias, contractile dysfunction, and the underlying changes in gene regulation. They demonstrated the conservation of toxicant routes and cellular targets among vertebrates, enabling the establishment of a broad framework for the worldwide issue of cardiotoxic pollution.

It's crucial to remember that the health impacts from exposure to PAHs can differ depending on the particular type of PAH, the exposure route (inhalation, ingestion, or skin contact), and individual susceptibility. To reduce potential health hazards, regulatory bodies from all over the world monitor and establish permitted amounts of PAHs in various environmental media. Furthermore, the risk of harmful health consequences can be decreased by minimizing exposure through appropriate safety precautions, such as wearing protective gear and avoiding tobacco smoke and other sources of combustion.

2.4.2.2 Health Effects from Exposure to Particulate Matters

One significant part of ambient air pollution is comprised of particulate particles or PMs. Different sizes of PMs can be found in the environment, such as $PM_{2.5}$ and PM_{10}. While the latter cannot be seen with the unaided eye, the former can. Along with other environmental pollutants, the dispersion of PMs could shift dramatically because of climate change. In addition to being physically present in the environment, PMs can carry various substances, move them across different geographic locations, and introduce them internally into living things. PM exposure, primarily through inhalation, can have a variety of negative health effects. These particles' sizes are very important. Smaller particles that have absorbed chemicals penetrate the respiratory system more deeply and are mostly responsible for global illness load and mortality.

Environmental PM and O_3 concentrations were determined using several models, and the ensuing health effects under various climate scenarios were assessed (Orru et al. 2017). Mortalities from O_3 and fine particle exposure were predicted to rise because of climate change. The estimated outcomes varied by region, though. Furthermore, there was a fair amount of uncertainty in the estimates of future air pollution emissions and health effects because of the reliance on and uncertainty around the climate change scenario. Therefore, additional work is required to slow down climate change and learn more about how it affects human health.

It has been demonstrated that atmospheric conditions significantly affect the temporal and spatial distribution of O_3 and other pollutants (Kinney 2018a, b). Higher outside temperatures resulted in an increase in O_3 and $PM_{2.5}$ precursor emissions. Higher temperatures and more sunshine accelerated the processes that create O_3. The effect of precipitation, winds, vertical mixing, and weather systems will influence the dispersion and movement of air pollutants in the atmosphere. Additional research suggests that future climate scenarios may result in higher O_3 concentrations in populated areas, even with constant emissions of anthropogenic air pollution.

Owing to the wide range of properties and chemical makeup of PMs, human exposure can result in a few health issues. The immediate- and short-term consequences include eye, nose, and throat discomfort, along with symptoms including coughing, sneezing, and dyspnoea. Long-term effects, such as the aggravation of respiratory diseases including asthma and chronic bronchitis, may arise with additional exposure (Pope et al. 2006). Reduced lung function would result from these disorders, especially in older adults and children. In addition, brief exposure increased the risk of various cardiovascular illnesses (CVD), heart attacks, and irregular heartbeats. Heart disease and associated CVD may occur because of extended exposure (Brook et al. 2010).

In fact, the connection between air pollution and its effects on health has been the subject of numerous studies. The most researched health outcomes in a systematic mapping review (from the screening of 3400 articles) were respiratory (mostly asthma and COPD) and cardiovascular disease (primarily stroke) issues (Dominski et al. 2021). Most air pollution was caused by incomplete combustion, which mixed gases and particulates. In fact, the pollutants that have been studied the most extensively were NO_2, $PM_{2.5}$, and PM_{10}. The largest sample size was the general population, which was followed by adults and children. Among the contaminants, $PM_{2.5}$'s impacts on health—especially CVDs—were the subject of frequent research.

Jeroen de Bont et al. (2022) conducted a thorough review and meta-analysis of the relationship between ambient air pollution and CVDs, emphasizing regional variations and susceptible subpopulations. Stroke (22 reviews), all-cause CVD mortality, and morbidity (19) were the most extensively researched outcomes. Higher exposure to ambient air pollution was found to be most strongly associated with all-cause CVD mortality and morbidity, stroke, blood pressure, and ischaemic heart disorders (IHDs). There is consistent evidence linking brief exposure to PMs to higher risks of hypertension, myocardial infarction (MI), and stroke (both fatal and nonfatal). Long-term $PM_{2.5}$ exposure was primarily linked to a higher risk of incident MI, atherosclerosis, hypertension, stroke, and stroke death.

Stronger correlations between these reports and susceptible subpopulations—particularly the elderly, cardiac patients, and those with higher weight status—were discovered in Asian countries. These findings highlight how critical it is to lower air pollution levels—particularly PMs—both globally and among populations that are particularly sensitive.

Children are more affected by pollution's negative effects on health than adults, especially in low- and middle-income nations. Furthermore, there is an unbreakable two-way relationship between climate change's effects on children's respiratory health and air pollution (Sly 2021). It is important to highlight that genetic, epigenetic, and metagenomics variables might alter an individual's susceptibility to diseases connected to exposure.

The collapse of the two World Trade Centre (WTC) towers following the terrorist attacks serves as an example of exposure to substantial PMs. First responders (FR; firefighters, medical personnel, and police officers), as well as thousands of civilians, were affected by the emission of dust, smoke, and vapours containing volatile chemicals in New York City. Studies conducted on the survivors' progeny reveal that they were progressively more vulnerable to age-related disorders affecting the neurological and cardiovascular systems (Mears et al. 2022).

The effects of PM exposure on the central nervous system (CNS) and brain health are a topic receiving more research (Hahad et al. 2020). It has been demonstrated that such an influence raises the risk of depression, neurodevelopmental disorders, stroke, dementia, Parkinson's disease, cognitive impairment, and other related problems. In fact, direct exposure to chemical and/or nonchemical stressors, including but not limited to PMs, can result in neurological problems. Positive initiatives towards enhancing human health include raising knowledge of the brain illnesses that are common in low- and middle-income countries and investing in the capacity for additional research connected to environmental health (Tshala-Katumbay et al. 2015).

Given that pregnancy is a time of rapid human growth and development, children's health may be especially compromised by a mother's exposure to PM. A study conducted by Johnson et al. (2021) examined the health consequences experienced by children following their prenatal exposure to ultrafine $PM_{0.1}$ and tiny $PM_{2.5}$. The study discovered correlations with disorders of the immune system, brain development, respiratory systems, and cardiometabolic health in youngsters. The biological mechanisms behind these deleterious effects, according to the authors, could be attributed to direct placental translocation of ultrafine particles, oxidative stress, and inflammation in the mother's body caused by both fine and ultrafine PM, epigenetic modifications, and possible endocrine effects that affect long-term health.

Particle size, source, and chemical makeup are some of the variables that must be considered in order to fully understand PM toxicity (Kelly and Fussell 2012). Furthermore, little is known about the underlying biological pathways of disease and susceptibility. However, new research indicates that inflammation and oxidative stress play a significant role in the pathophysiology of illnesses brought on by air pollution. This is because exposure to different air pollutants increases the generation of proinflammatory mediators and reactive oxygen species (ROS).

Regulatory bodies around the world have implemented air quality regulations to reduce PM levels and protect human health considering the known health hazards. Policies aimed at lowering maternal exposure and the health effects on kids and other vulnerable groups ought to be given top emphasis. Although $PM_{2.5}$ levels are controlled, it is known that people in lower socio-economic classes and the impoverished regions frequently face disproportionate exposures. Consequently, to reduce exposure for the public, and especially for vulnerable populations, preventative efforts that inform clinical and dietary advice, as well as neighbourhood/regional planning, are required.

People can then take steps to reduce their exposure, such as using air purifiers, remaining inside during times of high pollution, and supporting legislation that aims to improve the quality of the air. Global reduction of air pollution and its health effects depends on improved air quality management, which includes enhanced air quality monitoring and international partnerships backed by the US National Ambient Air Quality Standards (NAAQS) (Faridi et al. 2023).

2.4.2.3 Health Effects from Exposure to Volatile Organic Chemicals

As may be predicted, a class of organic compounds known as VOCs are those that readily evaporate into the atmosphere. Among VOCs, benzene, toluene,

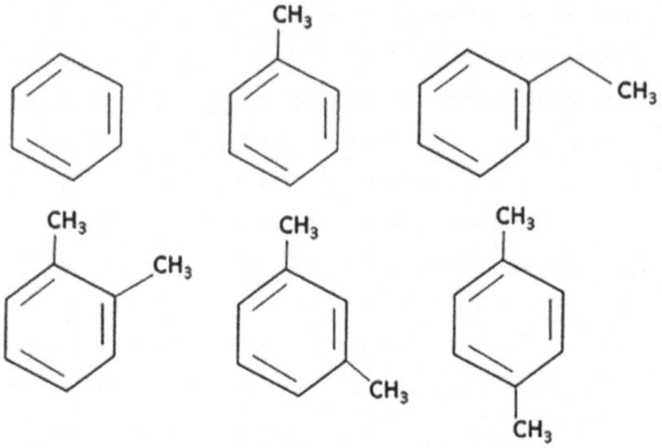

Fig. 2.6 Chemical structures of benzene, toluene, and ethylbenzene (upper three figures), and ortho-, meta-, and para-xylene. (Source: Montero-Montoya et al. 2018)

ethylbenzene, and xylenes are just a few examples of common and harmful environmental pollutants. The group was given the acronym BTEX because these four compounds are frequently found together, particularly in the vicinity of industrial regions (Montero-Montoya et al. 2018). The primary sources of VOCs and BTEX emissions are the combustion of fossil fuels and municipal trash. Nonetheless, mobile sources accounted for most VOC emissions in large cities. Since VOCs are so common, it is impossible to avoid human exposure to them, and harmful VOCs pose a major risk to health. The health effects of VOC pollution will worsen because of global warming (Fig. 2.6).

Because of the usage of goods like paints, cleansers, and furniture polishes, VOCs are also frequently discovered in indoor environments. Buildings with more insulation and sealing are expected to be more energy-efficient, which means there will be more indoor pollution. In addition, a greater amount of time is spent indoors by people engaging in various activities. Pollutant exposure inside is increased by such combinations. On the other hand, several health problems can arise from poor indoor air quality (USEPA 2018; https://www.epa.gov/indoor-air-quality-iaq).

The influence of indoor air pollution on populations with respiratory disorders was the subject of a recent review. There were 69 published studies in the review. According to the analysis, VOCs and PMs were the primary indoor air contaminants. Measurements of personal exposure revealed exposure to contaminants both indoors and outdoors, emphasizing the importance of home features and air exchange rates. There was a correlation between human exposure levels and temperature, humidity, educational attainment, air purifier use, and time spent near sources. Increased upper respiratory tract and asthma symptoms, as well as cancer, were linked to high VOC exposures (Maung et al. 2022).

A recent systematic review indicated a link between indoor VOC exposure and detrimental health effects (Liu et al. 2022). As per the review, the combined relative risk (RR) for leukaemia was 1.03 (95% CI: 1.01–1.05) for every 1 $\mu g/m^3$ increase in benzene and 1.25 (95% CI: 1.14–1.37) for every 0.1 $\mu g/m^3$ increase in butadiene. The pooled asthma relative risks were 1.08 (95% confidence interval: 1.02–1.14), 1.02 (95% confidence interval: 1.00–1.04), and 1.04 (95% confidence interval: 1.02–1.06) for every 1 $\mu g/m^3$ rise in benzene, toluene, and p-dichlorobenzene, in that order. When benzene concentration was increased by 1 μg/m^3, the pooled risk ratio for low birth weight was 1.12 (95% CI: 1.05–1.19).

Irritation of the respiratory system is a typical sign of VOC exposure. Breathing difficulties and coughing are examples of more severe side effects that might result from persistent discomfort. Furthermore, some VOCs can irritate the eyes and nose, resulting in symptoms like watering, redness, and congestion.

There are some VOCs that are known to induce nausea, headaches, and dizziness. Extended exposure could make these symptoms worse. Prolonged exposure to specific VOCs has been linked to effects on the central nervous system, such as neurological complaints and cognitive decline. Some VOCs are known or suspected carcinogens, including formaldehyde and benzene. Long-term exposure to certain substances at high concentrations may raise the risk of cancer.

Certain VOCs can have adverse effects on the liver and kidneys, particularly when exposure is chronic and at high concentrations. Moreover, pregnant women and developing foetuses may be particularly vulnerable to the effects of certain VOCs, which can impact foetal development and reproductive health (ATSDR, 2020) (https://www.atsdr.cdc.gov/toxprofiles/tp.asp?id=352&tid=62).

Exposure to VOCs can cause neurotoxicity. A recent report indicates that nonoccupational exposures were associated with adult depression, anxiety, and suicide (Dickerson et al. 2020).

The kind and severity of the health impacts depend on the exposure duration, as well as the quantities of volatile organic compounds. As a result, lowering VOC exposure can be a useful strategy for preventing diseases. Effective methods that use nanomaterials to remove VOCs include adsorption, catalysis, and photocatalysis. Furthermore, it is critical to advance understanding and expertise regarding the removal of VOCs to improve both indoor and outdoor environments, as well as global water supplies (David et al. 2021).

Additionally, it is advised to carry out research to close knowledge gaps that are crucial for addressing environmental injustice and health inequities in post-industrial cities, as well as safeguarding and maintaining vital freshwater resources (Miller et al. 2020). Furthermore, additional research is required for life course, longitudinal studies, studies that focus on neurotoxic exposures, studies that can quantify cumulative exposures in addition to shorter term exposures, and studies that lessen the risk of reverse causation.

WHO response to climate change and impact on health (12 October 2023: https://www.who.int/news-room/fact-sheets/detail/climate-change-and-health)

- WHO's Response to the Challenges Centred Around Three Main Objectives:
  - To promote actions that both reduce carbon emissions and improve health, supporting a rapid and equitable transition to a clean energy economy, ensuring that health is central to climate change mitigation policy, accelerating mitigation actions that bring the greatest health gains, and mobilizing the strength of the health community to drive policy change and build public support.

- To build better, more climate-resilient and environmentally sustainable health systems, ensuring core services, environmental sustainability, and climate resilience as central components of UHC and primary health care (PHC), supporting health systems to leapfrog to cheaper, more reliable, and cleaner solutions, while decarbonizing high-emitting health systems, and mainstreaming climate resilience and environmental sustainability into health service investments, including the capacity of the health workforce.
- To protect health from the wide range of impacts of climate change, assessing health vulnerabilities and developing health plans, integrating climate risk and implementing climate-informed surveillance and response systems for key risks, such as extreme heat and infectious disease, supporting resilience and adaptation in health-determining sectors such as water and food, and closing the financing gap for health adaptation and resilience.

• Future Prospects:
  - Climate change and the consequences of chemical toxins and human health belong to an eminent global problem. However, the two interacting components which are major contributors towards human health problems have seldom been investigated. Since the extent of the problem crosses national barriers, solutions to the problem need international effort. Such an international effort requires collaborations that are based on shared resources, transparency, accountability, and dedication. The effort can be initiated with a focus on the three recommendations from WHO, as described above. The effort should also be supported by research in mechanisms on causation. A research direction can be focused on how climate change affects the release and alteration of chemical toxins, and the extent of human exposure to the toxins. Another direction can be on how reduction from the use of fossil fuel affects climate change and the release of toxins. Effective management of the climate change problem will require better determination of exposure and health effects, as well as the effect of mitigation and adaptation efforts (Weber et al. 2023).

2.4.3 Biological Toxins

2.4.3.1 Introduction to Airborne Mycotoxins

Mycotoxins are fungal secondary metabolites, which pose a major threat to both human and animal health. These toxins contaminate agricultural products, including crops grown and stored for consumption, as well as processed foods, and feeds *Aspergillus*, *Fusarium*, *Penicillium*, and *Alternaria* genera are home to most toxigenic species of the 200 identified filamentous fungi found as contaminants (Tsitsigiannis et al. 2012). The presence of mycotoxins in food and feed is affected by factors such as climate, farming methods, and breaks in the agricultural-commercial supply chain. Mycotoxins are highly resistant to degradation caused by heat, physical, and chemical methods during processing (Patriarca and Pinto 2017). Mycotoxins can become airborne and pose a threat in indoor and agricultural environments. Exposure to airborne mycotoxins can occur in various occupational settings, and there is a growing interest in understanding their impact on human health (Robbins et al. 2000).

Exposure to airborne mycotoxins through inhalation, with activities related to animal care and management, shows the highest points of exposure. Symptoms and consequences associated with inhaling mycotoxins include irritation of mucous membranes, damage to epithelial tissues, systemic reactions such as fever, nausea, and fatigue, immunotoxicity, nephrotoxicity, liver damage, damage to the central nervous system, and an increased risk of cancer development (Marcelloni et al. 2024). Several studies have indicated that mycotoxins may contribute to the manifestation of sick building syndrome, characterized by symptoms like fever, headache, and asthenia, posing a potential risk of exposure to indoor occupants such as office workers (Straus 2009; Li and Yang 2004; Sarvtin 2023). Furthermore, there are reports suggesting that exposure to airborne mycotoxins could increase the risk of developing allergies (Jarvis and Miller 2005). Hospitalization complications related to biofilm-associated infections in medical devices, such as dialysis machines, are becoming more common and have raised rates of morbidity and mortality (Cooper et al. 2014). Filamentous moulds like *Fusarium* and *Aspergillus* can cause a wide range of illnesses, such as superficial, locally invasive, and disseminated infections. Their conidia are found in soil, water, and the air, and they are widely distributed across the environment (Dananche et al. 2017), including haemodialysis machines in the hospital environment. Another environment conducive to airborne mycotoxins is that of indoor damp surfaces and water-damaged areas. Many mycotoxins, including aflatoxins, T2 toxin, DON, ochratoxin A, gliotoxins, and sterigmatocystin, isolated from indoor environments have been linked to human health concerns (Viljoen and Claassen 2023). Another recent study reported on occupant morbidity and the prevalence of mycotoxins in indoor air and suggested that these are related. The indoor air condensate cytotoxicity test is a potentially useful method for evaluating building risks when it comes to moisture damage (Vaali et al. 2022).

The matter of damp or mouldy indoor spaces possibly causing and/or exacerbating various respiratory ailments has been addressed by the World Health Organisation in the form of guidelines on indoor air quality (WHO 2009).

Agricultural practices during harvest and post-harvest activities also contribute to airborne mycotoxin exposure. Furthermore, feed and grain handling in animal husbandry is significant sources of airborne mycotoxins. In healthcare settings, mycotoxin concentrations are generally low, but heating, ventilation, and air conditioning (HVAC) systems can act as a source of contamination (Haque et al. 2020; Marcelloni et al. 2024); this was also shown to be true for the air conditioning system of modern cars (Gołofit-Szymczak et al. 2023). The primary fungus linked to invasive lung infections was found to be *Aspergillus* (Gow et al. 2022). Exposure to airborne mycotoxins has also been identified in workers employed at waste sorting and treatment plants (Schlosser et al. 2020). Dust inhalation in the workplace has been recognized as an occupational exposure route for mycotoxins, with grain elevator workers as an example (Ndaw et al. 2021). Recently, consideration was also given to the air quality in public spaces (such as selected rooms in zoological gardens) (Plewa-Tutaj et al. 2024). Inhalation exposure to fungi outside of occupational settings can be generally categorized into fungi commonly found in outdoor air and building-associated fungi that thrive on damp building materials. Most fungi present in outdoor air consist primarily of two species belonging to the genus *Cladosporium*, alongside other *phylloplane* species, mushroom spores, and various pathogens responsible for plant diseases, while building-associated fungi constitute a narrower group of species that grow on the nutrients present in building materials and contents (Jarvis and Miller 2005). These fungi are often adapted to survive the specific levels of available water within the environment, but these will not be further discussed in this section of the chapter as the focus will remain on the airborne mycotoxins.

The various sources of exposure can result in the inhalation of several mycotoxins in variable concentrations and combinations. These factors have driven comprehensive sampling and analysis protocols to accurately assess the level of airborne mycotoxin exposure (Ndaw et al. 2021; Buiarelli et al. 2015). Exposure assessment of the aflatoxins, trichothecenes, ochratoxin A, and fusarium mycotoxins via inhalation has been described (Burg et al. 1981; Brera et al. 2002; Brasel et al. 2005; Smoragiewicz et al. 1993; Halstensen et al. 2006). When food and feed contaminated with the same mycotoxins are consumed, the levels of exposure through the inhalation route seem to be much lower than those through oral routes of exposure. However, mycotoxin inhalation is still considered an important route of exposure beyond the consumption of contaminated foods, as these toxins can be found in the environment even when the fungus is not visible, making it crucial to understand the mechanism underlying their induced toxicities.

2.4.3.2 Molecular Toxicity of Airborne Mycotoxins

Airborne mycotoxins have been linked to various adverse effects, posing a significant threat to human health. Understanding the mechanisms of toxicity is crucial for effective risk assessment and intervention strategies. Airborne mycotoxins have the potential to induce both chronic and acute health effects through inhalation and subsequent entry into the lymphatic system and bloodstream where they affect the innate immune response by dysregulating macrophages (Cleary 2022), disrupting protein synthesis, heightening the response to bacterial endotoxins, and inhibiting the clearance of particles from the lungs (Awuchi et al. 2022). Trichothecenes, fumonisins, and zearalenone are the three major and frequently occurring fusariotoxins (Escrivá et al. 2017). The following mycotoxins have been classified by IARC based on their carcinogenicity to humans, with aflatoxins B1, B2, G1, G2, M1 classified a carcinogenic (Group 1), ochratoxin A, fumonisin B1, fumonisin B2, fusarin C, sterigmatocystin classified as possibly carcinogenic (Group 2B), and deoxynivalenol, patulin, citrinin, zearalenone, fusarenon X classified as nonclassifiable as carcinogenic (Group 3) (IARC 1987, 1993, 2006, 2012, 2019).

Trichothecenes T-2 toxin and deoxynivalenol (DON)
Trichoderma, Myrothecium, Stachybotrys, Fusarium, and *Spicellum* are a few examples of the several filamentous fungus species that generate a structurally similar group of mycotoxins known as trichothecenes (TCT). Trichothecenes have been detected in air samples obtained during various agricultural activities, in ventilation systems of private residences and office buildings, and on the walls of houses in high-humidity areas. Some reports suggest a link between trichothecenes and the onset of sick building syndrome (Assouline-Dayan et al. 2002; Shoemaker and House 2006).

The most toxic TCT is T-2 toxin with both in vitro and in vivo investigations substantiating this claim. The lipophilic nature of this toxin allows it to be easily absorbed through the pulmonary mucosa, based on aerosol deposition data in mice, while inhalation of T-2 mycotoxin is at least ten times more toxic than systemic administration and at least twenty times more toxic than dermal administration (Creasia et al. 1987). Numerous studies conducted over recent decades have focused on elucidating the cytotoxic and genotoxic effects of T-2 toxin. At the cellular level, T-2 primarily inhibits protein synthesis, consequently leading to secondary disruptions in DNA and RNA synthesis. It is postulated that T-2 binds and deactivates peptidyl transferase activity at the transcription site, thereby halting protein synthesis. The principal molecular target of T-2 is the 60S ribosomal unit, where it impedes polypeptide chain initiation. Additionally, T-2 toxin intoxication is associated with oxidative stress, characterized by heightened lipid peroxidation, damage to nuclear and mitochondrial DNA, disruptions in cellular signalling, and inflammatory pathways. Furthermore, the T-2 disrupts the cell cycle and triggers apoptosis (Janik et al. 2021). These toxic effects were demonstrated to inhibit alveolar macrophage function with authors concluding that T-2 toxin likely inhibits macrophage activation by disrupting protein synthesis. Inhibition of macrophage monokines such as lymphocyte activating factor (interleukin (IL)-1), which plays a role in regulating the immune response to foreign antigens, is detrimental to the individual's immune status. Additionally, inhibition of macrophage phagocytosis would render an individual increasingly susceptible to opportunistic infections (Gerberick et al. 1984).

Deoxynivalenol, a member of the Trichothecene family of toxins, was also found in air samples (Ngundi et al. 2006). The biological toxicity of DON has been studied in both in vitro and in vivo settings. DON induces several physiological abnormalities, including digestive disorders, reproductive issues, and disruptions to the endocrine system. The mechanism of DON toxicity involves inhibition of protein synthesis, where the toxin interacts with the peptidyl transferase enzyme, binding to the 60S ribosomal subunit and inhibiting translation, leading to ribotoxic stress. Additionally, the dysregulation of several mitogen-activated protein kinases (MAPK), which mediate various effects of the toxin, such as oxidative stress, apoptosis, inflammatory responses, and endocrine disruptions, was reported. Furthermore, DON induces DNA damage through direct toxin action and via mechanisms like the formation of DNA adducts from free radicals. DON also triggers lipid peroxidation and malondialdehyde (MDA) formation, which reacts with deoxyguanosine and deoxyadenosine in DNA, resulting in the formation of DNA adducts, notably the mutagenic M1G (pyrimidol(1,2-a]purin-10(3H)-one) (Janik et al. 2020). Studies using primary human airway epithelial cells of nasal (hAECN) and bronchial origin (hAECB), and immortalized human bronchial (16HBE14o-) and alveolar (A549) epithelial cell lines showed that DON induces disruption to cytokines IL-6 and IL-8 and some degree of cytotoxicity (Ferreira Lopes et al. 2017). A study by Ndaw et al. (2021) could not show a correlation between the DON levels as measured in air samples and the levels in urine samples of grain elevator workers, emphasizing that for biomonitoring to be more accurate, inhalation together with dermal intake should be considered. A study by Limay-Rios and Schaafsma (2021) showed a linear correlation between the mycotoxin content in grain and grain dust, with DON showing the strongest correlation. The results suggest DON in grain bulk can be estimated by modelling levels in grain based upon levels in grain dust. These findings have practical implications for agricultural practices and occupational health, contributing to efforts to ensure food safety and protect human health.

Ochratoxin (OTs)

Penicillium and *Aspergillus* species produce ochratoxins, and these mycotoxins have three forms of metabolites: ochratoxin A (OTA), ochratoxin B (OTB), and ochratoxin C (OTC). Ochratoxin A (OTA) is a toxic mycotoxin produced by *A. niger*, *A. ochraceus*, *Penicillium verrucosum*, and *A. carbonarius*. Exposure to the mycotoxin, OTA, through inhalation of dust has been confirmed in agricultural and household environments (Skaug et al. 2001; Wontner-Smith et al. 2014; Richard et al. 1999).

Numerous studies have highlighted the adverse effects of OTA on both human and animal health. OTA induces various toxic effects, including nephrotoxicity, hepatotoxicity, genotoxicity, teratogenicity, immunotoxicity, and neurotoxicity. The mechanisms of OTA action are diverse, including G2/M phase cell cycle arrest, apoptosis, necrosis, inhibition of cell division processes, and protein synthesis. OTA has been shown to induce excessive generation of reactive oxygen species (ROS), leading to oxidative stress and ROS-mediated apoptosis, as well as DNA adducts and DNA single-strand breaks. Additionally, OTA treatment decreases glutathione (GSH), an important antioxidant, and increases lipid peroxides and nitric oxide (NO) generation (Janik et al. 2020).

Eosinophilic chronic rhinosinusitis (ECRS) encompasses various subtypes, including allergic fungal sinusitis, nonallergic fungal rhinosinusitis, aspirin-exacerbated ECRS, and superantigen-induced ECRS. A recent study aimed to assess the impact of ochratoxin A (OTA), one of the most prevalent mycotoxins, on the release of proinflammatory cytokines such as interleukin-6 (IL-6) and the chemokine interleukin-8 IL-8 from nasal epithelial cell cultures obtained from individuals with and without ECRS. The findings indicate that OTA, a widely distributed mycotoxin, exhibits a potent proinflammatory effect on nasal epithelial cell cultures, leading to the release of IL-6 and IL-8 (Cremer et al. 2012). Rats treated with OTA exhibited various pulmonary abnormalities, including alveolar congestion, alveolar cell hyperplasia, prominent alveolar septal vessels, variable intensity loss of alveolar architecture, intraparenchymal inflammatory infiltration, intraparenchymal hyperaemic vessels, respiratory epithelial proliferation, perivascular and peribronchial inflammation, pneumonic infiltration, distorted appearance of lung parenchyma, and emphysematous areas (Okutan et al. 2004). In a recent study, Kharboush et al. (2024) provided novel insights into the molecular mechanisms underpinning the cytotoxic and potential carcinogenic effects of OTA in foetal lung fibroblast cells. This study revealed the upregulation of onco-miR-155, accompanied by a correlation with widespread DNA hypermethylation, previously linked to upregulation in individuals diagnosed with lung cancer (Kharboush et al. 2024).

Aflatoxins

Among the 18 different types of aflatoxins, four are commonly encountered: B1 (AFB1), B2 (AFB2), G1 (AFG1), and G2 (AFG2). AFB1 is notably recognized as the most prevalent contaminant of foods and possesses the highest carcinogenic and mutagenic potential of its family. Airborne aflatoxin was shown to be generated during agricultural processes, particularly grain handling across different continents (Ghosh et al. 1997; Viegas et al. 2012). Data suggests that inhalation of AFB1 poses an additional risk in this occupational setting. Studies in rat lung tissue after exposure to AFB1 highlighted the increased presence of inflammatory cells, along with dilation of the alveolar sacs filled with exudates as a pathological consequence (Awuchi et al. 2022). Additionally, AFB1 was shown to impair motile and chemosensory functions of airway cilia, contributing to pathogenesis of fungal airway diseases (Lee et al. 2016).

The inhalation of dust particles contaminated with AFB1 can lead to significant local exposure of the nasal mucosa. The bioactivation and toxicity of AFB1 in the nasal mucosa following intranasal administration of the mycotoxin along with translocation of the mycotoxin from the nasal mucosa to the brain via olfactory neurons were studied in rats. The findings indicated that intranasal administration of AFB1 resulted in the formation of tissue-bound metabolites in sustentacular cells, some cells of Bowman's glands, and a subset of neuronal cells in the olfactory mucosa (Larsson and Tjälve 2000). Conversely, in the respiratory nasal mucosa, AFB1 bioactivation was selectively observed in mucous cells. Intranasal instillation of AFB1 led to disorganized undulating olfactory epithelium, characterized by injured neuronal and sustentacular cells. In the respiratory epithelium, selective destruction of mucous cells was observed. Further analysis using β-spectrometry and autoradiography with tape sections of the rat heads administered with 3H-AFB1 intranasally revealed the transport of AFB1 and/or AFB1 metabolites along the axons of primary olfactory neurons to their terminations in the glomeruli of the olfactory bulb. These results suggest that the materials transported in the olfactory nerves represent AFB1 and/or some of its nonreactive metabolites. The authors concluded that the application of AFB1 on the nasal mucosa in rats leads to significant local bioactivation of the mycotoxin in this tissue and translocation of AFB1 and/or its metabolites to the olfactory bulb via olfactory neurons (Larsson and Tjälve 2000). Recently, the exposure to airborne mycotoxins, OTA, gliotoxin, sterigmatocystin, and AFB1 was measured at various waste management facilities, with quantifiable results obtained in only 11% of the 94 air samples collected. While gliotoxin and ochratoxin A levels could not be quantified, the levels of sterigmatocystin and AFB1 were all below 1 ng/m^3 (Schlosser et al. 2020), suggesting that no additional prevention strategies were required at these waste management facilities against the mentioned mycotoxins.

Within liver cells, the toxin undergoes metabolism by microsomal-mixed function oxidase (MFO) enzymes, which belong to the cytochrome P450 (CYP450) superfamily. AFB1 is converted into more toxic, highly reactive 8,9-exo-epoxide and 8,9-endo-epoxide metabolites. The 8,9-exo-epoxide exhibits a strong affinity for binding to DNA, resulting in the formation of the 8,9-dihydro-8-(N′-guanyl)-9-hydroxy-AFB1 (AFB1-N-Gua) adduct, leading to DNA mutations. Moreover, the 8,9-exo-epoxide binds to other macromolecules such as RNA and proteins, resulting in the inhibition of RNA, DNA, and protein syntheses, ultimately disrupting cellular functions. This epoxide can undergo enzymatic and nonenzymatic conversion into AFB1-8,9-dihydrodiol, which can bind to proteins like albumin, or it can be converted into aflatoxin dialdehyde and excreted via urine through the action of aflatoxin aldehyde reductase. Additionally, in the endoplasmic reticulum of liver cells, AFB1 is hydroxylated to form less toxic metabolites, including aflatoxin M1, aflatoxin Q1, and aflatoxin P1 (Janik et al. 2020). The presence of CYP450 enzymes in the lungs (Hukkanen et al. 2002) suggests the potential for similar metabolic reactions as those observed in the liver. A study by Kelly et al. (1997) provided evidence that human lung microsomes have the capability to activate AFB1, leading to the formation of the exo-AFB1-8,9-epoxide, and it is likely that CYP enzymes belonging to the 3A subfamily are responsible for this activity. The lower level of AFB1 activation observed in human lung compared to that in human liver can be attributed to the lower abundance of CYP-containing cells in the lung. However, the potential for in situ AFB1 activation and the resulting carcinogenic risk cannot be overlooked, particularly in occupational settings where inhalation of AFB1-contaminated dusts is prevalent (Kelly et al. 1997). This raises concerns regarding the possibility of contributing to the development of lung cancer. The human respiratory tract possesses high activity of CYP2A13 and CYP1A2 and was shown to exert an influential role in human lung tumorigenesis upon exposure to airborne AFB1 and subsequent bioactivation. These findings offer insights into the aetiology of airborne AFB1 contamination and may aid in identifying biomarkers associated with human lung cancer caused by AFB1 exposure (Zhang et al. 2014; Van Vleet et al. 2002). Further evidence supporting the role of CYP2A13 in the respiratory transformation of AFB1 showed that bronchial epithelial cells overexpressing this family member of CYPs showed elevated levels of apoptosis and cytotoxicity. This heightened effect was linked to the increased expression of proapoptotic proteins like cleaved PARP, activated caspase-3, and Bax, along with the decreased expression of anti-apoptotic proteins such as Bcl-2 and phosphorylated-Bad. Moreover, an escalation in DNA adduct formation was observed in CYP2A13 cells following AFB1 treatment (Yang et al. 2012). These collective findings suggest that AFB1 exerts adverse effects in respiratory diseases mediated by CYP2A13.

Fusarium Mycotoxin: Fumonisins and Zearalenone (ZEA)

Fumonisins (FBs) are a diverse group of mycotoxins produced by Fusarium spp. and are commonly found as natural contaminants of agricultural produce. FBB1 is the best studied and most toxic of the family. In maize-growing areas where FB fumonisin contamination is endemic, grain dust was shown to contain FBB1.

The mode of action of FBB1 relies on its structural similarity to sphinganine. This similarity leads to the accumulation of sphingoid bases and disrupts the biosynthesis or function of sphingolipids through competitive inhibition of the enzyme. Consequently, a range of toxic effects and diseases have been associated with FBB1 exposure, including hepatocarcinogenicity, hepatotoxicity, nephrotoxicity, and cytotoxicity. Additionally, diseases such as leucoencephalomalacia in horses and pulmonary oedema in horses and swine have been documented as outcomes of FBB1 toxicity (Chen et al. 2021).

Fumonisin B1 was shown to accumulate over time in lung where an increase in sphinganine-1-phosphate and a decrease in glucosylceramides were observed (Guerre et al. 2022). The pulmonary effects of FBB1 primarily manifest as pulmonary oedema in animals, with pigs being particularly susceptible to the severe lung effects of FBB1. The predominant pathological findings include the accumulation of serous or serofibrinous exudate in interlobular tissue or alveolar lumina, as well as thickening of interalveolar septa due to epithelial hyperplasia and/or fibrin accumulation. These effects are attributed to FBB1-induced cardiovascular disturbances, wherein FBB1 blocks sphingosine-mediated L-type calcium channels, leading to heart failure and subsequent pulmonary oedema. FBB1 also leads to reductions in heart rate, myocardial contractility, and arterial blood pressure. Disturbances in the sphingolipid biosynthetic pathway caused by FBB1 affect the cardiovascular system of pigs, resulting in elevated concentrations of sphinganine (Sa) and sphingosine (So), which induce systemic hypotension. Additionally, increased levels of sphingosine-1-phosphate contribute to pulmonary hypertension (Chen et al. 2021).

Zearalenone (ZEA) is a known myco-oestrogen and endocrine disruptor and has been associated with various other detrimental impacts, including developmental toxicity, immunotoxicity, and genotoxicity. Emerging evidence suggests that these effects cannot solely be attributed to ZEA's oestrogenic potency; oxidative stress may play a significant role as a mediator of these observed toxic effects (So et al. 2014). A study by So et al., demonstrated that exposure to ZEA leads to significant alterations in gene expression profiles in lung cells. Pathway analysis revealed various cellular processes to be affected, including impaired DNA damage responses, cell cycle arrest, down-regulation of inflammatory responses, and alterations in epigenetic profiles. Experiments showed ZEA treatment at a concentration of 40μM decreased cell viability, induced apoptosis, and promoted the generation of reactive oxygen species (ROS) in a time-dependent manner. The immunosuppressive effects of ZEA were also demonstrated through the suppression of lipopolysaccharide (LPS)-induced expression of proinflammatory cytokines (IL-6, IL-8, and IL-1β). At the epigenetic level, the study found a marked decrease in global DNA methylation levels after exposure to ZEA. These findings collectively suggest that ZEA induces a wide range of toxic effects, with ROS potentially playing a central role in ZEA-induced cell death (So et al. 2014).

2.4.4 Mycotoxins and Toxicity in Respiratory Diseases

Several case studies have supported the pathologic effects of the toxigenic fungi *Stachybotrys* implicated in pulmonary haemorrhage (Elidemir et al. 1999; Novotny and Dixit 2000; Flappan et al. 1999; Hodgson et al. 1998). T-2 commonly produced by *Stachybotrys* impacts the vascular system, leading to dilation and swelling of microvessels, as well as damage to the plasma membrane and the blood vessel wall. The mechanism behind mycotoxin-induced lung injury remains unclear. While trichothecenes produced by *Stachybotrys* are potent inhibitors of protein synthesis, this mechanism alone cannot fully account for pulmonary haemorrhage. Another hypothesis suggests that trichothecenes inhibit collagen synthesis in rapidly growing young lungs, leading to increased capillary fragility. However, this hypothesis lacks support from animal models. Some animal studies with *Stachybotrys* have demonstrated haemorrhagic changes in the lungs, but these exposures involved intratracheal or intranasal instillation of large quantities of spores or toxins. It remains uncertain whether the haemorrhagic exudes result from the toxic effects of the mycotoxin or compensatory inflammatory mechanisms (Fung and Clark 2004).

Viral infections represent a major risk factor for upper respiratory symptoms, impacting areas such as the nasal cavity, paranasal sinuses, pharynx, and larynx (Jain et al. 2001), while bacterial infections often develop after viral infections and affect upper and lower respiratory systems (Khan et al. 2015; Bosch et al. 2013). *Bordetella bronchiseptica* (*B. bronchiseptica*) and *Pasteurella multocida* (*P. multocida*) are pathogenic bacteria commonly found in the airways, contributing to various respiratory diseases such as pneumonia, an inflammatory condition affecting the lung's alveoli (Mattoo and Cherry 2005; Wilkie et al. 2012). In pigs infected with *B. bronchiseptica* and exposed to FBB_1 (10 mg/kg, in vivo), clinical signs such as mild serous nasal discharge, sneezing, panting, and hoarseness were observed, unlike those not exposed to FBB_1 (Pósa et al. 2011). Additionally, *P. multocida* infection exacerbates the risk of pneumonia and the severity of associated pathological changes. Although neither FBB1 exposure (0.5 mg/kg, in vivo) nor *P. multocida* infection alone significantly affected weight gain or caused serious clinical signs or lung lesions, co-treatment with FBB1 and *P. multocida* resulted in delayed growth, induced cough, increased production of bronchoalveolar lavage fluid (BALF) cells (such as macrophages and lymphocytes), and prominent lung lesions accompanied by elevated levels of proinflammatory cytokines, such as TNF-α, IFN-γ, and IL-18 (Halloy et al. 2005; Pósa et al. 2011; Kovács et al. 2016). Although these experiments were conducted in pigs, there is a pressing need for additional research to comprehensively understand the interactions between mycotoxins and pathogen infections, particularly focusing on host defence mechanisms and disease progression.

The trichothecene mycotoxin DON is known to dysregulate mucosal and systemic immunity. DON interferes with the murine immune response to viral respiratory infection. DON enhanced reovirus-induced interferon (IFN)-β and IFN-γ mRNA responses in the lung, while concurrently suppressing mRNA expression for IFN-α, IFN-αβ receptor (IFNAR), and IFN-γ receptor (IFNGR). Additionally, DON impaired the induction of two type 1 IFN-dependent antiviral genes, namely, double-stranded RNA–activated protein kinase R (PKR) and oligoadenylate synthase 2 (OAS2). In mice, respiratory reovirus infection resulted in mild bronchopneumonia, which was significantly exacerbated by DON, as evidenced by pronounced inflammatory cell infiltration, marked alveolar damage, and a higher volume density of intraepithelial mucosubstances in pulmonary airways. DON markedly intensified viral-induced elevations of protein, MCP-1, TNF-α, and inflammatory cells in the bronchoalveolar lavage fluid (BALF). Furthermore, DON exposure upregulated the induction of reovirus-specific immunoglobulin A (IgA) in BALF, faecal pellets, and serum. The effect of DON on BALF IgA was preceded by elevated IL-6 expression and secretion in the lung. Overall, these findings suggest that DON compromises resistance to respiratory viral infection. The reduced expression of IFNAR and type 1 IFN-mediated genes in the lung may contribute to DON impairment of pulmonary reovirus clearance, while exacerbation of bronchopneumonia and IgA responses corresponds to increased MCP-1, TNF-α, and IL-6 expression (Li et al. 2007). A separate study on T-2 pretreatment markedly exacerbated bronchiolitis with alveolitis during respiratory reovirus infection. T-2 toxin inhibited reovirus-induced elevation of interferon-gamma (IFN-γ) in BALF, but increased the production of interleukin-6 (IL-6) and monocyte chemoattractant protein-1 (MCP-1). Pretreatment with T-2 toxin also suppressed reovirus-specific mucosal immunoglobulin A (IgA) responses in the lung and enteric tract, while enhancing serum IgA and IgG responses. Overall, T-2 toxin increased lung viral burden, bronchopneumonia, and pulmonary cellular infiltration in reovirus-infected mice. These effects may be attributed to reduced levels of alveolar macrophages and altered cytokine and mucosal Ig responses (Li et al. 2006).

Aflatoxin B1 is known to modulate immune responses in mammals and has a significant impact on swine influenza virus (SIV), a major pathogen affecting both animals and humans. In vitro studies have revealed that even low concentrations of AFB1 (ranging from 0.01 to 0.25μg/ml) markedly enhance SIV replication, as evidenced by increased viral titres, matrix protein (M) mRNA, and nucleoprotein (NP) levels in MDCK cells, A549 cells, and porcine alveolar macrophages (PAMs). In vivo experiments demonstrated that exposure to AFB1 at doses ranging from 10 to 40μg/kg exacerbates SIV infection in mice, resulting in significantly higher lung virus titres, viral M mRNA levels, NP levels, lung indexes, and more severe lung damage. Further investigations revealed that AFB1 upregulates Toll-like receptor 4 (TLR4), but not other Toll-like receptors, in SIV-infected PAMs. Additionally, AFB1 activates TLR4 signalling, leading to increased phosphorylation of NFκB p65 and release of TNF-α in PAMs and mice. Conversely, TLR4 knockdown or treatment with BAY 11-7082, a specific inhibitor of NFκB, inhibited AFB1-induced SIV replication and inflammatory responses in PAMs. Moreover, administration of a TLR4-specific antagonist, TAK242, or genetic knockout of TLR4 attenuated AFB1-mediated promotion of SIV replication, inflammation, and lung damage in mice. These authors concluded that exposure to AFB1 exacerbates SIV replication, inflammation, and lung damage by activating TLR4-NFκB signalling pathway (Sun et al. 2018).

Such investigations are essential not only for optimizing agricultural production but also for shedding light on the effects of respiratory pathogens and their interactions in humans. These crucial findings underscore the importance of monitoring and addressing the potential health risks associated with mycotoxin exposure and concurrent pathogen infections in both agricultural settings and human populations.

2.5 Conclusion

Therefore, environmental pollutions are found to be increasing due to climate change resulting in the reduction in the dispersion of air pollutants from the interference with convection currents, in addition to the enhancement in the formation of photochemical pollutants, and seasonal variation that increases the release of pollens and fungal spores. Acid precipitation damages freshwater ecosystems and increases chemical reactions of the pollutants, and the increase in water evaporation rates will also lead to an increase in aquatic chemical pollutants, which leads to eutrophication and formation of toxic algal bloom.

Moreover, numerous studies have consistently detected mycotoxins in both indoor and outdoor samples, and the type of toxins and the extent of exposure vary depending on factors such as the working environment, products handled, or the nature of the work performed. The exposure to multiple mycotoxins at different concentrations and combinations is common, emphasizing the importance of employing comprehensive sampling and analytical protocols to accurately assess and characterize occupational risks.

Agricultural settings where post-harvest practices and animal husbandry techniques are carried out are associated with the highest exposure level of airborne mycotoxins, yet compliance with personal protective equipment usage remains a challenge. Further field investigations are imperative to corroborate existing findings and identify additional sectors or tasks susceptible to mycotoxin inhalation exposure. Such information is pivotal for occupational hygienists and technicians to develop and implement effective prevention and control measures. Determining the real exposure to these mycotoxins is difficult, and with climate change, mycotoxins are predicted to become a very real health concern. The diversity of sources results in the inhalation of several mycotoxins in variable concentrations and combinations, necessitating comprehensive sampling and analysis protocols for accurate risk assessment. The consistent use of personal protective equipment, notably respiratory protective devices, should be encouraged under these conditions given health risks posed. In certain environments, additional prevention strategies need to be implemented against these airborne mycotoxins to mitigate the health risks.

Understanding the molecular toxicity of airborne mycotoxins is crucial for effective risk assessment and mitigation strategies. Airborne mycotoxins most extensively studied include AFB1, DON, ZEA, FBB1, and OTA. A wide array of techniques was employed to explore mycotoxin toxicity in different models of the respiratory system that investigated oxidative stress, cytotoxicity, and cell death mechanisms and transcriptomic analyses including miRNA profiles. AFB1 stands out as the major toxin investigated with its biotransformation implicated in lung cancer.

Mycotoxins were also found to exacerbate infections through dysregulation of the immune system in respiratory organs. Continued research is needed to comprehensively understand the interactions between mycotoxins and pathogens, host defence mechanisms, and disease progression, highlighting the importance of monitoring and addressing the potential health risks associated with mycotoxin exposure.

Despite advancements in understanding the mechanisms involved in airborne mycotoxin toxicity, their pervasive and significant health impact persists globally. As regulatory approaches to mycotoxins vary worldwide, strategic initiatives to prevent occupational exposure and focused research should be prioritized to identify ways to mitigate the levels of airborne mycotoxins present.

References

Abbafati C et al (2020) Global burden of 369 diseases and injuries in 204 countries and territories, 1990–2019: a systematic analysis for the Global Burden of Disease Study 2019. Lancet 396(10258):1204–1222. https://doi.org/10.1016/S0140-6736(20)30925-9

Abdelhak S, Menard Y, Artigas J (2023) Effects of global change on the ability of stream biofilm to dissipate the herbicide glyphosate. Environ Pollut 324:121406. https://doi.org/10.1016/j.envpol.2023.121406

Agency for Toxic Substances and Disease Registry (ATSDR) (2020) Toxicological profile for volatile organic compounds (VOCs). Retrieved from https://www.atsdr.cdc.gov/toxprofiles/tp.asp?id=352&tid=62

Ahmady-Birgani H, Agahi E, Ahmadi SJ, Erfanian M (2018a) Sediment source fingerprinting of the Lake Urmia Sand Dunes. Sci Rep 8(1):206. Published 2018 Jan 9. https://doi.org/10.1038/s41598-017-18027-0

Ahmady-Birgani H et al (2018b) Sediment source fingerprinting of the lake Urmia Sand Dunes. Scientific Report 2017 8(1):1–15. https://doi.org/10.1038/s41598-017-18027-0

Akin J (2011) Homeostatic processes for thermoregulation. Nat Educ Knowl 3(10):7. https://www.nature.com/scitable/knowledge/library/homeostatic-processes-for-thermoregulation-23592046/

Al-Delaimy W (2020) Vulnerable populations and regions: Middle East as a case study. In: Al-Delaimy W, Ramanathan V, Sanchez Sorondo M (eds) Health of people, health of planet and our responsibility: climate change, air pollution and health. Springer Open, Cham, pp 121–133. https://doi.org/10.1007/978-3-030-31125-4

Allen MD et al (2019) Suboptimal hydration remodels metabolism, promotes degenerative diseases, and shortens life. JCI Insight 4(17). https://doi.org/10.1172/JCI.INSIGHT.130949

Altizer S, Bartel R, Han BA (2011) Animal migration and infectious disease risk. Science 331:296–302. https://doi.org/10.1126/science.1194694

Aminharati F et al (2018) The effect of environmental parameters on the incidence of Shigella outbreaks in Yazd province, Iran. Water Supply 18(4):1388–1395. https://doi.org/10.2166/ws.2017.205

Amster E (2021) Public health impact of coal-fired power plants: a critical systematic review of the epidemiological literature. Int J Environ Health Res 31(5):558–580. https://doi.org/10.1080/09603123.2019.1674256

António DC et al (2018) Toward climate change impact: vectors carrying viral infection What we should know. JRC Tech Report. https://doi.org/10.2760/067022

Archer VE (1990) Air pollution and fatal lung disease in three Utah counties. Arch Environ Health 45(6):325–334. https://doi.org/10.1080/00039896.1990.10118751

Armstrong LE et al (2024) Low daily water intake profile-is it a contributor to disease? Nutr Health 30:435. https://doi.org/10.1177/02601060241238826

Assouline-Dayan Y, Leong A, Shoenfeld Y, Gershwin ME (2002) Studies of sick building syndrome. Iv mycotoxicosis. J Asthma 39:191–201

Awuchi CG, Ondari EN, Nwozo S, Odongo GA, Eseoghene IJ, Twinomuhwezi H, Ogbonna CU, Upadhyay AK, Adeleye AO, Okpala COR (2022) Mycotoxins' toxicological mechanisms involving humans, livestock and their associated health concerns: a review. Toxins 14:167

Ayotte JD, Szabo Z, Focazio MJ, Eberts SM (2011) Effects of human-induced alteration of groundwater flow on concentrations of naturally-occurring trace elements at water-supply wells. Appl Geochem 26:747–762

Baccini M et al (2011) Impact of heat on mortality in 15 European cities: attributable deaths under different weather scenarios. J Epidemiol Community Health 65(1):64–70. https://doi.org/10.1136/jech.2008.085639

Ballester J et al (2023) Heat-related mortality in Europe during the summer of 2022. Nat Med 29(7):1857–1866. https://doi.org/10.1038/s41591-023-02419-z

Bănăduc D et al (2022) Freshwater as a sustainable resource and generator of secondary resources in the 21st century: stressors, threats, risks, management and protection strategies, and conservation approaches. Int J Environ Res Public Health 19(24). https://doi.org/10.3390/IJERPH192416570

Barcellos C, Sabroza PC (2001) The place behind the case: leptospirosis risks and associated environmental conditions in a flood-related outbreak in Rio de Janeiro. Cad Saude Publica 17 Suppl:59–67. https://doi.org/10.1590/s0102-311x2001000700014

Basu R (2009) High ambient temperature and mortality: a review of epidemiologic studies from 2001 to 2008. Environ Health 8(1). https://doi.org/10.1186/1476-069X-8-40

Bharti AR et al (2003) 'Leptospirosis: a zoonotic disease of global importance', Lancet Infectious Diseases, vol 3. Elsevier, pp 757–771. https://doi.org/10.1016/S1473-3099(03)00830-2

Bolan S, Padhye LP, Jasemizad T, Govarthanan M, Karmegam N, Wijesekara H et al (2024) Impacts of climate change on the fate of contaminants through extreme weather events. Sci Total Environ 909:168388

Bonyadi Z et al (2020) Quantification of mortality and morbidity attributed to the ambient air criteria pollutants in Shiraz city, Iran. Chemosphere 257:127233. https://doi.org/10.1016/J.CHEMOSPHERE.2020.127233

Borgomeo E et al (2018) The water-energy-food Nexus in the Middle East and North Africa Scenarios for a Sustainable Future. The World Bank, Washington, D.C.. Available at: www.worldbank.org/water. Accessed: 5 Apr 2021

Bosch AA, Biesbroek G, Trzcinski K, Sanders EA, Bogaert D (2013) Viral and bacterial interactions in the upper respiratory tract. PLoS Pathog 9:e1003057

Bourque F, Van Der Ven E, Malla A (2011) A meta-analysis of the risk for psychotic disorders among first- and second-generation immigrants. Psychol Med 41(5):897–910. https://doi.org/10.1017/S0033291710001406

Boxall ABA, Hardy A, Beulke S, Boucard T, Burgin L, Falloon PD et al (2009) Impacts of climate change on indirect human exposure to pathogens and chemicals from agriculture. Environ Health Perspect 117(4):508–514. https://doi.org/10.1289/ehp.0800084

Brasel TL, Martin JM, Carriker CG, Wilson SC, Straus DC (2005) Detection of airborne Stachybotrys chartarum macrocyclic trichothecene mycotoxins in the indoor environment. Appl Environ Microbiol 71(11):7376–7388. https://doi.org/10.1128/AEM.71.11.7376-7388.2005

Brera C, Caputi R, Miraglia M, Iavicoli I, Salerno A, Carelli G (2002) Exposure assessment to mycotoxins in workplaces: Aflatoxins and ochratoxin A occurrence in airborne dusts and human sera. Microchem J 73:167–173

Brook RD, Rajagopalan S, Pope CA 3rd et al (2010) Particulate matter air pollution and cardiovascular disease: an update to the scientific statement from the American Heart Association. Circulation 121(21):2331–2378. https://doi.org/10.1161/CIR.0b013e3181dbece1

Buiarelli F, Di Filippo P, Riccardi C, Pomata D, Rumolo E, Giannetti L, Neri B (2015) Analytical method for the determination of mycotoxins in indoor/outdoor airborne particulate matter by HPLC-MS-MS. Int J Environ Anal Chem 95:713–729

Burg WR, Shotwell OL, Saltzman BE (1981) Measurements of airborne aflatoxins during the handling of contaminated corn. Am Ind Hyg Assoc J 42:1–11

Burnett RT et al (2014) An integrated risk function for estimating the global burden of disease attributable to ambient fine particulate matter exposure. Environ Health Perspect 122(4):397–403. https://doi.org/10.1289/ehp.1307049

Burnett R et al (2018) Global estimates of mortality associated with long-term exposure to outdoor fine particulate matter. Proc Natl Acad Sci USA 115(38):9592–9597. https://doi.org/10.1073/pnas.1803222115

Byrne C et al (2006) Continuous thermoregulatory responses to mass-participation distance running in heat. Med Sci Sports Exerc 38(5):803–810. https://doi.org/10.1249/01.MSS.0000218134.74238.6A

Camargo JA, Alonso Á (2006) Ecological and toxicological effects of inorganic nitrogen pollution in aquatic ecosystems: a global assessment. Environ Int 32:831–849

Caminade C, McIntyre KM, Jones AE (2019) 'Impact of recent and future climate change on vector-borne diseases', Annals of the New York Academy of Sciences, vol 1436. Blackwell Publishing Inc., pp 157–173. https://doi.org/10.1111/nyas.13950

Casas G, Martinez-Varela A, Vila-Costa M, Jim'enez B, Dachs J (2021) Rain amplification of persistent organic pollutants. Environ Sci Technol 55(19):12961–12972. https://doi.org/10.1021/acs.est.1c03295

Chen CY, Driscoll CT (2018) Integrating mercury research and policy in a changing world. Ambio 47(2):111–115. https://doi.org/10.1007/s13280-017-1010-y

Chen J, Wen J, Tang Y, Shi J, Mu G, Yan R, Cai J, Long M (2021) Research progress on fumonisin b1 contamination and toxicity: a review. Molecules 26:5238

Chowell G et al (2019) Assessing the potential impact of vector-borne disease transmission following heavy rainfall events: a mathematical framework. Philos Trans R Soc Lond Ser B Biol Sci 374(1775):20180272. https://doi.org/10.1098/RSTB.2018.0272

Ciota AT, Keyel AC (2019) The role of temperature in transmission of zoonotic arboviruses, vol 11. MDPI AG, Viruses. https://doi.org/10.3390/v11111013

Cleary D (2022) Effects of mycotoxins on macrophages and their possible clinical implications. Macrophage targeted delivery systems: basic concepts and therapeutic applications. Springer

Cohen AJ et al (2017) Estimates and 25-year trends of the global burden of disease attributable to ambient air pollution: an analysis of data from the Global Burden of Diseases Study 2015. Lancet 389(10082):1907–1918. https://doi.org/10.1016/S0140-6736(17)30505-6

Colley DG et al (2014) Human schistosomiasis. Lancet (London) 383(9936):2253. https://doi.org/10.1016/S0140-6736(13)61949-2

Comair F et al (2022) Report of the task force on water resources, Eastern Mediterranean and Middle East climate change initiative. The Cyprus Institute. https://emme-cci.org/scientific-task-forces/task-force-water-resources/

Cooper VS, Staples RK, Traverse CC, Ellis CN (2014) Parallel evolution of small colony variants in Burkholderia cenocepacia biofilms. Genomics 104:447–452. https://doi.org/10.1016/j.ygeno.2014.09.007, PMID: 25263109

Creasia D, Thurman J, Jones L III, Nealley M, York C, Wannemacher JRR, Bunner D (1987) Acute inhalation toxicity of t-2 mycotoxin in mice. Fundam Appl Toxicol 8:230–235
Cremer B, Soja A, Sauer J-A, Damm M (2012) Pro-inflammatory effects of ochratoxin a on nasal epithelial cells. Eur Arch Otorrinolaringol 269:1155–1161
D'Amato G (2008) Outdoor air pollution and allergic airway disease. In: Kay AB, Kaplan AP, Bousquet J, Holt PG (eds) Allergy and allergic diseases. Wiley-Blackwell, pp 1266–1278
D'Amato G, Liccardi G, D'Amato M, Holgate ST (2005) Environmental risk factors and allergic bronchial asthma. Clin Exp Allergy 35:1113–1124
D'Amato G, Cecchi L, D'Amato M, Liccardi G (2010) Urban air pollution and climate change as environmental risk factors of respiratory allergy: an update. J Investig Allergol Clin Immunol 20(2):95–102
D'Amato G, Cecchi L, D'Amato M, Annesi-Maesano I (2014) Climate change and respiratory diseases. Eur Respir Rev 23:161
D'Amato G, Pawankar R, Vitale C et al (2016) Climate change and air pollution: effects on respiratory allergy. Allergy Asthma Immunol Res 8(5):391–395. https://doi.org/10.4168/aair.2016.8.5.391
David E, Niculescu VC (2021) Volatile organic compounds (VOCs) as environmental pollutants: occurrence and mitigation using nanomaterials. Int J Environ Res Public Health 18(24):13147. https://doi.org/10.3390/ijerph182413147
de Bont J, Jaganathan S, Dahlquist M, Persson Å, Stafoggia M, Ljungman P (2022) Ambient air pollution and cardiovascular diseases: an umbrella review of systematic reviews and meta-analyses. J Intern Med 291(6):779–800. https://doi.org/10.1111/joim.13467
de Souza WM, Weaver SC (2024) Effects of climate change and human activities on vector-borne diseases. Nat Rev Microbiol 22:476. https://doi.org/10.1038/s41579-024-01026-0
Dickerson AS, Wu AC, Liew Z, Weisskopf M (2020) A scoping review of non-occupational exposures to environmental pollutants and adult depression, anxiety, and suicide. Curr Environ Health Rep 7(3):256–271. https://doi.org/10.1007/s40572-020-00280-7
Diggs DL, Huderson AC, Harris KL et al (2011) Polycyclic aromatic hydrocarbons and digestive tract cancers: a perspective. J Environ Sci Health C Environ Carcinog Ecotoxicol Rev 29(4):324–357. https://doi.org/10.1080/10590501.2011.629974
Dominski FH, Lorenzetti Branco JH, Buonanno G, Stabile L, Gameiro da Silva M, Andrade A (2021) Effects of air pollution on health: a mapping review of systematic reviews and meta-analyses. Environ Res 201:111487. https://doi.org/10.1016/j.envres.2021.111487
Dudley JP, Hoberg EP, Jenkins EJ et al (2015) Climate change in the North American Arctic: a One Health perspective. EcoHealth 12:713–725
East PL, Gahagan S, Al-Delaimy WK (2018) 'The Impact of Refugee Mothers' trauma, post-traumatic stress, and depression on their children's adjustment. J Immigr Minor Health 20(2):271–282. https://doi.org/10.1007/s10903-017-0624-2
Edelson PJ et al (2023) Climate change and the epidemiology of infectious diseases in the United States. Clin Infect Dis:950–956. https://doi.org/10.1093/cid/ciac697
Elidemir O, Colasurdo GN, Rossmann SN, Fan LL (1999) Isolation of stachybotrys from the lung of a child with pulmonary hemosiderosis. Pediatrics 104:964–966
Elmorshedy H et al (2020) Elimination of schistosomiasis requires multifactorial diagnostics: evidence from high- and low-prevalence areas in the Nile Delta, Egypt. Infect Dis Poverty 9(1):31. https://doi.org/10.1186/S40249-020-00648-9
Ernst KC et al (2023) Differences in longevity and temperature-driven extrinsic incubation period correlate with varying dengue risk in the Arizona-Sonora desert region. Viruses 15(4). https://doi.org/10.3390/V15040851
Escrivá L, Oueslati S, Font G, Manyes L (2017) Alternaria mycotoxins in food and feed: an overview. J Food Qual Article ID 1569748. https://doi.org/10.1155/2017/1569748
Falkovich AH, Rudich Y (2001) Analysis of semivolatile organic compounds in atmospheric aerosols by direct sample introduction thermal desorption GC/MS. Environ Sci Technol 35(11):2326–2333. https://doi.org/10.1021/es000280i

Faridi S, Krzyzanowski M, Cohen AJ et al (2023) Ambient air quality standards and policies in Eastern Mediterranean countries: a review. Int J Public Health 68:1605352. https://doi.org/10.3389/ijph.2023.1605352

Ferreira Lopes S, Vacher G, Ciarlo E, Savova-Bianchi D, Roger T, Niculita-Hirzel H (2017) Primary and immortalized human respiratory cells display different patterns of cytotoxicity and cytokine release upon exposure to deoxynivalenol, nivalenol and fusarenon-x. Toxins 9:337

Flappan SM, Portnoy J, Jones P, Barnes C (1999) Infant pulmonary hemorrhage in a suburban home with water damage and mold (stachybotrys atra). Environ Health Perspect 107:927–930

Frogner-Kockum P, Goransson G, Haeger-Eugensson M (2020) Impact of climate change on metal and suspended sediment concentrations in urban waters. Front Environ Sci 8. https://doi.org/10.3389/fenvs.2020.588335

Fung F, Clark RF (2004) Health effects of mycotoxins: a toxicological overview. J Toxicol Clin Toxicol 42:217–234

Gat D et al (2017) Origin-dependent variations in the atmospheric microbiome community in Eastern Mediterranean dust storms. Environ Sci Technol 51(12):6709–6718. https://doi.org/10.1021/acs.est.7b00362

Gauer R, Meyers BK (2019) Heat-related illnesses. Am Fam Physician 99(8):482–489

Gerberick GF, Sorenson WG, Lewis DM (1984) The effects of t-2 toxin on alveolar macrophage function in vitro. Environ Res 33:246–260

Ghosh S, Desai MR, Pandya G, Venkaiah K (1997) Airborne aflatoxin in the grain processing industries in India. Am Ind Hyg Assoc J 58:583–586

Ginexi EM et al (2000) Natural disaster and depression: a prospective investigation of reactions to the 1993 Midwest Floods. Am J Community Psychol 28(4):495–518. https://doi.org/10.1023/A:1005188515149

Gołofit-Szymczak M, Wójcik-Fatla A, Stobnicka-Kupiec A, Górny RL (2023) Filters of automobile air conditioning systems as in-car source of exposure to infections and toxic moulds. Environ Sci Pollut Res 30:108188–108200

Gow NAR, Johnson C, Berman J et al (2022) The importance of antimicrobial resistance in medical mycology. Nat Commun 13(1):5352. Published 2022 Sep 12. https://doi.org/10.1038/s41467-022-32249-5

Guerre P, Matard-Mann M, Collén PN (2022) Targeted sphingolipid analysis in chickens suggests different mechanisms of fumonisin toxicity in kidney, lung, and brain. Food Chem Toxicol 170:113467

Guo Y et al (2017) Heat wave and mortality: a multicountry, multicommunity study. Environ Health Perspect 125(8):087006. https://doi.org/10.1289/EHP1026

Hahad O, Lelieveld J, Birklein F, Lieb K, Daiber A, Münzel T (2020) Ambient air pollution increases the risk of cerebrovascular and neuropsychiatric disorders through induction of inflammation and oxidative stress. Int J Mol Sci 21(12):4306. Published 2020 Jun 17. https://doi.org/10.3390/ijms21124306

Halloy DJ, Gustin PG, Bouhet S, Oswald IP (2005) Oral exposure to culture material extract containing fumonisins predisposes swine to the development of pneumonitis caused by pasteurella multocida. Toxicology 213:34–44

Halstensen AS, Nordby K-C, Eduard W, Klemsdal SS (2006) Real-time pcr detection of toxigenic fusarium in airborne and settled grain dust and associations with trichothecene mycotoxins. J Environ Monit 8:1235–1241

Hamidianpour M et al (2021) Climatology of the Sistan Levar wind: atmospheric dynamics driving its onset, duration and withdrawal. Atmos Res 260:105711. https://doi.org/10.1016/J.ATMOSRES.2021.105711

Haque MA, Wang Y, Shen Z, Li X, Saleemi MK, He C (2020) Mycotoxin contamination and control strategy in human, domestic animal and poultry: a review. Microb Pathog 142:104095

Hayes K et al (2018) Climate change and mental health: risks, impacts and priority actions. Int J Ment Heal Syst 12(1):28. https://doi.org/10.1186/S13033-018-0210-6

Heydari N et al (2017) Household dengue prevention interventions, expenditures, and barriers to Aedes aegypti control in Machala, Ecuador. Int J Environ Res Public Health 14(2). https://doi.org/10.3390/ijerph14020196

Hodgson MJ, Morey P, Leung W-Y, Morrow L, Miller D, Jarvis BB, Robbins H, Halsey JF, Storey E (1998) Building-associated pulmonary disease from exposure to stachybotrys chartarum and aspergillus versicolor. J Occup Environ Med 40:241–249

Horstick O, Runge-Ranzinger S (2019) Multisectoral approaches for the control of vector-borne diseases, with particular emphasis on dengue and housing. In: Transactions of the royal society of tropical medicine and hygiene, vol 113. Oxford University Press, pp 822–827. https://doi.org/10.1093/trstmh/trz020

Hukkanen J, Pelkonen O, Hakkola J, Raunio H (2002) Expression and regulation of xenobiotic-metabolizing cytochrome p450 (cyp) enzymes in human lung. Crit Rev Toxicol 32:391–411

Hunter P (1997) Waterborne disease: epidemiology and ecology. Wiley, Chichester. Available at: https://books.google.com/books?hl=en&lr=&id=UB03WJlbOMkC&oi=fnd&pg=PR7&dq=Waterborne+disease:+epidemiology+and+ecology&ots=ZwNhkjMZA-&sig=nPK5-dhYal2daC5Qc_PoCCA7mC4. Accessed: 9 Apr 2024

Hunter PR, Macdonald AM, Carter RC (2010) Water supply and health. PLoS Med 7(11):e1000361. https://doi.org/10.1371/journal.pmed.1000361

IPCC (2019) In: Shukla PR, Skea J, Buendia EC, Masson-Delmotte V, Portner H-O, Roberts DC, Zhai P, Slade R, Connors S, van Diemen R, Ferrat M, Haughey E, Luz S, Neogi S, Pathak M, Petzold J, Pereira JP, Vyas P, Huntley E, Kissick K, Belkacemi M, Malley J (eds) Climate change and land: an IPCC special report on climate change. Desertification, Land Degradation, Sustainable Land Management, Food Security, and Greenhouse Gas Fluxes in Terrestrial Ecosystems. www.ipcc.ch

Jain N, Lodha R, Kabra S (2001) Upper respiratory tract infections. Ind J Pediatr 68:1135–1138

Janik E, Niemcewicz M, Ceremuga M, Stela M, Saluk-Bijak J, Siadkowski A, Bijak M (2020) Molecular aspects of mycotoxins—a serious problem for human health. Int J Mol Sci 21:8187

Janik E, Niemcewicz M, Podogrocki M, Ceremuga M, Stela M, Bijak M (2021) T-2 toxin—the most toxic trichothecene mycotoxin: metabolism, toxicity, and decontamination strategies. Molecules 26:6868

Jarvis BB, Miller JD (2005) Mycotoxins as harmful indoor air contaminants. Appl Microbiol Biotechnol 66:367–372

Jéquier E, Constant F (2010) Water as an essential nutrient: the physiological basis of hydration. Eur J Clin Nutr 64(2):115–123. https://doi.org/10.1038/EJCN.2009.111

Johnson NM, Hoffmann AR, Behlen JC et al (2021) Air pollution and children's health—a review of adverse effects associated with prenatal exposure from fine to ultrafine particulate matter. Environ Health Prev Med 26(1):72. https://doi.org/10.1186/s12199-021-00995-5

Karande S et al (2003) An observational study to detect leptospirosis in Mumbai, India, 2000. Arch Dis Child 88(12):1070–1075. https://doi.org/10.1136/adc.88.12.1070

Kelly FJ, Fussell JC (2012) Size, source and chemical composition as determinants of toxicity attributable to ambient particulate matter. Atmos Environ 60:504–526. https://doi.org/10.1016/j.atmosenv.2012.06.039

Kelly JD, Eaton DL, Guengerich FP, Coulombe RA Jr (1997) Aflatoxin b1 activation in human lung. Toxicol Appl Pharmacol 144:88–95

Kenneson A et al (2017) Social-ecological factors and preventive actions decrease the risk of dengue infection at the household-level: results from a prospective dengue surveillance study in Machala, Ecuador. PLoS Negl Trop Dis 11(12):e0006150. https://doi.org/10.1371/journal.pntd.0006150

Keramat A, Marivani B, Samsami M (2011) Climatic change, drought and dust crisis in Iran. World Acad Sci Eng Technol 81(9):10–13

Keyantash J, Dracup JA (2002) The quantification of drought: an evaluation of drought indices. Bull Am Meteorol Soc 83(8):1167–1180. https://doi.org/10.1175/1520-0477-83.8.1167

Khan S, Priti S, Ankit S (2015) Bacteria etiological agents causing lower respiratory tract infections and their resistance patterns. Iran Biomed J 19:240

Kharboush TG, Ahmed IA, Farag AA, Kharboush T, Sayed AE-DH, Abdel-Kareim AM, Attia H, Eid RA, Zaki MSA, Al-Tabbakh A-SM (2024) Epigenetic alterations of mir-155 and global DNA methylation as potential mediators of ochratoxin a cytotoxicity and carcinogenicity in human lung fibroblasts. Environ Sci Pollut Res 31:5473–5483

Kinney PL (2018a) Interactions of climate change, air pollution, and human health. Curr Environ Health Rep 5(1):179–186. https://doi.org/10.1007/s40572-018-0188-x

Kinney PL (2018b) Interactions of climate change, air pollution, and human health. Curr Environ Health Rep 5:179–186. https://doi.org/10.1007/s40572-018-0188-x

Kovács M, Pósa R, Tuboly T, Donkó T, Repa I, Tossenberger J, Szabó-Fodor J, Stoev S, Magyar T (2016) Feed exposure to fb1 can aggravate pneumonic damages in pigs provoked by P. Multocida. Res Veter Sci 108:38–46

Kumar M, Bolan NS, Hoang SA, Sawarkar AD, Jasemizad T, Gao B, Keerthanan S, Padhye LP, Singh L, Kumar S, Vithanage M, Li Y, Zhang M, Kirkham MB, Vinu A, Rinklebe J (2021) Remediation of soils and sediments polluted with polycyclic aromatic hydrocarbons: to immobilize, mobilize, or degrade? J Hazard Mater 420:126534. https://doi.org/10.1016/j.jhazmat.2021.126534

Lake IR, Foxall CD, Lovett AA et al (2005) Effects of river flooding on PCDD/F and PCB levels in cows' milk, soil, and grass. Environ Sci Technol 39(23):9033–9038. https://doi.org/10.1021/es051433a

Lamon L, Von Waldow H, Macleod M, Scheringer M, Marcomini A, Hungerbühler K (2009) Modeling the global levels and distribution of polychlorinated biphenyls in air under a climate change scenario. Environ Sci Technol 43(15):5818–5824. https://doi.org/10.1021/es900438j

Lang-Yona N et al (2020) Links between airborne microbiome, meteorology, and chemical composition in northwestern Turkey. Sci Total Environ 725:138227. https://doi.org/10.1016/j.scitotenv.2020.138227

Larsson P, Tjälve H (2000) Intranasal instillation of aflatoxin b1 in rats: bioactivation in the nasal mucosa and neuronal transport to the olfactory bulb. Toxicol Sci 55:383–391

Lawrance EL et al (2022) The impact of climate change on mental health and emotional wellbeing: a narrative review of current evidence, and its implications, vol 34. International Review of Psychiatry. Taylor and Francis Ltd, pp 443–498. https://doi.org/10.1080/09540261.2022.2128725

Lee RJ, Workman AD, Carey RM, Chen B, Rosen PL, Doghramji L, Adappa ND, Palmer JN, Kennedy DW, Cohen NA (2016) Fungal aflatoxins reduce respiratory mucosal ciliary function. Sci Rep 6:33221

Lelieveld J et al (2015) The contribution of outdoor air pollution sources to premature mortality on a global scale. Nature 525(7569):367–371. https://doi.org/10.1038/nature15371

Lenka SP, Jasemizad T, Balaneji IR, Huang B, Campbell B, Whittaker C, Padhye LP (2021) The fate of microplastics in natural and engineered aquatic systems: a case study of unplanned indirect potable reuse. Curr Opin Environ Sci Health 24:100302. https://doi.org/10.1016/j.coesh.2021.100302

Levy BS, Sidel VW, Patz JA (2017) Climate change and collective violence. Annu Rev Public Health 38(1):241–257. https://doi.org/10.1146/annurev-publhealth-031816-044232

Li D-W, Yang CS (2004) Fungal contamination as a major contributor to sick building syndrome. Adv Appl Microbiol 55:31–112

Li M, Harkema JR, Islam Z, Cuff CF, Pestka JJ (2006) T-2 toxin impairs murine immune response to respiratory reovirus and exacerbates viral bronchiolitis. Toxicol Appl Pharmacol 217:76–85

Li M, Harkema JR, Cuff CF, Pestka JJ (2007) Deoxynivalenol exacerbates viral bronchopneumonia induced by respiratory reovirus infection. Toxicol Sci 95:412–426

Lim CL (2020) Fundamental concepts of human thermoregulation and adaptation to heat: a review in the context of global warming. Int J Environ Res Public Health 17(21):1–33. https://doi.org/10.3390/ijerph17217795

Lim CL, Byrne C, Lee JK (2008) Human thermoregulation and measurement of body temperature in exercise and clinical settings. Ann Acad Med Singap 37(4):347–353

Limay-Rios V, Schaafsma AW (2021) Relationship between mycotoxin content in winter wheat grain and aspirated dust collected during harvest and after storage. ACS Omega 6:1857–1871. https://doi.org/10.1021/acsomega.0c04256

Linares C et al (2020) Impacts of climate change on the public health of the Mediterranean Basin population—current situation, projections, preparedness and adaptation. Environmental Research Academic Press Inc 182:109107. https://doi.org/10.1016/j.envres.2019.109107

Liu C, Chen R, Sera F et al (2019) Ambient particulate air pollution and daily mortality in 652 cities. N Engl J Med 381(8):705–715. https://doi.org/10.1056/NEJMoa1817364

Liu N, Bu Z, Liu W et al (2022) Health effects of exposure to indoor volatile organic compounds from 1980 to 2017: a systematic review and meta-analysis. Indoor Air 32(5):e13038. https://doi.org/10.1111/ina.13038

Longo SB, York R (2009) Structural influences on water withdrawals: an exploratory macro-comparative analysis. Human Ecol Rev:75–83. Available at: https://www.jstor.org/stable/24707738. Accessed: 9 Apr 2024

Lorentzou C et al (2019) Extreme desert dust storms and COPD morbidity on the Island of Crete. Int J COPD 14:1763–1768. https://doi.org/10.2147/COPD.S208108

Louis S, Carlson AK, Suresh A et al (2023) Impacts of climate change and air pollution on neurologic health, disease, and practice: a scoping review. Neurology 100(10):474–483. https://doi.org/10.1212/WNL.0000000000201630

Manisalidis I et al (2020) Environmental and health impacts of air pollution: a review, vol 8. Frontiers in Public Health. Frontiers Media SA, p 14. https://doi.org/10.3389/fpubh.2020.00014

Marcantonio R, Javeline D, Field S, Fuentes A (2021) Global distribution and coincidence of pollution, climate impacts, and health risk in the anthropocene. PLoS One 16:e0254060

Marcelloni AM, Pigini D, Chiominto A, Gioffrè A, Paba E (2024) Exposure to airborne mycotoxins: the riskiest working environments and tasks. Ann Work Expos Health 68:19–35

Marris CR, Kompella SN, Miller MR et al (2020) Polyaromatic hydrocarbons in pollution: a heart-breaking matter. J Physiol 598(2):227–247. https://doi.org/10.1113/JP278885

Mattila T, Santonen T, Andersen HR et al (2021) Scoping review—the association between asthma and environmental chemicals. Int J Environ Res Public Health 18(3):1323. Published 2021 Feb 1. https://doi.org/10.3390/ijerph18031323

Mattoo S, Cherry JD (2005) Molecular pathogenesis, epidemiology, and clinical manifestations of respiratory infections due to bordetella pertussis and other bordetella subspecies. Clin Microbiol Rev 18:326–382

Maung TZ, Bishop JE, Holt E, Turner AM, Pfrang C (2022) Indoor air pollution and the health of vulnerable groups: a systematic review focused on particulate matter (PM), volatile organic compounds (VOCs) and their effects on children and people with pre-existing lung disease. Int J Environ Res Public Health 19(14):8752. https://doi.org/10.3390/ijerph19148752

Mazar Y et al (2016) Effect of dust storms on the Atmospheric Microbiome in the Eastern Mediterranean. Environ Sci Technol 50(8):4194–4202. https://doi.org/10.1021/acs.est.5b06348

Meade RD et al (2020) Physiological factors characterizing heat-vulnerable older adults: a narrative review. Environment International Elsevier Ltd 144:105909. https://doi.org/10.1016/j.envint.2020.105909

Mears MJ, Aslaner DM, Barson CT, Cohen MD, Gorr MW, Wold LE (2022) Health effects following exposure to dust from the World Trade Center disaster: an update. Life Sci 289:120147. https://doi.org/10.1016/j.lfs.2021.120147

Medina-Ramón M, Schwartz J (2007) Temperature, temperature extremes, and mortality: a study of acclimatisation and effect modification in 50 US cities. Occup Environ Med 64(12):827–833. https://doi.org/10.1136/oem.2007.033175

Miller CJ, Runge-Morris M, Cassidy-Bushrow AE et al (2020) A review of volatile organic compound contamination in post-industrial urban centers: reproductive health implications using a detroit lens. Int J Environ Res Public Health 17(23):8755. https://doi.org/10.3390/ijerph17238755

Millyard A et al (2020) Impairments to thermoregulation in the elderly during heat exposure events, vol 6. Gerontology and Geriatric Medicine, p 233372142093243. https://doi.org/10.1177/2333721420932432

Moller L, Schuetzle D, Autrup H (1994) Future research needs associated with the assessment of potential human health risks from exposure to toxic ambient air pollutants. Environmental Health Perspectives. National Institute of Environmental Health Sciences, pp 193–210. https://doi.org/10.1289/ehp.102-1566924

Montero-Montoya R, López-Vargas R, Arellano-Aguilar O (2018) Volatile organic compounds in air: sources, distribution, exposure and associated illnesses in children. Ann Glob Health 84(2):225–238. https://doi.org/10.29024/aogh.910

Mordecai EA et al (2020) Climate change could shift disease burden from malaria to arboviruses in Africa. In: The Lancet Planetary Health, vol 4. Elsevier B.V, pp e416–e423. https://doi.org/10.1016/S2542-5196(20)30178-9

Mousavi A et al (2020) Climate change and health in Iran: a narrative review. J Environ Health Sci Eng 18(1):367–378. https://doi.org/10.1007/s40201-020-00462-3

Murdock CC, Sternberg ED, Thomas MB (2016) Malaria transmission potential could be reduced with current and future climate change. Sci Rep 6(1):1–7. https://doi.org/10.1038/srep27771

National Institute of Environmental Health Sciences, Human health impacts of climate change. Last Reviewed: November 07, 2022. Accessed: 7 May 2024. https://www.niehs.nih.gov/research/programs/climatechange/health_impacts/index.cfm

Ndaw S, Jargot D, Antoine G, Denis F, Melin S, Robert A (2021) Investigating multi-mycotoxin exposure in occupational settings: a biomonitoring and airborne measurement approach. Toxins 13:54

Negev M et al (2015) Impacts of climate change on vector borne diseases in the Mediterranean basin—implications for preparedness and adaptation policy. Int J Environ Res Public Health 12(6):6745–6770. https://doi.org/10.3390/ijerph120606745

Neira M et al (2023) Climate change and human health in the Eastern Mediterranean and Middle East: Literature review, research priorities and policy suggestions. Environ Res 216:114537. https://doi.org/10.1016/j.envres.2022.114537

Ngowi HA (2020) Prevalence and pattern of waterborne parasitic infections in Eastern Africa: A systematic scoping review. Food Waterb Parasit 20:e00089. https://doi.org/10.1016/j.fawpar.2020.e00089

Ngundi MM, Qadri SA, Wallace EV, Moore MH, Lassman ME, Shriver-Lake LC, Ligler FS, Taitt CR (2006) Detection of deoxynivalenol in foods and indoor air using an array biosensor. Environ Sci Technol 40:2352–2356

North Atlantic Treaty Organization (NATO) (2019) Water scarcity in the Middle East. NATO Strategic Direction South Hub, Naples. Available at: www.TheSouthernHub.org. Accessed: 23 Mar 2021

Novotny WE, Dixit A (2000) Pulmonary hemorrhage in an infant following 2 weeks of fungal exposure. Arch Pediatr Adolesc Med 154:271–275

Obradovich N et al (2018) Empirical evidence of mental health risks posed by climate change. Proc Natl Acad Sci USA 115(43):10953–10958. https://doi.org/10.1073/PNAS.1801528115

Oke TR (1982) The energetic basis of the urban heat Island. Q J R Meteorol Soc 108(455):1–24. https://doi.org/10.1002/qj.49710845502

Okutan H, Aydin G, Ozcelik N (2004) Protective role of melatonin in ochratoxin a toxicity in rat heart and lung. J Appl Toxicol 24:505–512

Orru H, Ebi KL, Forsberg B (2017) The interplay of climate change and air pollution on health. Curr Environ Health Rep. 4(4):504–513. https://doi.org/10.1007/s40572-017-0168-6

Oviedo MV, Romoser WS, James CB, Mahmood F, Reisen WK (2011) Infection dynamics of western equine encephalomyelitis virus (Togaviridae: Alphavirus) in four strains of Culex tarsalis (Diptera: Culicidae): an immunocytochemical study. Res Rep Trop Med 2011(2):65–77. https://doi.org/10.2147/RRTM.S13946

Padhy SK et al (2015) Mental health effects of climate change. Ind J Occup Environ Med 19:3–7. https://doi.org/10.4103/0019-5278.156997

Padmanabha H et al (2010) Ecological links between water storage behaviors and Aedes aegypti production: Implications for dengue vector control in variable climates. EcoHealth 7(1):78–90. https://doi.org/10.1007/s10393-010-0301-6
Pal JS, Eltahir EAB (2016) Future temperature in southwest Asia projected to exceed a threshold for human adaptability. Nat Clim Chang 6(2):197–200. https://doi.org/10.1038/nclimate2833
Palinkas LA, Wong M (2020) Global climate change and mental health, current opinion in psychology. Elsevier BV, pp 12–16. https://doi.org/10.1016/j.copsyc.2019.06.023
Pardo M et al (2017) Mitochondria-mediated oxidative stress induced by desert dust in rat alveolar macrophages. GeoHealth 1(1):4–16. https://doi.org/10.1002/2016GH000017
Parker CL, Wellbery CE, Mueller M (2019) The changing climate: managing health impacts. Am Fam Physician 100(10):618–626
Patriarca A, Pinto VF (2017) Prevalence of mycotoxins in foods and decontamination. Curr Opin Food Sci 14:50–60
Plewa-Tutaj K, Krzyściak P, Dobrzycka A (2024) Mycological air contamination level and biodiversity of airborne fungi isolated from the zoological garden air – preliminary research. Environ Sci Pollut Res Int 31(30):43066–43079. https://doi.org/10.1007/s11356-024-33926-2
Pope CA 3rd, Dockery DW (2006) Health effects of fine particulate air pollution: lines that connect. J Air Waste Manag Assoc 56(6):709–742. https://doi.org/10.1080/10473289.2006.10464485
Pope CA, Dockery DW (1992) Acute health effects of PM10 pollution on symptomatic and asymptomatic children. Am Rev Respir Dis 145(5):1123–1128. https://doi.org/10.1164/ajrccm/145.5.1123
Pósa R, Donkó T, Bogner P, Kovács M, Repa I, Magyar T (2011) Interaction of bordetella bronchiseptica, pasteurella multocida, and fumonisin b1 in the porcine respiratory tract as studied by computed tomography. Can J Vet Res 75:176–182
Qadri F et al (2005) Enterotoxigenic Escherichia coli and Vibrio cholerae diarrhea, Bangladesh, 2004. Emerg Infect Dis 11(7):1104–1107. https://doi.org/10.3201/eid1107.041266
Racinais S et al (2019) Core temperature up to 41.5oC during the UCI road cycling world championships in the heat. Br J Sports Med 53(7):426–429. https://doi.org/10.1136/BJSPORTS-2018-099881
Reiter P (2001) Climate change and mosquito-borne disease. Environ Health Perspect 109(SUPPL. 1):141–161. https://doi.org/10.1289/ehp.01109s1141
Richard JL, Plattner RD, May J, Liska SL (1999) The occurrence of ochratoxin a in dust collected from a problem household. Mycopathologia 146:99–103
Robbins CA, Swenson LJ, Nealley ML, Kelman BJ, Gots RE (2000) Health effects of mycotoxins in indoor air: a critical review. Appl Occup Environ Hyg 15:773–784
Robine JM, Cheung SL, Le Roy S et al (2008) Death toll exceeded 70,000 in Europe during the summer of 2003. C R Biol 331(2):171–178. https://doi.org/10.1016/j.crvi.2007.12.001
Romanello M et al (2022) The 2022 report of the Lancet Countdown on health and climate change: health at the mercy of fossil fuels. Lancet (London) 400(10363):1619–1654. https://doi.org/10.1016/S0140-6736(22)01540-9
Prathumratana L, Kim R, Kim KW (2008) Heavy metal contamination of the mining and smelting district in Mitrovica, Kosovo. In: Proceedings of the International Symposia on Geoscience Resources and Environments of Asian Terranes (GREAT 2008). pp 24–26. https://citeseerx.ist.psu.edu/document?repid=rep1&type=pdf&doi=6ea560bac2d31762054b0d06950aee139b3d3c4d
Saad-Hussein A, Ramadan HKA (2023) Impacts of climate change on environmental toxins and pollutants causing liver health problems. In: Saad-Hussein A, Elwakil R, Khomsi K (eds) Impact of climate change on health in Africa. Springer, Cham. https://doi.org/10.1007/978-3-031-39466-9_3
Saad-Hussein A, El-Mofty HM, Hassanien MA (2011) Climate change and fungal keratitis trend: in Egypt. Eastern Med Health J 17(6):468–473
Saad-Hussein A, Elserougy S, Beshir S et al (2012) Work-related airborne fungi and the biological levels of mycotoxin in textile workers. J Appl Sci Res 8:719–726

Saad-Hussein A, Taha M, Fadl N et al (2016) Effects of airborne aspergillus on serum aflatoxin B1 and liver enzymes in workers handling wheat flour. Hum Exp Toxicol 35:3–9
Saad-Hussein A, Soliman KM, Moubarz G (2022) 18SrRNA gene sequencing for environmental aflatoxigenic fungi and risk of hepatic carcinoma among exposed workers. J Environ Sci Health A 57:174–182
Saha KC (2003) Review of arsenicosis in West Bengal, India—a clinical perspective. Crit Rev Environ Sci Technol 33:127–163. https://doi.org/10.1080/10643380390814514
Sahoo BM, Ravi Kumar BVV, Banik BK, Borah P (2020) Polyaromatic Hydrocarbons (PAHs): structures, synthesis and their biological profile. Curr Org Synth 17(8):625–640. https://doi.org/10.2174/1570179417666200713182441
Samoli E et al (2011) Acute effects of air pollution on pediatric asthma exacerbation: evidence of association and effect modification. Environ Res 111(3):418–424. https://doi.org/10.1016/j.envres.2011.01.014
Sarvtin MT (2023) A review of the role of indoor fungi in sick building syndrome. Int J Med Lab
Schiedek D, Sundelin B, Readman JW, Macdonald RW (2007) Interactions between climate change and contaminants. Mar Pollut Bull 54(12):1845–1856. https://doi.org/10.1016/j.marpolbul.2007.09.020
Schlosser O, Robert S, Noyon N (2020) Airborne mycotoxins in waste recycling and recovery facilities: occupational exposure and health risk assessment. Waste Manag 105:395–404
Schmeltz MT, Gamble JL (2017) Risk characterization of hospitalizations for mental illness and/or behavioral disorders with concurrent heat-related illness. PLoS One 12(10):e0186509. https://doi.org/10.1371/JOURNAL.PONE.0186509
Schmidt WP et al (2011) Population density, water supply, and the risk of dengue fever in Vietnam: Cohort study and spatial analysis. PLoS Med 8(8):e1001082. https://doi.org/10.1371/journal.pmed.1001082
Schweitzer MD et al (2018) Lung health in era of climate change and dust storms. Environ Res 163(February):36–42. https://doi.org/10.1016/j.envres.2018.02.001
Semenza JC, Paz S (2021) Climate change and infectious disease in Europe: impact, projection and adaptation. The Lancet Regional Health—Europe. Lancet Reg Health Eur. https://doi.org/10.1016/j.lanepe.2021.100230
Sherwood SC, Huber M (2010) An adaptability limit to climate change due to heat stress. Proc Natl Acad Sci USA 107(21):9552–9555. https://doi.org/10.1073/PNAS.0913352107/-/DCSUPPLEMENTAL/PNAS.0913352107_SI.PDF
Shoemaker RC, House DE (2006) Sick building syndrome (SBS) and exposure to water-damaged buildings: time series study, clinical trial and mechanisms. Neurotoxicol Teratol 28:573–588
Sips GJ et al (2023) Climate change and infectious diseases. Ned Tijdschr Geneeskd 167:D7503–D7503. Available at: https://europepmc.org/article/med/37289898. Accessed: 11 Apr 2024
Siraj AS, Oidtman RJ, Huber JH et al (2017) Temperature modulates dengue virus epidemic growth rates through its effects on reproduction numbers and generation intervals. PLoS Negl Trop Dis 11(7):e0005797. Published 2017 Jul 19. https://doi.org/10.1371/journal.pntd.0005797
Skaug MA, Wignand E, Stormer FC (2001) Ochratoxin A in airborne dust and fungal conidia. Mycopathologia 151:93–98. https://doi.org/10.1023/a:1010953401173
Sly PD (2021) Adverse environmental exposure and respiratory health in children. Pediatr Clin N Am 68(1):277–291. https://doi.org/10.1016/j.pcl.2020.09.018
Smoragiewicz W, Cossette B, Boutard A, Krzystyniak K (1993) Trichothecene mycotoxins in the dust of ventilation systems in office buildings. Int Arch Occup Environ Health 65:113–117
So MY, Tian Z, Phoon YS, Sha S, Antoniou MN, Zhang J, Wu RS, Tan-un KC (2014) Gene expression profile and toxic effects in human bronchial epithelial cells exposed to zearalenone. PLoS One 9:e96404
Son JY, Lee JT, Anderson GB, Bell ML (2012) The impact of heat waves on mortality in seven major cities in Korea. Environ Health Perspect 120(4):566–571. https://doi.org/10.1289/ehp.1103759
Straus DC (2009) Molds, mycotoxins, and sick building syndrome. Toxicol Ind Health 25(9–10):617–635. https://doi.org/10.1177/0748233709348287

Sun Y, Su J, Liu Z et al (2018) Aflatoxin B1 promotes influenza replication and increases virus related lung damage via activation of TLR4 signaling. Front Immunol 9:2297. Published 2018 Oct 4. https://doi.org/10.3389/fimmu.2018.02297
Sur D, Dutta P, Nair GB, Bhattacharya SK (2000) Severe cholera outbreak following floods in a northern district of West Bengal. Indian J Med Res 112:178–182
Talley SM, Coley PD, Kursar TA (2002) The effects of weather on fungal abundance and richness among 25 communities in the Intermountain West. BMC Ecol 2:7. https://doi.org/10.1186/1472-6785-2-7
Tang X, Wilson SR, Solomon KR, Shao M, Madronich S (2011) Changes in air quality and tropospheric composition due to depletion of stratospheric ozone and interactions with climate. Photochem Photobiol Sci 10(2):280–291. https://doi.org/10.1039/c0pp90039g
Tarrass F, Benjelloun M (2012) The effects of water shortages on health and human development. Perspect Public Health 132:240–244. https://doi.org/10.1177/1757913910391040
Thomson MC, Jeanne I, Djingarey M (2009) Dust and epidemic meningitis in the Sahel: a public health and operational research perspective. IOP Conf Ser: Earth Environ Sci 7(1):012017. https://doi.org/10.1088/1755-1307/7/1/012017
Tozan Y et al (2020) Transmission dynamics of dengue and chikungunya in a changing climate: do we understand the eco-evolutionary response? Expert Rev Anti-Infect Ther 18(12):1187–1193. https://doi.org/10.1080/14787210.2020.1794814
Tshala-Katumbay D, Mwanza JC, Rohlman DS, Maestre G, Oriá RB (2015) A global perspective on the influence of environmental exposures on the nervous system. Nature 527(7578):S187–S192. https://doi.org/10.1038/nature16034
Tsitsigiannis DI, Dimakopoulou M, Antoniou PP, Tjamos EC (2012) Biological control strategies of mycotoxigenic fungi and associated mycotoxins in Mediterranean basin crops. Phytopathol Mediterr 51:158–174. https://oajournals.fupress.net/index.php/pm/article/view/5471
United States Environmental Protection Agency (EPA) (2018) Volatile organic compounds' impact on indoor air quality. Retrieved from https://www.epa.gov/indoor-air-quality-iaq/volatile-organic-compounds-impact-indoor-air-quality
van Vleet TR, Macé K, Coulombe RA Jr (2002) Comparative aflatoxin b1 activation and cytotoxicity in human bronchial cells expressing cytochromes p450 1a2 and 3a4. Cancer Res 62:105–112
Vaali K, Tuomela M, Mannerström M, Heinonen T, Tuuminen T (2022) Toxic indoor air is a potential risk of causing immuno suppression and morbidity – a pilot study. J Fungi 8(2):104. https://doi.org/10.3390/jof8020104
Viegas S, Veiga L, Malta-Vacas J, Sabino R, Figueredo P, Almeida A, Viegas C, Carolino E (2012) Occupational exposure to aflatoxin (afb1) in poultry production. J Toxic Environ Health A 75:1330–1340
Viljoen M, Claassen N (2023) Pathophysiological aspects of exposure to dampness-associated indoor mould and mycotoxins: a mini-overview. J Hazard Mater Advance 9:100228. https://doi.org/10.1016/j.hazadv.2022.100228
Vörösmarty CJ et al (2000) Global water resources: vulnerability from climate change and population growth. Science (New York) 289(5477):284–288. https://doi.org/10.1126/SCIENCE.289.5477.284
Walinski A et al (2023) The effects of climate change on mental health. Dtsch Arztebl Int 120(8):117. https://doi.org/10.3238/ARZTEBL.M2022.0403
Wang XY et al (2015) The impacts of heatwaves on mortality differ with different study periods: a multi-city time series investigation. PLoS One 10(7):e0134233. https://doi.org/10.1371/JOURNAL.PONE.0134233
Watson JT, Gayer M, Connolly MA (2007) Epidemics after natural disasters. Emerg Infect Dis 13(1):1. https://doi.org/10.3201/EID1301.060779
Watts N et al (2019) The 2019 report of The Lancet Countdown on health and climate change: ensuring that the health of a child born today is not defined by a changing climate, vol 394. Lancet Publishing Group, p 1836. https://doi.org/10.1016/S0140-6736(19)32596-6

Watts N et al (2021) The 2020 report of The Lancet Countdown on health and climate change: responding to converging crises. Lancet Publishing Group, pp 397:129–170. https://doi.org/10.1016/S0140-6736(20)32290-X

Weber E, Downward GS, Ebi KL, Lucas PL, van Vuuren D (2023) The use of environmental scenarios to project future health effects: a scoping review. Lancet Planet Health 7:e611–e621

WHO (World Health Organization) (2023) Climate change: Key facts. Accessed: 7 May 2024. https://www.who.int/news-room/fact-sheets/detail/climate-change-and-health

WHO Regional Office for Europe (2021) Human health effects of polycyclic aromatic hydrocarbons as ambient air pollutants: report of the Working Group on Polycyclic Aromatic Hydrocarbons of the Joint Task Force on the Health Aspects of Air Pollution. License: CC BY-NC-SA 3.0 IGO, Copenhagen. https://iris.who.int/bitstream/handle/10665/350636/9789289056533-eng.pdf

Wikimedia Commons Contributors (2025) 'File:Benzo-a- pyrene.svg', Wikimedia Commons, 13 November 2023, 10:15 UTC, [accessed 24 February 2025]

Wilkie IW, Harper M, Boyce JD, Adler B (2012) Pasteurella multocida: diseases and pathogenesis. Pasteurella multocida: molecular biology, toxins and infection. Curr Top Microbiol Immunol 36:1–22

Wontner-Smith T, Bruce D, Cardwell S, Armitage D, Jennings P (2014) Ochratoxin a production during ambient-air drying. J Stored Prod Res 56:1–7

Yang HY et al (2005) Clinical distinction and evaluation of leptospirosis in Taiwan—a case-control study. J Nephrol 18(1):45–53. Available at: https://europepmc.org/article/med/15772922. Accessed: 12 Apr 2024

Yang X-J, Lu H-Y, Li Z-Y, Bian Q, Qiu L-L, Li Z, Liu Q, Li J, Wang X, Wang S-L (2012) Cytochrome p450 2a13 mediates aflatoxin b1-induced cytotoxicity and apoptosis in human bronchial epithelial cells. Toxicology 300:138–148

Yang Z et al (2022) 'Health Effects of Long-Term Exposure to Ambient PM2.5 in Asia-Pacific: a Systematic Review of Cohort Studies. Curr Environ Health Rep 9:130–151. https://doi.org/10.1007/s40572-022-00344-w

Yu YY, Jin H, Lu Q (2022) Effect of polycyclic aromatic hydrocarbons on immunity. J Transl Autoimmun 5:100177. https://doi.org/10.1016/j.jtauto.2022.100177

Zhang Z, Lu H, Huan F, Meghan C, Yang X, Wang Y, Wang X, Wang X, Wang SL (2014) Cytochrome p450 2a13 mediates the neoplastic transformation of human bronchial epithelial cells at a low concentration of aflatoxin b1. Int J Cancer 134:1539–1548

Zhang W, Yi M, Xiao G, Ma J, Wu G, Xu C (2018) Al2o3-coated h-bn composite powders and as-prepared si3n4-based self-lubricating ceramic cutting tool material. Int J Refract Met Hard Mater 71:1–7

Zhang W et al (2019) Projected changes in maternal heat exposure during early pregnancy and the associated congenital heart defect burden in the United States. J Am Heart Assoc 8(3):e010995. https://doi.org/10.1161/JAHA.118.010995

Zittis G et al (2016) Projected changes in heat wave characteristics in the eastern Mediterranean and the Middle East. Reg Environ Chang 16(7):1863–1876. https://doi.org/10.1007/s10113-014-0753-2

Zittis G et al (2022) Climate change and weather extremes in the Eastern Mediterranean and Middle East. Rev Geophys 60(3):e2021RG000762. https://doi.org/10.1029/2021RG000762

Climate Change Impacts on Occupational Toxins, and Its Health Effects

3

Amal Saad-Hussein, Aisha M. Samir, Nermin H. Zawilla, Mahmoud M. Amr, Mohamed M. AlBatanony, Hussein H. Zayet, Heba Mahdy-Abdallah, and Marwa M. Fouad

Abstract

Climate change faces an immediate threat to the expansion of the real gross domestic product and to the working conditions and productivity of labor. Addressing these issues could strengthen economies and raise working standards, according to the International Labour Organization, published in 2023. The impact of climate change on the workers' health will be largely determined by personal, socioeconomic, and environmental pollutants in the workplaces, as well as the behavioral changes made to lessen the whole spectrum of health dan-

A. Saad-Hussein (✉)
Department of Environmental & Occupational Medicine, Institute of Environment and Climate Change Research, National Research Centre, Giza, Egypt

National Committee of Toxicology, Academy of Scientific Research and Technology (ASRT), Ministry of Higher Education and Scientific Research, Cairo, Egypt
e-mail: al.hussein@nrc.sci.eg

A. M. Samir · N. H. Zawilla · M. M. AlBatanony · H. H. Zayet · M. M. Fouad
Occupational and Environmental Medicine Department, Faculty of Medicine, Cairo University, Cairo, Egypt
e-mail: aishasamir@kasralainy.edu.eg; zawilla@kasralainy.edu.eg; marwa.fouad@kasralainy.edu.eg

M. M. Amr
National Committee of Toxicology, Academy of Scientific Research and Technology (ASRT), Ministry of Higher Education and Scientific Research, Cairo, Egypt

Occupational and Environmental Medicine Department, Faculty of Medicine, Cairo University, Cairo, Egypt

H. Mahdy-Abdallah
Environmental & Occupational Medicine Department, Environment & Climate Change Research Institute, National Research Centre, Giza, Egypt
e-mail: hm.mahdy@nrc.sc.eg

A. Saad-Hussein et al. (eds.), *Climate Change Impacts on Toxins and Health Effects*,
https://doi.org/10.1007/978-981-96-3416-3_3

gers. Will these factors, circumstances, and adjustments, together with climate change, have an impact on workers' occupational safety and health? This is the question that will be answered in this chapter. This chapter has underscored the intricate interplay between climate change, occupational toxins, and the health of the exposed workers, with specific emphasis on the possible ways of adaptation and mitigation.

Keywords

Climate change · Occupational exposures · Chemical pollutants · Biological pollutants · Occupational health problems

Abbreviations

A1AT	alpha-1-anti-trypsin gene
AFB_1	aflatoxin B1
APEX1	apurinic/apyrimidinic endonuclease gene
BPDE-Alb	benzo(a)pyrene diol epoxide albumin adduct
C	carbon
CFCs	Chlorofluorocarbons
CH_4	methane
CHIP	climate-health-infrastructure-pathways
Cl	chlorine
CO_2	carbon dioxide
F	fluorine
GDP	gross domestic product
GHGs	greenhouse gases
GST	glutathione S-transferase
$GSTM_1$	glutathione S-transferase M1 gene polymorphism
$GSTT_1$	glutathione S-transferase T1 gene polymorphism
H	hydrogen
HCFCs	hydrochlorofluorocarbons
ILO	International Labour Organization
IPCC	Intergovernmental Panel on Climate Change
N_2O	nitrous oxide
NOx	nitrogen oxides
OEM	Occupational and environmental medicine
PAHs	Polycyclic aromatic hydrocarbons
PFAS	Perfluorinated compounds
PM	particulate matter
SLCP	Short-lived climate pollutants
SLR	Sea level raise
VOCs	volatile organic compounds

3.1 Introduction

Climate change is gaining attention worldwide, according to the scenarios studied by the Intergovernmental Panel on Climate Change (IPCC) and mentioned in the report of 2021. Global surface temperature was increased by 1.1 °C or more over the next decade, above the surface temperature before the industrial era, and there is more than a 50% chance that the temperature could reach 1.5 °C more quickly under a high-emission scenario between 2021 and 2040 (IPCC 2021). But there are new estimates of the probability of exceeding this level (IPCC 2023).

Climate change refers to a wider range of alterations to the planet, such as variations in precipitation, wind patterns, and plant blooming, in addition to temperature variation. These climate hazards increase the risk of deaths, noncommunicable diseases, emergence, and remerging of infectious diseases, besides the variation of spread of these diseases, and health emergencies (WHO 2023).

Occupational poisons could be linked to environmental effects of worldwide climate change. Effective prevention, monitoring, and response to potential exposures that may affect human health due to climate factors require an understanding of these exposures (Cailee 2022). It has an impact on the physical environment in addition to all aspects of humans and ecosystems. For many workers, heat and humidity exposure at work may be more dangerous than exposure in the community since workers typically have less control over exposures during work-related tasks (Dunne et al. 2013).

Impacts to workers will include the direct effects of climate change-potentiated occupational hazards, especially the increase in ambient temperatures, or indirect climate change-associated occupational hazards that are likely to occur from industrial transitions, new industrial technologies, and emitted environmental pollutants (Schulte et al. 2016). Workers might be exposed to known occupational hazards for longer occupational time and at higher intensities, which could eventually lead to an increase in the severity and frequency of these exposures, as well as the emergence of new ones (Moda et al. 2019).

The Middle East is among the most vulnerable regions to the accelerating impacts of human-caused climate changes, mainly extreme heatwaves, decreasing precipitation, protracted droughts, more powerful sandstorms and floods, and rising sea levels (Neira et al. 2023). However, the effects will not be felt equally in the whole area. Resource-poor nations will be particularly hard hit since they lack the infrastructure, technology, human, and physical capital necessary for adaptation. This is true when it comes to how global warming is affecting rural livelihoods and endangering water and food security. The consequences will exacerbate long-standing injustices and decades of unviable public policies, especially those pertaining to land and water. The general population is already experiencing extra morbidity and mortality because of exposure to extremely high heat levels. The effect of climate change on workers' health has not received much attention, even though workers in hot areas are also susceptible to heat exposure and climate change (Levi et al. 2018).

At-risk areas, guidelines, and heat management systems to combat rising heat exposure in work environments are still far from being properly implemented. Furthermore, numerous researches have verified a noteworthy decline in productivity as a result of prolonged exposure to high temperatures. The workplace is greatly impacted by climate change, particularly in terms of safe chemical handling, storage, and production. The production and use of numerous chemicals in the workplace might have an important impact on the environment, as well as on climate (ILO 2023).

Climate change is affecting all parts of the planet; changing weather patterns create a threat to water and food production, and rising sea levels raise the possibility of catastrophic flooding (IPCC 2021).

3.2 Impact of Occupational Toxins on Climate Change

Occupational toxins are substances released into the environment in quantities, concentrations, or under circumstances that have harmful effects on the environment and human life in the short or long term. The synergistic effects of climate change with occupational exposure to toxins will have potentially serious adverse consequences for human health, as well as on the ecosystem.

There are two main causes of climate change: natural and human. Human activities include deforestation, agriculture, and industry. These activities are to blame for the human-induced climatic change. The second largest production sector in the world is the chemical industry. It has a long history of consistent growth; between 2000 and 2017, the industry's production capacity almost doubled to 2.3 billion tons, averaging between 4% and 4.5% annually, suggesting possible future increases in the amount of production of chemicals to the environment (UNEP 2019).

Industrial emissions have a major role in the chemical production of greenhouse gases (GHGs), that is, the main cause of climate change, as well as the production of huge amounts of chemical pollutants in the environment. This load has been greatly increased by industrial sources of dangerous substances, such as pesticides, fertilizers, heavy metals, plastics, and dyes that result in currently represented huge environmental risks for human health (Naidu et al. 2021). The industrial sector is proved to be attributed to more than one-third of the world's direct and indirect GHG emissions (Climate Action 2020; https://www.climateaction100.org/wp-content/uploads/2020/12/CA100-Progress-Report.pdf).

The GHGs, such as carbon dioxide, are emitted at every stage of a chemical's lifecycle, including production, use, and waste. Industries, such as iron and steel, cement, chemicals, electronics, plastics, clothing, and aluminum, are the main contributors to climate change worldwide. Also, manufacturing industries, such as food processing, mining, and construction industries, account for about a quarter (23%) of GHG emissions. These direct emissions come from a variety of sources, not only from the on-site burning of fossil fuels for electricity and heat but also from the industrial processes and chemical reactions involved in these industries. These include carbon dioxide (CO_2), methane (CH_4), nitrous oxide (N_2O), and fluorinated gases. As these gases envelop the earth, they trap more solar energy, causing global warming (Davda 2024; EPA 2024).

Chemical pollutants have the ability to spread around the world via the air, water, waste materials, human and animal vectors, and nanoparticles. Certain harmful compounds can be biodegraded by the environment, while some are resistant to these processes, causing long-term environmental harm and building up in food chains. Therefore, chemical emissions have negative effects on ecosystems, environment, and climate that are severe and long-lasting (Naidu et al. 2021).

The use of fossil fuels for power generation, industry, and polluting transportation are all major sources of both particulate matter (PM) and CO_2. The use of fossil fuels to produce heat and electricity increases global emissions dramatically, because most of the energy used is still produced by burning gas, coal, or oil, which releases strong GHGs like N_2O and CO_2 that cover the earth and decrease the natural retention of solar radiation. The petrochemical sector is believed to be one of the primary forces behind the rise in the demand for fossil fuels during the last decade till now. Manufacturing equipment often runs on coal, oil, or gas, and certain goods like plastics are made using chemicals that come from fossil fuels (Fagbohunka 2015).

The unintended and undesirable escape, leakage, or emission of gases or vapors from pressure-containing facilities or equipment, as well as plant components including valves, piping flanges, pumps, storage tanks, valves, compressors, and so forth, is known as fugitive emission. Fugitive emissions may be a more urgent concern, even though this environmental harm is undesirable. Workers at chemical factories or oil refineries run the danger of serious health implications from prolonged exposure to volatile organic compounds (VOCs), such as benzene, which can be extremely hazardous to humans. Similar to this, nearby communities are in danger due to these plants (Khajeh Hoseini et al. 2022).

Pollutant emissions into the atmosphere have the potential to alter the climate. The term "climate forcers" is frequently used to describe these pollutants, which include GHGs. The climate is warmed by ozone in the atmosphere, but it can also be warmed or cooled by various PM components. For instance, particulate sulfates chill the earth's atmosphere, yet black carbon, a particulate pollutant from burning, warms the planet (Forster et al. 2007).

The ground-level ozone occurs in the atmospheric air closer to earth, usually, and is produced chemically through photochemical reactions under sunlight by interactions between nitrogen oxides (NOx) released in car exhaust and VOCs from industrial emissions. VOCs enter the atmosphere from industrial emissions, such as paints, solvents, evaporated gasoline, and natural sources. Therefore, pollution from car exhaust, refineries, industrial boilers, power plants, and chemical industries is the source of the elevation of the ground levels of ozone (Bloomer et al. 2009).

The manufacturing sector releases toxic waste vapors that are hazardous to health as well as the ecosystem, which is considered to be the main source of global air pollution. The health and safety of nearby neighbors as well as plant workers may be impacted by these emissions, which can also cause environmental degradation, offensive odors, and air pollution. A wide range of sectors are linked to the release of VOCs, including flexographic printing, food and beverage, biogas, medicines, oil, and gas. Nonetheless, there are great effects of the industrial emissions on climate change (Khajeh Hoseini et al. 2022).

Chlorofluorocarbons (CFCs) and hydrochlorofluorocarbons (HCFCs) are wholly or partially halogenated hydrocarbons containing carbon (C), hydrogen (H), chlorine (Cl), and fluorine (F). Perfluorinated compounds (PFAS) release HCFC-22, a GHG that is 5000 times more powerful than CO_2. Numerous CFCs have found widespread use as solvents, gaseous fire suppression systems, propellants (in aerosol applications), refrigerants, air conditioners, and foam insulation. CFCs contribute to ozone depletion in the upper atmosphere. Unfortunately, in many cases, CFCs were replaced by HFCs, which do not damage the ozone layer but are similarly effective in global warming phenomena (Darby 2014).

The main cause of the waste industry's climate impact is methane, which is created during the anaerobic breakdown of organic materials in landfills. One of the unique greenhouse gases referred to as short-lived climate pollutants (SLCPs) is methane. As their name suggests, these gases do not last long, but they have a profound effect. For instance, although soot is a short-lived secondary light-colored particle, over a 20-year period it can have an impact 4000 times greater than carbon dioxide (Bailey 2017).

Air pollutants, such as methane and black carbon, are powerful SLCPs that contribute to climate change and ill health. Although SLCPs persist in the atmosphere for short lifetimes, their global warming potential is often much greater than CO_2. Black carbon, a component of fine particulate matter, is one of the largest contributors to global warming after CO_2. Black carbon warms the earth's atmosphere by absorbing sunlight, thereby accelerating the melting of snow and ice (Klimont et al. 2017).

Although methane is likewise transient, it leaves a significant impression. Methane will be 84 times more potent than CO_2 during the next 20 years, meaning that every ton of methane will trap the same amount of heat in the atmosphere as 84 tons of carbon dioxide (Bailey 2017).

Polycyclic aromatic hydrocarbons (PAHs) are organic pollutants made up of two or more fused aromatic rings of hydrogen and carbon atoms, which are usually colorless, white, or light yellow-colored solid substances. The primary cause of PAH emissions is from a variety of industrial processes, including waste incineration, the production of iron and steel, aluminum, cement, coal-tar pitch, dyes, asphalt, rubber tire manufacturing, the production of fungicides and insecticides, refinery exhaust, and the generation of electricity. Coal gasification, electric arc furnaces, oxygen furnaces, diesel engines, and gasoline-powered engines of big machinery are further industrial sources of PAH emissions. Open burning of biomass and burning of agricultural wastes are agricultural sources of PAH emissions (Ravindra et al. 2008).

The two main processes by which PAHs are changed in the atmosphere are ozone-induced oxidation and hydroxylation, both of which are triggered by sunshine. UVA and visible light can be absorbed by PAHs. UVA light absorption causes the cells to create reactive intermediates when excited PAHs combine with oxygen or other molecules, as well as reactive species when electrons or energy are transferred from the excited PAHs. Proteins, nucleic acids, and cell membranes are among the biological components that are harmed by these reactive species (Yan et al. 2004).

3.3 Impact of Climate Change on the Bioavailability of Occupational Toxins

There are many researches considering the impacts of climate change on the chemical and biological toxins. Climate change is likely to increase human exposures to air chemical pollutants through several mechanisms. Directly, an increase in the atmospheric temperature may enhance the release and degradation of chemicals, such as agricultural contaminants particularly pesticides and fertilizers, and may increase its toxicity (Quiroga et al. 2019). Indirectly, seasonal and quantity variation in precipitation found to have a significant role in the transportation and mobilization of chemical pollutants to contaminate underground water, as well as mobilization to unpolluted areas (Bloomfield et al. 2006). Therefore, there is increasing risk of exposure to these agricultural contaminants among agricultural workers especially among pesticide sprayers.

The increases in exposure and the magnitude of toxicity of chemical pollutants are highly dependent on the chemical composition of the contaminant and its physical criteria (Boxall et al. 2009). Transportation, dispersion, and deposition of air chemical pollutants are mainly dependent on the wind speeds and directions, as stagnation of the wind speed is suspected to cause increases in the concentration of pollutants (Saad-Hussein and Ramadan, 2023). Moreover, there are seasonal variations detected in ambient air pollution levels of some chemicals, such as VOCs, that were commonly polluting water bodies and underground waters, soils and sediments, and biological systems, in addition to the ambient air (Egbuna et al. 2021).

Global warming especially in humid ill-ventilated areas was found to cause increased exposure to high concentrations of fungi and a possibility of elevation in the mycotoxin levels in the biological fluids of the exposed workers (Saad-Hussein et al. 2012a, b, 2016a, b). Therefore, climate changes have proved to have the possibility to increase the toxicity and bioaccumulation of the pollutants in the environment and in living creatures.

3.4 Impacts of Climate Change on Workers' Health

3.4.1 Susceptibility and Vulnerability

Mostly all over the world, workers are underlying low socioeconomic communities, especially in poor nations, which increase the risk of being the most vulnerable population to climate change impacts (Nerbass et al. 2017). Always these low economy' communities are living in crowded inadequate ventilated houses, and the majority of them perform physical activities with insufficient nutritional requirements, with irregular pay-by-output occupations with little or limited healthcare providence.

Climate change has potentially enhanced the frequency and severity of certain current occupational hazards, as outdoor workers are directly exposed to the increased temperature, polluted air, and extreme weather, as well as introducing new and unforeseen dangers (Ramadan and Mourad 2022). The increase in the environmental pollution is found to be due to reduction in the dispersion of air pollutants due to interference with convection currents and enhancement in the formation of photochemical pollutants such as smog and increase in the release of pollens and fungal spores (Tang et al. 2011).

Moreover, the urban heat island effect is a situation that causes temperatures to be somewhat higher in the urban areas than in the nearby suburban and rural areas, due to the low green areas, so air pollution concentrations may increase, which could explain why death rates are higher (Le Tertre et al. 2006).

Humidity and heat waves exacerbate the urban heat island affect, which leads to an increase in mortality rates due to the higher heat load, the longer the duration, and the higher rate of change and less acclimatization (Read 2004). Humidity rises may worsen the effects of intense heat on human health. In addition, humidity with warming conditions in ill-ventilated workplaces could increase the possibility of fungal growth and increase the hazardous effects of respiratory fungal infections among the exposed workers (Saad et al. 2006; Saad-Hussein et al. 2012b; Morcos et al. 2019), as well as increase the rate of their exposure to mycotoxins (Saad-Hussein et al. 2012a, 2016a, b).

Climate change has indirect impacts on indoor workers, especially those who work in hot indoor workplaces; e.g., in steel mills, dry cleaners, manufacturing plants, warehouses, and other industries without air conditioning, they are exposed to excessive heat and/or indoor air pollution (Ramadan and Mourad 2022). Indoor workers are vulnerable to climate change consequences, such as heat-related stress, heat exhaustion, heat stroke, reduced productivity, increased accident rates, acute myocardial events, and cerebrovascular events, as well as the development of acute exacerbations of asthma and COPD due to their exposure to the worsening air pollution, besides the increase in their exposure to hospitable habitats vector-borne diseases (Sherman et al. 2023).

Aside from these hazardous impacts of climate change and exposure to air pollutants, whether outdoor or indoor, there are various variables that form risk factors that enhance workers' vulnerability to climate-related occupational risks; these factors include individual, institutional, and environmental factors (Habibi et al. 2021). Individual variation is mainly dependent on the personal data of the worker, in the form of age, gender, nutritional status, and medical history of chronic diseases; these factors have an important role in their adjustment to climate change impacts through their physiological, behavioral, and occupational daily activities, in addition to technological improvements (Amoadu et al. 2023).

Although total physiological acclimatization to hot weather may take several years, it might happen in a matter of days in younger ages, less physically active persons, and according to their behavioral clothes, frequency of drinking water, as well as the ventilation of the working places. It was detected that elevation of UV radiation exposure may also influence human immunity and change the body's capacity to eliminate the first mutated cells that trigger the start of cancer (Toyooka et al. 2006). Moreover, the rate of changes in infrastructure and technologies used will also significantly affect the tolerance with climatic change variations.

Additionally, the increase in the incidence of malnutrition and starvation, as well as the increase in the incidences of the food-borne diseases and food poisoning due to rapid multiplication of microorganisms due to weather warming, leads to an increase in the vulnerability of the workers to climate change impacts (Stanke et al. 2013). Also, there is a possibility of an increase in the incidences of fascioliasis and food-borne trematodes due to the effects of increasing atmospheric temperature on snails, the intermediate host of these trematodes (Zanaty et al. accepted 2024). Drought also increases the risk to the development of water-borne diseases and poor hygiene diseases due to drinking unsafe water and the securing of safe water (Stanke et al. 2013). Food insecurity, water scarcity, a delicate marine ecosystem, dependence on seafood, and small island communities are increasing factors to vulnerability to climate change (Begum et al. 2022). Therefore, all these factors increase the vulnerability of the workers to the impacts of climate change.

In addition to the previous factors, environment–gene interaction plays a very important role in increasing personal susceptibility to occupational and environmental health problems. Examples of this interaction, *GST* gene polymorphisms, appeared to have a significant role in increasing susceptibility to hepatotoxic effects of occupational exposure to toxic chemicals. It was found that pesticide sprayers with *GSTT1* were the most sensitive workers to hepatotoxicity (Saad-Hussein et al. 2022a, b), as well as to liver cancers (Saad-Hussein et al. 2019).

Occupational exposure to organic dust, such as cotton, flour, and wood dust, is proved to cause the elevation of aflatoxin levels (AFB_1) in the biological fluids of exposed workers (Saad-Hussein et al. 2012a, b; 2013, 2016a, b). It was found that workers with elevated AFB_1 were susceptible to hepatic parenchymal cell injury, except the workers with *GSTT1 and GSTM1* null genotypes (Saad-Hussein et al. 2016a, b, 2022a, b).

The genes' polymorphism for human apurinic/apyrimidinic endonuclease (*APEX1*) and alpha-1-anti-trypsin (*A1AT*) gene mutation also plays an important role in susceptibility to lung cancers in workers occupationally exposed to PAHs. It was found that lung cancer tumor biomarkers were significantly elevated in the aluminum workers with high levels of benzo(a)pyrene diol epoxide albumin adduct (BPDE-Alb) carrying homozygous *APEX1* genotype *Glu/Glu,* as well as in those carrying *A1AT* mutant allele (Moubarz et al. 2023).

Therefore, it is important to study gene polymorphisms of the workers during selection of their type of work to decrease the risk of developing occupational health problems due to the interaction phenomena between environmental exposures and gene susceptibility.

3.4.2 HealthCare Infrastructure Role

The increase in poverty and citizens of the poor countries may increase the vulnerability to climate change; due to the unfavorable environments, and low socioeconomic status (Hamidi et al. 2020). This could be because of the lack of infrastructure, technology, and physical capital necessary to adapt and prepare to deal with the risks associated with climate change events (Sarkodie and Strezov 2019). Climate change extreme events can be threatening risks to infrastructure, such as healthcare services, roads, bridges, and buildings (Maxwell et al. 2018). For example, sea level raise, due to melting ice sheets and thermal expansion, will force some people to migrate from their areas to other areas, which could increase the development of slums, homelessness, and crime rates. In addition, there will be an increased demand for special healthcare services, protective programs, and favorable housing. Moreover, the developing slums will suffer from overcrowding, increased pollution, overheating, unimagined stress on buildings, roads, and unseen infrastructure (Clarke 2018), as well as the disturbance in the transportation networks and food and water sources. All these factors increase the stress on the population in general, especially on the productive categories to overcome these prepending stresses.

Moreover, these extreme events are straining healthcare workers, disrupting physical infrastructure, and interrupting supply chains, making it more difficult to reach those in need of medical interventions, providing medical care for patients, and threatening healthcare quality and safety due to the shortages of medications and hospital facilities. Increasing in the weather temperature may lead to the destruction of the medications, as well as pharmaceutical raw materials, and with worsening medication shortages for drugs, the risk of unpreventable harm and deaths raised (Kelley et al. 2015).

Therefore, healthcare infrastructure must be resilient to the climate change effects through climate-health-infrastructure-pathways (CHIP) model presented by the University of Cambridge, UK. The main pathways of the CHIP model that must be taken into consideration are climate hazards, vulnerability characteristics, physical infrastructure design that pose threats to health, and risks (direct or indirect). The direct risks include infrastructure damage and exposure to extreme weather conditions like droughts, floods, fires, or elevated temperatures. Indirect risks include things like contaminated water supplies, air pollution, and difficulty accessing healthcare. The importance of infrastructure readiness in promoting health outcomes is very essential to minimize the climate change risks, and it should be taken into account when managing and developing infrastructure systems (Wake 2023).

3.4.3 Competencies and Capacities Needed

Climate change threatens human health, especially in developing countries with large physical workforce sectors carrying out their tasks under hot temperatures with a lack of safety regulations that require global efforts to combat the increase in global temperatures and the health effects of climate change (Ansah et al. 2021).

There are several published evidences linking climate change and the increase in the frequency of health problems among workers, such as injuries, fatigue, heat exhaustion and stress, psychological stress, and acceleration of the attacks from chronic diseases, such as cardiovascular, respiratory, cancers, and kidney diseases, that may lead to death (Lundgren et al. 2013; Berhane et al. 2016; Moda et al. 2019; Heydari et al. 2021). Therefore, there are essential requirements for developing economies that prioritize cutting-edge researches to provide needed information, occupational health and safety, policies, campaigns, and interventions that protect the ecosystem and the workforce (Ansah et al. 2021).

In the Caribbean, longitudinal surveys revealed potential changes in health professional behaviors, communications, engagement, and applications toward climate change impacts on health after their implementation in 10 weeks' live-virtual, regional-specific, and evidence and competency courses targeting health risks sensitive to climate change. The three essential items covered in the courses were increasing knowledge about the communication between climate and health, incorporating this knowledge and the skills of the health professional practices, and engaging them with climate and health initiatives within their communities (Sorensen et al. 2023).

Globally, understanding the perceptions of occupational health and safety (OHS) professionals regarding climate-related hazards on workers' health and performances is an essential step to implement future educational and training programming, researches, and translating activities worldwide (Pedersen et al. 2021). The findings of the study done by Pedersen et al. (2021) revealed that the impacts of OHS in specific region are influenced by social and economic factors. So, it is important to train OHS workforce about identify and mitigate climate-related hazards. Trained OHS can have important role in mitigation of the emissions causing global warming and its impacts on the workers' health through designing and implementing workplace controls measures.

Moreover, Jagals and Ebi (2021) designed cross-sectoral subcompetencies and learning framework content that can be used for training health workers to function at local, regional, and global levels to increase their responses to climate and environmental risks and impacts. This framework contained six domains: knowledge related to climate and environmental sciences, drivers of climate change, iterative risk management, evidence, projections and assessments of climate and environmental risks and impacts, and methods for mitigation, adaptation, and health cobenefits, in addition to the global strategies and the international/regional/local agreements and frameworks.

Therefore, the development of an educational reaching program, maintaining and improving the different levels of competence of the health workforce, will mitigate climate change risks and impacts on the general population as well as in occupational health of the workers in different sectors.

3.5 Interaction Between Climate Change and Occupational Toxins on Health

Numerous chemicals used and produced in the workplace have the potential to affect the climate and the environment; on the other hand, climate change can have an impact on how safe it is to use, store, and transport chemicals. Environmental degradation puts 1.2 billion jobs, or 40% of the world's workforce, at risk, and 23 million working-life years were lost annually because of various environment-related hazards. As global warming gets worse, it will destroy employment and livelihoods, harm infrastructure, and disrupt business activity (ILO 2023).

It poses an immediate threat to the expansion of the real gross domestic product and to the working conditions and productivity of labor. On the other hand, addressing these issues could strengthen economies and raise working standards (ILO 2023).

3.5.1 Occupational Carcinogens Toxicity

Burning fossil fuels releases particulate matter (PM) and other carcinogenic substances including sulfur and nitrogen dioxide in addition to greenhouse gases. The International Agency for Research on Cancer has classified fine particles PM2.5, or particles less than 2.5 microns, as a group 1 carcinogen because they are very dangerous (IARC 2013).

Numerous factors, such as increased UV exposure, the possibility of being exposed to air pollution and hazardous chemicals, extreme weather effects, and heat waves, are linked to an increased risk of cancer due to climate change. Air pollution is believed to be responsible for approximately 14% of lung cancer cases worldwide (Hiatt et al. 2020).

Increased exposure to harmful chemicals that are known or suspected of causing cancer after periods of heavy rainfall and increased chemical volatilization in response to rising temperatures are two potential direct effects of climate change on cancer. When there is a lot of rain or flooding, there may be more leaching of heavy metals and harmful chemicals from storage sites, as well as more runoff contaminating water with pollutants that are already persistent in the ecosystem. Mammals that live in the sea can also suffer direct signs of cancer that are associated with long-term or persistent exposure to chemical pollutants in the marine environment (MacDonald et al. 2002).

The use of pesticides, as an example, is found to be increasing to overcome the rapid growth and maturation of the different insects, as well as the increase in its frequencies, due to climate change. This increase in exposure to pesticides was found to cause an increase in hepatic tumor biomarkers, the DNA damage detected by comet assay, shortening of telomere length, and decrease in telomerase enzyme activity among pesticide sprayers compared to their controls, especially among those carrying $GSTT_1$ genotype (Saad-Hussein et al. 2019). The authors recommend to use telomere relative length and comet assay with *GST* gene polymorphisms for early prediction of liver cancer susceptibility in workers exposed to pesticides.

Moreover, epigenetics studies will give a full picture of predisposition to gene mutations due to multiple environmental pollutants. Epigenetic changes were found to be associated with pesticide exposure among urban-exposed workers compared to rural workers, and that could be attributed to urbanization quality of life including dietary habits, working and living in closed areas, and exposure to extra pollutions emitted from urbanization sources (Saad-Hussein and Taha 2020).

UV radiation exposure will rise because of the depletion of stratospheric ozone, another direct consequence of climate change. Exposure to UV radiation raises the risk of both skin cancer and cataracts. There are other variables including temperature and exposure to other compounds that can amplify the carcinogenic potential of UV radiation (Tucker 2009).

One of the most carcinogenic PAHs is benzo(a)pyrene, which is usually used as an exposure marker for risk evaluations. PAH exposure cannot be avoided. Three important ways of exposure to PAHs include ingestion, inhalation, and skin contact. Most of the exposure happens at work, such as in food processing and coke manufacturing industries, or for traffic who inhale PAH-containing road dust and vehicle exhaust. It is extremely toxic, immune toxicogenic, mutagenic, carcinogenic, and teratogenic to all living forms (Abdel-Shafy and Mansour 2016).

PAHs can be more phototoxic when paired with the increase in UV radiation exposure, which can harm DNA (Toyooka et al. 2006). It was found that although occupational exposure to PAHs was for a long time, even if its concentrations in the workplace were within the permissible limits, DNA damage was detected in the exposed workers and was expressed by high BPDE-albumin adduct levels (Moubarz et al. 2023). The long-term exposure to PAHs can cause tumors in the skin, lung, esophagus, colon, pancreas, bladder, and breast (Lee, 2010).

Climate change is also linked with the increase in exposure to fungi especially among workers exposed to organic dusts, which were found to cause elevation in the levels of aflatoxin B1 (AFB_1) and its metabolite AFM_1 in the biological fluids of the exposed workers, and these elevations were associated with the elevation of liver cancer biomarkers (Saad-Hussein et al. 2013, 2014).

3.5.2 Occupational Cardiotoxicity

Numerous studies (Jerrett et al. 2009; Thurston et al. 2016; Pinault et al. 2017) have demonstrated that exposure to $PM_{2.5}$ and other pollutants such as sulfur dioxide (SO_2) and nitrogen oxides such as nitrogen dioxide (NO_2), photochemical oxidants such as O_3, both short and long term, raise the risk of hospital admissions and cardiovascular mortalities, especially ischemic heart diseases.

Particulate matter is linked to numerous pathophysiological alterations, such as pulmonary embolism, deep vein thromboses, blood vessel dysfunction, atherosclerotic disease, systemic inflammation, and abnormal coagulation and thrombosis (Brook et al. 2008). The short-term exposure to PM2.5 is associated with increases in the incidence of MI (Rajagopalan et al. 2018).

Numerous studies (Shah et al. 2013; Bai et al. 2019) have demonstrated a relation between hospitalizations for heart failure (HF) and exposure to air pollution. Cardiovascular problems were found to be negatively impacted by exposure to fuel mixtures, which were linked to exposure to high emissions of particulates or other pollutants like carbon monoxide and nitrous oxides during the industrial processes (Hill et al. 2009).

Extreme heat can directly cause exacerbations of cardiovascular disease. Also, there is evidence that heat exacerbates the harmful effects of ozone and other particles on heart disease.

A systemic inflammatory response can be triggered by severe cases of heatstroke and hyperthermia. This reaction is mediated by circulating mRNAs that activate inflammatory cytokines, which cause an excessive activation of endothelial cells and leukocytes. Electrolyte imbalances include hypokalemia and hyperkalemia, as well as hypomagnesemia, which raises the risk of arrhythmias, which can be triggered by hyperthermia. On the other hand, bradycardia can result from hypothermia (Heidari et al. 2016).

Dehydration from hyperthermia leads to hemoconcentration, leading to thrombocytosis and leukocytosis in those exposed to high temperatures. A hypercoagulable state brought on by hemoconcentration may raise the risk of thrombosis and myocardial infarction (Stewart et al. 2017).

3.5.3 Occupational Hepatotoxicity

The liver is regarded as a key organ in the detoxification process when a significant quantity of foreign substances or xenobiotics enters the body. Therefore, prolonged exposure to chemicals in the environment will result in chronic intrahepatic exposure to these substances, which will show up as liver disease and decreased liver function (Christensen et al. 2013).

Climate change found to cause an increase of environmental exposure to persistent chemical pollutants, such as persistent organic compounds (Saad-Hussein and Ramadan 2023). Workers highly occupationally exposed to persistent chemicals are at risk to develop toxicant-associated steatohepatitis (TASH), which includes hepatocyte necrosis, higher proinflammatory cytokine, lowered antioxidants, insulin resistance with altered adipokines, and mitochondrial dysfunction, such as workers exposed to vinyl chloride (Cave et al. 2010), polychlorinated biphenyls (Shi et al. 2012).

Occupational exposures to high concentrations of VOCs were found to be associated also with the development of TASH (Brautbar and Williams 2002). Human exposure to VOCs occurs by either inhalation or dermal absorption (Agency 2017).

3.6 Climate Change Impact on Occupational Health and Safety

Detection and prevention of climate change impacts on occupational health are necessary steps to decrease the risk of work-related diseases and injuries through the politicians, governments, and employers and the health safety staff, as well as the workers to safeguard employees against the long-term health effects of the synergistic effects of climate change and occupational exposures (Ramadan and Mourad 2022).

Reduction of combustion of fossil fuels, whatever its source is, results in reduction of the emissions of CO_2, NOx, and PM. Reduction of NOx emissions leads to restriction in the formation of ozone and nitrate aerosols, as well as other secondary pollutants (Pinho-Gomes et al. 2023). This reduction can be achieved by switching to physically active travel as an example; therefore, physical activities will improve physical and mental health as well as reduce the risk of obesity (Quam et al. 2017).

3.7 Responsibilities of Occupational Medicine Provider in Management of Climate Change-Related Health Problems

Workers are susceptible to the health hazards due to their exposure to environmental changes, both environmental pollution and climate changes. Direct climate change hazards include increasing atmospheric temperature in the form of heatwaves, reduction in air quality due to wind stagnation, and allergen sensitization due to seasonal variation of the agricultural products, while indirect climate change hazards include disaster zone exposures such as floods and dryness, stress and mental health impacts from the disasters, and waterborne and vector-borne disease due to variation of the seasonal growth and the changes in their habitats (Perkison et al. 2018).

Occupational and environmental medicine (OEM) provider must be qualified and have the needed competencies to recognize occupational risks and how to mitigate hazards, recognize, and respond to health effects in the occupational environment. Cloeren et al. (2014) mentioned that OEM provider should be competent to address individual and organizational factors needed to control the climate change health impacts on the workers' health and initiate programs to manage the productivity of workers and create policies for healthy workforce. In addition, OEM providers must be trained and provided with the required knowledge about the direct and indirect impacts of climate change on the working areas to detect, respond, and manage climate change-related health hazards on the exposed workers and react within the organizations to mitigate health risks and respond to health crises according to the combined effects of hazards in the workplace and climate change (Applebaum et al. 2016).

Thus, the most important responsibility of OEM could be summarized in the following issues:

1. Prepare suitable guidance to be provided to the employers explaining ways for protection of outdoor workers potentially exposed to the extreme temperatures (Balbus et al. 2016; Perkison et al. 2018).
2. Reorganization of the workers with acute and chronic health problems, such as cardiovascular and respiratory diseases, sensitive to the increasing temperature and worsening air quality due to chemical or biological pollutants, inform of high concentrations of ozone, particulate matter, and pollen count (Crimmins et al. 2016; Fann et al. 2016; Perkison et al. 2018).
3. Creating a protective action plan with the employer to decrease the potential adverse effects of extreme events before, during, and after the events and be sure of the availability of the essential infrastructure for emergencies (Crimmins et al. 2016; Bell et al. 2016; Perkison et al. 2018).
4. Planning to provide effective guidance to the employers and the managers about seasonal variations and possibility of increasing risks of vector-borne disease among the workers (Crimmins et al. 2016; Beard et al. 2016; Perkison et al. 2018).
5. Provide emergencies clinical management and isolation of employees affected by food-borne illnesses and work with the employers and the mangers in protecting the healthcare providers for those individuals (Crimmins et al. 2016; Ziska et al. 2016; Perkison et al. 2018).
6. Provide the employees at risk for mental health problems and support them to overcome the facing stressors due to disasters, losses, and migrations by providing them more comprehensive programs through their employment (Crimmins et al. 2016; Dodgen et al., 2016; Perkison et al. 2018).
7. Creating communication strategies during climate change extreme events related to climatic scenarios, inform exposure to extreme heat, poor air quality, reduce food and water quantity, concomitant UV exposure, changes in infectious agents, and population displacement (Crimmins et al. 2016; Perkison et al. 2018).

3.8 Role of Mitigation and Adaptation

The essential strategy to minimize the hazardous effects of climate change is the combination of mitigation and adaptation strategies. According to IPCC definition: "*Mitigation strategy is an intervention to reduce the emissions sources or enhance the sinks of GHGs,*" and "*Adaptation strategy is an adjustment in natural or human systems in response to actual or expected climatic stimuli or their effects, which moderates harm or exploits beneficial opportunities*" (IPCC 2001).

Mostly, the use of coal and fossil fuel for the production of energy increases the emission of GHGs. Therefore, according to IPCC-AR6 2022, the reduction of the usage of fossil fuel and switching to alternative energy sources depending on renewable energy, such as geothermal, small hydro, biomass cogeneration, and wind energy, will decrease GHG emissions and mitigate climate change (IPCC 2022).

World Bank (2010) reported that the global costs of adaptation in developing countries were estimated to be between 70 and 100 billion $US per year to the year 2050, as adaptation was found to be closely linked to poverty and development to reduce vulnerability to climate change. In African countries to adapt to climate change, simultaneously reducing vulnerability and increasing resilience are essential. Therefore, adaptation strategies and investment planning must be integrated into national and regional strategies and policies as part of multisectoral approaches (Connolly-Boutin and Smit 2016). But, in Africa, these adaptation strategies are not symmetrical, and only a few countries have documented their efforts. Egypt, as an example, has undertaken many activities to deal with the issue of climate change, starting with the signature of the Kyoto Agreement, and its confirmation of signing the Paris Agreement in 2015. According to these agreements, the first national reporting was in 1999, the second national reporting was in 2010, the third national reporting was in 2016, and the fourth national reporting was in 2022. Egypt also established the National Council of Climate Change (ENCCC), which represented all the Egyptian Ministries to discuss all the essential issues related to climate change and collaborate on the efforts for adaptation and mitigation between the different Ministries. Through the COP26 in the year 2021, Egypt launched the National Climate Change Strategy 2050, which was presented in COP27 held in Sharm El Sheikh, Egypt.

3.9 Research Gaps

While the influence of climate change on human health is undeniable, it is still difficult to precisely quantify the scope and gravity of many dangers associated with climate change, especially among occupational workers exposed to additional sources of health risks. As science progresses, it will be more precisely identify the dangers and scope of occupational health problems and to link them to global warming.

Climate change is an environmental issue essential to develop specific scientific researches to reach the best means of adaptation and mitigation and its effects. It is also linked indirectly with education and training, urban development, and foreign policies.

Finding national, regional, and international strategies for mitigation and adaptation to climate change risky problems is the most targeted objective of the policy-makers and stockholders. Vulnerability assessment of occupational health problems is important to provide information that can be used to develop management actions in response to climate change and inform mitigation and adaptation strategies. Tiepolo and Bacci (2017) denoted that freely accessible information is required to develop proper management actions for monitoring and evaluation programs, as well as resilience and adaptation projects for climate change, and that vulnerability tracking is a regional and national decision-making tool for designing suitable adaptation policies.

Therefore, future researches are needed to focus on three primary areas: tracking vulnerability to climate change, especially for occupational health, identification of the effects of climate change on the workers' health, and the collection of data supporting future projections of how climate change will affect occupational health.

Moreover, infrastructure solutions are also an important area that needs further research to help avoid or even minimize these health hazards.

References

Abdel-Shafy HI, Mansour MSM (2016) A review on polycyclic aromatic hydrocarbons: Source, environmental impact, effect on human health and remediation. Egypt J Pet 25:107–123. https://doi.org/10.1016/j.ejpe.2015.03.011

Agency EP (2017) Technical overview of volatile organic compounds. Indoor Air Quality, Washington, DC. https://www.epa.gov/indoor-air-quality-iaq/what-are-volatile-organic-compounds-vocs

Amoadu M, Ansah EW, Sarfo JO, Hormenu T (2023) Impact of climate change and heat stress on workers' health and productivity: a scoping review. J Clim Chang Health 12:100249

Ansah EW, Ankomah-Appiah E, Amoadu M, Sarfo JO (2021) Climate change, health and safety of workers in developing economies: a scoping review. J Clim Chang Health 3:100034

Applebaum KM, Graham J, Gray GM et al (2016) An overview of occupational risks from climate change. Curr Environ Health Rep 3:13–22

Bai L, Shin S, Burnett RT et al (2019) Exposure to ambient air pollution and the incidence of congestive heart failure and acute myocardial infarction: A population-based study of 5.1 million Canadian adults living in Ontario. Environ Int 132:105004. https://doi.org/10.1016/j.envint.2019.105004

Bailey K (2017) How the waste industry can lead on climate change. https://www.waste360.com/emissions/how-the-waste-industry-can-lead-on-climate-change

Balbus J, Crimmins A, Gamble JL et al (2016) Introduction: climate change and human health. In: The impacts of climate change on human health in the United States: a scientific assessment. US Global Change Research Program, Washington, DC, pp 26–42

Beard CB, Eisen RJ, Barker CM et al (2016) Vector-borne diseases. In: The impacts of climate change on human health in the United States: a scientific assessment. US Global Change Research Program, Washington, DC, pp 129–155

Begum RA, Lempert R et al (2022) "Chapter 1: point of departure and key concept" (PDF). In: Climate change 2022: impacts, adaptation and vulnerability. The sixth assessment report of the intergovernmental panel on climate change. Cambridge University Press, p 170

Bell JE, Herring SC, Jantarasami L et al (2016) Impacts of extreme events on human health. In: The impacts of climate change on human health in the United States: a scientific assessment. US Global Change Research Program, Washington, DC, pp 99–128

Berhane K, Kumie A, Samet J (2016) Health effects of environmental exposures, occupational hazards and climate change in Ethiopia: synthesis of situational analysis, needs assessment and the way forward. Ethiop J Health Dev 30(1 Spec Iss):50–56

Bloomer BJ, Stehr JW, Piety CA, Salawitch RJ, Dickerson RR (2009) Observed relationships of ozone air pollution with temperature and emissions. Geophys Res Lett 36(9). https://doi.org/10.1029/2009GL037308

Bloomfield J, Williams R, Gooddy D et al (2006) Impacts of climate change on the fate and behaviour of pesticides in surface and groundwater—a UK perspective. Sci Total Environ 369:163–177

Boxall AB, Hardy A, Beulke S et al (2009) Impacts of climate change on indirect human exposure to pathogens and chemicals from agriculture. Environ Health Perspect 117(4):508–514. https://doi.org/10.1289/ehp.0800084

Brautbar N, Williams J II (2002) Industrial solvents and liver toxicity: risk assessment, risk factors and mechanisms. Int J Hyg Environ Health 205:479–491

Brook BW, Sodhi NS, Bradshaw CJ (2008) Synergies among extinction drivers under global change. Trends Ecol Evol 23(8):453–460. https://doi.org/10.1016/j.tree.2008.03.011

Cailee H (2022) PCPHCoP (poison center & public health collaborations Community of practice) newsletter is Climate Change and Toxic Exposures; September 2022 edition

Cave M, Falkner KC, Ray M et al (2010) Toxicant-associated steatohepatitis in vinyl chloride workers. Hepatology 51:474–481

Christensen KLY, Carrico CK, Sanyal AJ, Gennings C (2013) Multiple classes of environmental chemicals are associated with liver disease: NHANES 2003–2004. Int J Hyg Environ Health 216(6):703–709. https://doi.org/10.1016/j.ijheh.2013.01.005

Clarke L (2018) Ch.17: sector interactions, multiple stressors, and complex systems. In: Impacts, risks, and adaptation in the United States: fourth national climate assessment, volume II. U.S. Global Change Research Program, Washington, DC, p 643

Climate Action (2020) 2020 Progress Report. Climate Action 100+ initiative. https://www.climate-action100.org/wp-content/uploads/2020/12/CA100-Progress-Report.pdf

Cloeren M, Gean C, Kesler D et al (2014) American College of Occupational and Environmental Medicine's occupational and environmental medicine Competencies-2014: ACOEM OEM competencies task force. J Occup Environ Med 56:e21–e40

Connolly-Boutin L, Smit B (2016) Climate change, food security, and livelihoods in sub-Saharan Africa. Reg Environ Chang 16:385–399. https://doi.org/10.1007/s10113-015-0761-x

Crimmins A, Balbus JL, Gamble CB et al (2016) The impacts of climate change on human health in the United States: a scientific assessment. US Global Change Research Program, Washington, DC. Available at: https://doi.org/10.7930/J0BZ63ZV

Darby M (2014) Ozone layer treaty could tackle super polluting HFCs. rtcc.org. Archived from the original on 19 August 2014. Retrieved April 11, 2022. https://web.archive.org/web/20140819125632/, http://www.rtcc.org/2014/07/15/ozone-layer-treaty-could-tackle-super-polluting-hfcs/

Davda K (2024) How do industrial emissions contribute to climate change? Published on March 4, 2024. https://oizom.com/how-do-industrial-emissions-contribute-to-climate-change/. Available 4th May 2024

Dodgen D, Donato D, Kelly N et al (2016) Mental health and Well-being. In: The impacts of climate change on human health in the United States: a scientific assessment. US Global Change Research Program, Washington, DC, pp 217–245

Dunne JP, Stouffer RJ, John JG (2013) Reductions in labour capacity from heat stress under climate warming. Nat Clim Chang 3(6):563–566

Egbuna C, Amadi CN, Patrick-Iwuanyanwu KC et al (2021) Emerging pollutants in Nigeria: a systematic review. Environ Toxicol Pharmacol 85:103638

EPA (United States-Environmental Protection Agency) (2024) Overview of greenhouse gases. https://www.epa.gov/ghgemissions/overview-greenhouse-gases. Available 4 May 2024

Fagbohunka A (2015) The influence of industrial clustering on climate change: an overview, economic and environmental studies (E&ES), ISSN 2081-8319, Opole University, Faculty of Economics, Opole, Vol. 15, Iss. 4, pp. 433–443

Fann N, Brennan T, Dolwick P et al (2016) Air quality impacts. In: The impacts of climate change on human health in the United States: a scientific assessment. U.S. Global Change Research Program, Washington, DC, pp 69–98

Forster P, Ramaswamy V, Artaxo P, Berntsen T, Betts R, Fahey DW et al (2007) Changes in atmospheric constituents and in radiative forcing. In: Climate Change 2007: the physical science basis contribution of Working Group I to the Fourth Assessment Report of the Intergovernmental Panel on Climate Change. Cambridge University Press, Cambridge

Habibi P, Moradi G, Moradi A, Heydari A (2021) The impacts of climate change on occupational heat strain in outdoor workers: a systematic review. Urban Clim 36:100770. https://doi.org/10.1016/j.uclim.2021.100770

Hamidi AR, Zeng Z, Khan MA (2020) Household vulnerability to floods and cyclones in Khyber Pakhtunkhwa, Pakistan. Int J Disaster Risk Reduct 46:101496

Heidari L, Winquist A, Klein M, O'Lenick C, Grundstein A, Ebelt Sarnat S (2016) Susceptibility to heat-related fluid and electrolyte imbalance emergency department visits in Atlanta,

Georgia, USA. Int J Environ Res Public Health 13(10):982. Published 2016 Oct 2. https://doi.org/10.3390/ijerph13100982
Heydari P, Moradi G, Moradi A, Heydari A (2021) The impacts of climate change on occupational heat strain in outdoor workers: a systematic review. Urban Clim:36. https://doi.org/10.1016/j.uclim.2021.100770
Hiatt RA, Beyeler N (2020) Cancer and climate change [published correction appears in Lancet Oncol. 2020 Dec;21(12):e553. https://doi.org/10.1016/S1470-2045(20)30698-7]. Lancet Oncol 21(11):e519–e527. https://doi.org/10.1016/S1470-2045(20)30448-4
Hill J, Polasky S, Nelson E, Tilman D, Huo H, Ludwig L, Neumann J, Zheng H, Bonta D (2009) Climate change and health costs of air emissions from biofuels and gasoline. Proceedings of the National academy of sciences of the United states of America 106(6):2077–2082. https://doi.org/10.1073/pnas.0812835106
IARC (International Agency for Research on Cancer) (2013) IARC: Outdoor air pollution a leading environmental cause of cancer deaths. Press Release No. 221. https://www.iarc.who.int/wpcontent/uploads/2018/07/pr221_E.pdf
ILO (2023) Chemicals and climate change in the world of work: impacts for occupational safety and health. Research report. https://www.ilo.org/wcmsp5/groups/public/%2D%2D-ed_dialogue/%2D%2D-lab_admin/documents/publication/wcms_887111.pdf
Intergovernmental Panel on Climate Change (IPCC) (2021) Ch. 11: weather and climate extreme events in a changing climate. In: Climate Change 2021: the physical science basis. Contribution of Working Group 1 to the Sixth Assessment Report of the Intergovernmental Panel on Climate Change. Cambridge University Press, pp 11–16. (In press)
IPCC (2001) Climate Change 2001: Mitigation contribution of working group III to the third assessment report of the Intergovernmental panel on climate change. Metz B, Davidson O, Swart R, Pan J eds Cambridge University Press, Cambridge, pp 760. https://www.ipcc.ch/report/ar3/wg3/
IPCC (2022) Summary for policymakers archived 2023-01-22 at the wayback machine [Pörtner H-O, Roberts DC, Poloczanska ES, Mintenbeck K, Tignor M, Alegría A, Craig M, Langsdorf S, Löschke S, Möller, Okem A (eds)]. In: climate change 2022: Impacts, adaptation and vulnerability. Contribution of Working Group II to the Sixth Assessment Report of the Intergovernmental Panel on Climate Change. Cambridge University Press, Cambridge, UK and New York, NY, USA, pp 3–33. https://doi.org/10.1017/9781009325844.001
IPCC (2023) Sections. In: Core Writing Team, Lee H, Romero J (eds) Climate Change 2023: synthesis report. Contribution of Working Groups I, II and III to the sixth assessment report of the Intergovernmental Panel on Climate Change. IPCC, Geneva, pp 35–115. https://doi.org/10.59327/IPCC/AR6-9789291691647
Jagals P, Ebi K (2021) Core competencies for health workers to deal with climate and environmental change. Int J Environ Res Public Health 18(8):3849. Published 2021 Apr 7. https://doi.org/10.3390/ijerph18083849
Jerrett M, Burnett RT, Pope CA III, Ito K, Thurston G, Krewski D, Shi Y, Calle E, Thun M (2009) Long-term ozone exposure and mortality. N Engl J Med 360(11):1085–1095. https://doi.org/10.1056/NEJMoa0803894
Kelley CP, Mohtadi S, Cane MA, Seager R, Kushnir Y (2015) Climate change in the Fertile Crescent and implications of the recent Syrian drought. Proc Natl Acad Sci USA 112(11):3241–3246. https://doi.org/10.1073/pnas.1421533112
Khajeh Hoseini L, Jalilzadeh Yengejeh R, Mohammadi Rouzbehani M, Sabzalipour S (2022) Health risk assessment of volatile organic compounds (VOCs) in a refinery in the southwest of Iran using SQRA method. Front Public Health 10:978354. https://doi.org/10.3389/fpubh.2022.978354
Klimont Z, Kupiainen K, Heyes C et al (2017) Global anthropogenic emissions of particulate matter including black carbon. Atmos Chem Phys 17(14):8681–8723. https://doi.org/10.5194/acp-17-8681-2017
Le Tertre A, Lefranc A, Eilstein D, Declercq C, Medina S, Blanchard M et al (2006) Impact of 2003 heat wave on all-cause mortality in 9 French cities. Epidemiology 17(1):75–79

Lee B-K (2010) Sources, distribution and toxicity of polyaromatic hydrocarbons (pahs) in particulate matter. In: Air pollution, (Villanyi V, ed). Rijeka: Intech, 99–122. https://doi.org/10.5772/10045

Levi M, Kjellstrom T, Baldasseroni A (2018) Impact of climate change on occupational health and productivity: a systematic literature review focusing on workplace heat. Med Lav 109(3):163–179. https://doi.org/10.23749/mdl.v109i3.6851

Lundgren K, Kuklane K, Gao C, Holmer I (2013) Effects of heat stress on working populations when facing climate change. Ind Health 51(3–15). https://doi.org/10.2486/indhealth.2012-0089

Macdonald R, Mackay D, Hickie B (2002) Contaminant amplification in the environment. Environ Sci Technol 36(23):456A–462A. https://doi.org/10.1021/es022470u

Maxwell K et al (2018) Ch. 11: Built environment, urban systems, and cities. In: Impacts, risks, and adaptation in the United States: fourth national climate assessment, volume II. U.S. Global Change Research Program, Washington, DC, p 442

Moda HM, Filho WL, Minhas A (2019) Impacts of climate change on outdoor workers and their safety: some research priorities. Int J Environ Res Public Health 16(18):3458. https://doi.org/10.3390/ijerph16183458

Morcos NYS, Saad-Hussein A, Ibrahim KS, Abou-ElMakarem SR, Abd El-Zaher N, Moubarz G (2019) Study of the immunological changes associated with *aspergillus* infection among ceramic workers. Aerobiologia 35(4):605–612

Moubarz G, Saad-Hussein A, Shahy EM et al (2023) Lung cancer risk in workers occupationally exposed to polycyclic aromatic hydrocarbons with emphasis on the role of DNA repair gene. Int Arch Occup Environ Health 96(2):313–329. https://doi.org/10.1007/s00420-022-01926-9

Naidu R, Biswas B, Willett IR, Cribb J, Kumar SB, Paul NC, Coulon F, Semple KT, Jones KC, Barclay A, Aitken RJ (2021) Chemical pollution: a growing peril and potential catastrophic risk to humanity. Environ Int 156:106616

Neira M, Erguler K, Ahmady-Birgani H et al (2023) Climate change and human health in the Eastern Mediterranean and Middle East: literature review, research priorities and policy suggestions. Environ Res 216(Pt 2):114537. https://doi.org/10.1016/j.envres.2022.114537

Nerbass FB, Pecoits-Filho R, Clark WF, Sontrop JM, McIntyre CW, Moist L (2017) Occupational heat stress and kidney health: From farms to factories. Kidney Int Rep 2(6):998–1008. https://doi.org/10.1016/j.ekir.2017.08.012

Pedersen KM, Busch Isaksen TM, Baker MG, Seixas N, Errett NA (2021) Climate change impacts and workforce development needs in federal region X: a qualitative study of occupational health and safety professionals' perceptions. Int J Environ Res Public Health 18(4):1513. https://doi.org/10.3390/ijerph18041513

Perkison WB, Kearney GD, Saberi P et al (2018) Responsibilities of the occupational and environmental medicine provider in the treatment and prevention of climate change-related health problems. J Occup Environ Med 60(2):e76–e81. https://doi.org/10.1097/JOM.0000000000001251

Pinault LL, Weichenthal S, Crouse DL, Brauer M, Erickson A, van Donkelaar A, Martin RV, Hystad P, Chen H, Finès P, Brook JR, Tjepkema M, Burnett RT (2017) Associations between fine particulate matter and mortality in the 2001 Canadian Census Health and Environment Cohort. Environ Res 159:406–415. https://doi.org/10.1016/j.envres.2017.08.037

Pinho-Gomes A-C, Roaf E, Fuller G, Fowler D, Lewis A, ApSimon H et al (2023) Air pollution and climate change. Lancet Planet Health 7:e727–e728. https://doi.org/10.1016/S2542-5196(23)00189-4

Quam VGM, Rocklöv J, Quam MBM, Lucas RAI (2017) Assessing greenhouse gas emissions and health co-benefits: a structured review of lifestyle-related climate change mitigation strategies. Int J Environ Res Public Health 14:468

Quiroga LB, Sanabria EA, Fornés MW et al (2019) Sublethal concentrations of chlorpyrifos induce changes in the thermal sensitivity and tolerance of anuran tadpoles in the toad Rhinella arenarum? Chemosphere 219:671–677

Rajagopalan S, Al-Kindi SG, Brook RD (2018) Air Pollution and Cardiovascular Disease: JACC State-of-the-Art Review. J Am Coll Cardiol 72(17):2054–2070. https://doi.org/10.1016/j.jacc.2018.07.099

Ramadan MA, Mourad BH (2022) Impact of climate change on occupational health and its possible preventive strategies (review article). Egypt J Occup Med 46(3):47–62. https://ejom.journals.ekb.eg/article_257813_e984daac16730c96d2bc3c533fda781e.pdf

Ravindra K, Sokhi R, Van Grieken R (2008) Atmospheric polycyclic aromatic hydrocarbons: source attribution, emission factors and regulation. Atmos Environ 42:2895–2921. https://doi.org/10.1016/j.atmosenv.2007.12.010

Read C (2004) Climate change and human health: risks and responses. BMJ 328(7451):1324

Saad A, Awad A-H, Aziz H (2006) Assessment of respiratory health problems due to exposure to airborne fungi in ceramics industry. Egypt J Occup Med 30(2):193–216

Saad-Hussein A, Ramadan HKA (2023) Impacts of climate change on environmental toxins and pollutants causing liver health problems. In: Saad-Hussein A, Elwakil R, Khomsi K (eds) Impact of climate change on health in Africa. Springer, Cham. https://doi.org/10.1007/978-3-031-39466-9_3

Saad-Hussein A, Taha MM (2020) Epigenetic study of global gene methylation in PON1, XRCC1 and GSTs different genotypes in rural and urban pesticide exposed workers. J Complement Integr Med. Published online June 11, 2020. https://doi.org/10.1515/jcim-2019-0166

Saad-Hussein A, Elserougy S, Beshir S, Ibrahim MIM, Awad A, Abdel-Wahhab MA (2012a) Work-related airborne fungi and the biological levels of mycotoxin in textile workers. J Appl Sci Res 8(2):719–726

Saad-Hussein A, Morcos NYS, Rizk SA, Ibrahim KS, Abd El-Zaher N, Moubarz G (2012b) *Aspergillus* hazardous problem in ceramics workers. Toxicol Ind Health 28(10):886–893

Saad-Hussein A, Beshir S, Moubarz G, Elserougy S, Ibrahim MIM (2013) Effect of occupational exposure to aflatoxins on some liver tumor markers in textile workers. Am J Ind Med 56(7):818–824

Saad-Hussein A, Taha MM, Beshir S, Shahy EM, Shaheen W, Elhamshary M (2014) Carcinogenic effects of aflatoxin B1 among wheat handlers. Int J Occup Environ Health 20(3):215–219

Saad-Hussein A, Shahy EM, Shaheen W, Taha MM, Mahdy-Abdallah H, Ibrahim KS, Hafez SF, Fadl NN, ElShamy KA (2016a) Comparative hepatotoxicity of Aflatoxin B1 among workers exposed to different organic dust with emphasis on polymorphism role of Glutathione S-transferase gene. Maced J Med Sci 4(2):312–318. https://doi.org/10.3889/oamjms.2016.051

Saad-Hussein A, Taha MM, Fadl NN, Awad AA, Mahdy-Abdallah H, Moubarz G, Aziz H, El-Shamy KA (2016b) Effects of Airborne *Aspergillus* on Serum Aflatoxin B1 and liver enzymes in workers handling wheat flour. Hum Exp Toxicol 35(1):3–9. https://doi.org/10.1177/0960327115573596

Saad-Hussein A, Beshir S, Taha MM, Shahy EM, Shaheen W, Abdel-Shafy EA, Thabet E (2019) Early prediction of liver carcinogenicity due to occupational exposure to pesticides. Mutat Res Genetic Toxicol Environ Mutagenesis 838:46–53

Saad-Hussein A, Shahy EM, Ibrahim KS, Mahdy-Abdallah H, Taha MM, Abdel-Shafy EA, Shaban EE (2022a) Influence of GSTM1, T1 genes polymorphisms on oxidative stress and liver enzymes in rural and urban pesticides-exposed workers. Arch Environ Occup Health 77(10):800–808. https://doi.org/10.1080/19338244.2021.2025024

Saad-Hussein A, Soliman KM, Moubarz G (2022b) 18S rRNA gene sequencing for environmental aflatoxigenic fungi and risk of hepatic carcinoma among exposed workers. J Environ Sci Health A 2:174–182. https://doi.org/10.1080/10934529.2022.2046428

Sarkodie SA, Strezov V (2019) Economic, social and governance adaptation readiness for mitigation of climate change vulnerability: evidence from 192 countries. Sci Total Environ 656:150–164. https://www.cdc.gov/nceh/hsb/chemicals/pdfs/pcphcop-newsletter_climate-change-and-toxic-exposures-508.pdf

Schulte PA, Bhattacharya A, Butler CR, Chun HK, Jacklitsch B, Jacobs T, Kiefer M, Lincoln J, Pendergrass S, Shire J, Watson J, Wagner GR (2016) Advancing the framework for considering the effects of climate change on worker safety and health. J Occup Environ Hyg 13(11):847–865. https://doi.org/10.1080/15459624.2016.1179388

Shah AS, Langrish JP, Nair H et al (2013) Global association of air pollution and heart failure: a systematic review and meta-analysis. Lancet 382(9897):1039–1048. https://doi.org/10.1016/S0140-6736(13)60898-3

Sherman JD, MacNeill AJ, Biddinger PD, Ergun O, Salas RN, Eckelman MJ (2023) Sustainable and resilient health care in the face of a changing climate. Annu Rev Public Health 44(1):255–277. https://doi.org/10.1146/annurev-publhealth-071421-051937

Shi X, Wahlang B, Wei X et al (2012) Metabolomic analysis of the effects of polychlorinated biphenyls in nonalcoholic fatty liver disease. J Proteome Res 11:3805–3815

Sorensen C, Hamacher N, Campbell H et al (2023) Climate and health capacity building for health professionals in the Caribbean: a pilot course. Front Public Health 11:1077306. Published 2023 Jan 26. https://doi.org/10.3389/fpubh.2023.1077306

Stanke C, Kerac M, Prudhomme C, Medlock J, Murray V (2013) Health effects of drought: a systematic review of the evidence. PLoS Curr 5:ecurrents.dis.7a2cee9e980f91ad7697b570bcc4b004. https://doi.org/10.1371/currents.dis.7a2cee9e980f91ad7697b570bcc4b004. PMID: 23787891; PMCID: PMC3682759

Stewart S, Keates AK, Redfern A, McMurray JJV (2017) Seasonal variations in cardiovascular disease. Nat Rev Cardiol 14(11):654–664. https://doi.org/10.1038/nrcardio.2017.76

Tang X, Wilson SR, Solomon KR, Shao M, Madronich S (2011) Changes in air quality and tropospheric composition due to depletion of stratospheric ozone and interactions with climate. Photochem Photobiol Sci 10(2):280–291. https://doi.org/10.1039/c0pp90039g

Thurston GD, Burnett RT, Turner MC et al (2016) Ischemic heart disease mortality and long-term exposure to source-related components of U.S. fine particle air pollution. Environ Health Perspect 124(6):785–794. https://doi.org/10.1289/ehp.1509777

Tiepolo M, Bacci M (2017) Tracking climate change vulnerability at municipal level in rural Haiti using open data. In: Tiepolo M, Pezzoli A, Tarchiani V (eds) Renewing local planning to face climate change in the tropics. Green energy and technology. Springer, Cham. https://doi.org/10.1007/978-3-319-59096-7_6/

Toyooka T, Ibuki Y, Takabayashi F, Goto R (2006) Coexposure to benzo[a]pyrene and UVA induces DNA damage: First proof of double-strand breaks in a cell-free system. Environ Mol Mutagen 47:38

Tucker MA (2009) Melanoma epidemiology. Hematol Oncol Clin North Am 23(3):383–vii. https://doi.org/10.1016/j.hoc.2009.03.010

UNEP (2019) UN environment global chemicals outlook II-from legacies to innovative solutions: implementing the 2030 agenda for sustainable development

Wake B (2023) Infrastructure and climate–health risks. Nat Clim Chang 13:1161. https://doi.org/10.1038/s41558-023-01858-3

WHO (2023) climate change. https://www.who.int/news-room/fact-sheets/detail/climate-change-and-health

World Bank (2010) Margulis, Sergio; Narain, Urvashi. The costs to developing countries of adapting to climate change: New methods and estimates - the global report of the economics of adaptation to climate change study (English). Washington, DC: World Bank. https://documents.worldbank.org/curated/en/667701468177537886

Yan J, Wang L, Fu PP, Yu H (2004) Photomutagenicity of 16 polycyclic aromatic hydrocarbons from the US EPA priority pollutant list. Mutat Res Gen Tox En 557:99–108. https://doi.org/10.1016/j.mrgentox.2003.10.004

Zanaty N, Faisal NI, Ramadan HK, Ahmad AA, Saad-Hussein A (accepted). Climate change plays a role in the emerging outbreak of human Fascioliasis in Upper Egypt. Tropical Diseases, Travel Medicine and Vaccines

Ziska L, Crimmins A, Auclair A et al (2016) Food safety, nutrition, and distribution. In: The impacts of climate change on human health in the United States: a scientific assessment. US Global Change Research Program, Washington, DC, pp 189–216

Climate Change Health Impacts of Food Toxins

4

Mosaad A. Abdel-Wahhab, Aziza A. El-Nekeety, and Yasser S. Sayed

Abstract

The challenges of climate change represent a major problem for food safety around the world. Long-term changes in patterns of temperature, precipitation, humidity, and extreme weather events are affecting agricultural activities, crop production, and the nutritional quality of most food crops. The allergy of germs, microorganisms, and pests to environmental factors has revealed that climate change is leading to the emergence of invasive stranger species that harm the health of plants and animals. Warming of surface seawater and increased input of nutrients lead to an abundance of toxin-producing algae causing rampant seafood contamination. Global efforts are continuing to reduce the emissions of greenhouse gases and mitigate the effect of climate change on feed safety related to plant, animal, human, and environmental health. This chapter discusses the impact of climate change on food safety and the efforts made internationally to decrease the microorganisms, chemicals, and biological contaminants that affect the health of plants, animals, and humans.

M. A. Abdel-Wahhab (✉) · A. A. El-Nekeety
Department of Food Toxicology & Contaminants, National Research Centre, Dokki, Giza, Egypt

National Committee of Toxicology, Academy of Scientific Research and Technology (ASRT), Ministry of Higher Education and Scientific Research, Cairo, Egypt
e-mail: ma.abdelwahab@nrc.sci.eg; aa.elnekteey@nrc.sci.eg

Y. S. Sayed
Department of Veterinary Forensic Medicine and Toxicology, Faculty of Veterinary Medicine, Damanhour University, Damanhour, Egypt
e-mail: elsayed-ys@vetmed.dmu.edu.eg

A. Saad-Hussein et al. (eds.), *Climate Change Impacts on Toxins and Health Effects*,
https://doi.org/10.1007/978-981-96-3416-3_4

Keywords

Climate change · Food safety · Mycotoxins · Heavy metals · Pesticide · Food toxins

Abbreviations

AFs	Aflatoxins
AFB_1	Aflatoxin B_1
As	Arsenic
Cd	Cadmium
Cr	Chromium
DNA	Deoxyribonucleic acid
DON	Deoxynivalenol
EFSA	European Food Safety Authority
FAO	Food and Agriculture Organization
FBs	Fumonisins
FDA	Food and Drug Administration
FSMA	The Food Safety Modernization Act
FSSAI	Food Safety and Standards Authority of India
GAP	Good agricultural practices
HACCP	Hazard Analysis and Critical Control Points
Hg	Mercury
IARC	International Agency for Research on Cancer
ISO	International Organization for Standardization
NADH	Nicotinamide Adenine Dinucleotide
NFSA	National Food Safety Authority
OTs	Ochratoxins
PAT	Patulin
Pb	Lead
TRC	Trichothecenes
WHO	World Health Organization
ZEN	Zearalenone

4.1 Introduction

There are several ongoing discussions about the effects of climate change, but its relationship to food safety issues is often overlooked. As shown in Table 4.1, climate change is an important driver of emerging risks to food safety and quality (Maggiore et al. 2020).

However, with the number of involved risks and factors, there is a great deal of uncertainty in the relationship between climate change and food safety. Climate change is believed to hurt humans, animals, plants, and ecosystems with the

Table 4.1 Summary of major effects of climate-dependent environmental factors on the behavior of foodborne pathogens and impact on food safety[a]

Climatic factor	Effect of climate on environmental factors involved in the growth, survival, and pathogenicity of foodborne pathogens	References
Temperature Increase	Increased occurrence of parasites in freshwater fish and plants	Herrera et al. (2016)
	Detection of new mycotoxin-producing fungal species in maize in Europe	Moretti et al. (2018)
	Increase in mastitis incidence in cows	Lacetera (2019)
	Increase in Salmonella in poultry	Herrera et al. (2016)
	Increased number of Vibrio cells in seafood	Marques et al. (2010)
Decrease	Increased contamination of berries by norovirus and hepatitis A	Calder et al. (2003), EFSA (2014)
Precipitation and humidity		
Increase in precipitation	Internalization of pathogenic E. coli and Salmonella in leafy green vegetables	Ge et al. (2012), Liu et al. 2013)
	Increased contamination of seafood by fecal indicator organisms due to water runoffs	Marques et al. (2010)
	Increased risk of splash dispersal and aerosolized Salmonella infecting tomatoes due to increased frequency of short periods of heavy rainfall	Cevallos-Cevallos et al. (2012)
Decrease in precipitation and humidity	Increased mycotoxin contamination by xerophilic fungi in maize at preharvest stage	Moretti et al. (2018)
pH and salinity		
Decrease in pH	Ocean acidification leads to increased HABs	Marques et al. (2010)
Decrease in salinity	Increases in bioaccumulation of toxic metals in molluscs	Wu et al. (2016)
Light		
Increase	Favors the growth of HABs	Marques et al. (2010)

[a]Adapted from Duchenne et al. (2021)

potential to worsen the frequency and severity of some foodborne diseases (Tirado et al. 2010). The fourth United Nations International Panel on Climate Change report noted that rising carbon dioxide levels, increasing temperatures, severe weather, and changes in patterns of rainfall are likely to jeopardize food safety (Van der Spiegel et al. 2012). The Climate Change and Emerging Risks to Food Safety Project concluded that climate change impacts the presence and behavior of freshwater and marine algal blooms, fungi, viruses, bacteria, and parasites, as well as disease vectors for plants and/or animals (Tirado et al. 2010; Uyttendaele et al. 2015; Wu et al. 2016). Thus, these climatic changes are likely to influence the sources and routes of transmission, growth, ecology, and survival of pathogens in different foods (Tirado et al. 2010; Wu et al. 2016). For example, temperature changes may affect the reproductive period of pathogens, thus altering their growth and distribution, leading to the emergence of unknown pathogens for particular hosts in some regions (Wu et al. 2016). The main consequences of climate change are the development of unknown new hazards or increased host exposure to existing and known risks (Maggiore et al. 2020). Overall, climate change is significantly influencing global food safety. Long-term changes in the frequency of severe weather events, rainfall patterns, temperature, and humidity will affect the production of crops and their nutritional quality, as well as entire agricultural practices. Thus, climate change is considered the driver of emerging risks to food safety and quality and identifies the potential impacts on a wide range of food safety-related issues.

4.2 Climate Change and Food Toxins

Food toxins can be presented naturally in plant or animal foods that have been exposed to toxins. It can also be caused by the activities of humans, such as environmental pollutants, pesticides, or veterinary medicines, and can be found in a variety of foods, including organic foods. Natural toxins are those synthesized by living organisms such as fungi, plants, algae, bacteria, and marine biotoxins (Tan et al. 2016). These chemical compounds differ in their composition, which causes them to vary in biological function and toxicity and cause severe health problems for both humans and animals (WHO 2018).

4.3 Climate Change and Food Contamination

Food contamination is a global safety concern due to the threat posed by several foodborne pathogens to human health. The food safety and food manufacturing sections face novel challenges due to the globalization of the food chain and changes in modern consumer preferences. Food contamination is divided into chemical, biological, and physical contamination according to the methods of spoiling food (Boa 2003). There are two main sources of contamination: primary contamination from an outside source (exogenous) and secondary contamination formed in food (endogenous). Food contamination causes food poisoning, which can lead to vomiting, diarrhea, fever, and dehydration. The contamination of food may happen at any step of the food production process, at harvesting, processing, transportation, and storage (Polo 2023).

4.3.1 Climate Change and Biological Contamination

Biological contamination occurs when fungi, bacteria, viruses, and parasites contaminate food along the production chain. These pathogenic microorganisms are ubiquitous and are coupled with climate change, so they can contaminate food at any stage, from farm to fork (Milićević et al. 2019). Some examples of usual food risk pairs are histamine in fish, *Salmonella* in poultry, *Vibrio* spp. in mollusks, *Campylobacter* spp. in dairy products, and noroviruses in shellfish or leafy green vegetables (Safavieh et al. 2014; Tirado et al. 2010). There are three mechanisms by which biological contamination causes human diseases, including intoxication, infection, and immunologic responses.

4.3.1.1 Bacteria

Bacteria are living microorganisms that play a vital role in food processing and safety (Gracias and McKillip 2004). Pathogenic bacteria are the main source of foodborne illnesses. Each type of bacteria needs optimum conditions to survive, such as temperature, oxygen, pH level, food (especially protein), and water (Multari et al. 2023). Bacteria distribution in a wide range includes air, soil, the human body, dust, raw food (meat, milk, etc.), food handlers, and clothes. The main bacteria causing food poisoning worldwide are *Escherichia coli* (causes traveler's diarrhea), *Salmonella* spp. (*Typhimurium, enteritis*), *Listeria monocytogenes, Shigella, Clostridium perfringens, Campylobacter, Bacillus anthracis,* and *Staphylococcus aureus.* The most effective way to protect against these pathogenic bacteria is to maintain the best food safety practices steps: clean, separate, cook, and chill (Callejón et al. 2015).

4.3.1.2 Mycotoxigenic Fungi and Mycotoxins

Mycotoxins

Mycotoxins are the secondary metabolites synthesized by diverse genera of filamentous fungi belonging mainly to the genera *Aspergillus, Penicillium, Fusarium,* and *Alternaria* (Barkai-Golan and Paster 2008; Nleya et al. 2018). There is increased concern about fungal diseases caused by *Fusarium* and *Aspergillus* spp. in staple crops, as they cause yield loss and produce mycotoxin in the food chain (Barea 2015). These mycotoxins pose serious risks to public health because of their carcinogenic and mutagenic properties (Abdel-Wahhab et al. 2020; Marin et al. 2013). The Food and Agriculture Organization (FAO) has estimated that mycotoxins affect 25% of the food crops worldwide, including many staple foods and animal feeds. More than 300 types of mycotoxins were isolated, and their chemical composition has been determined. The main mycotoxins that pose a potent health risk to animals and humans as feed and food contaminants are aflatoxins (AFs), ochratoxins (OTs), fumonisins (FBs), zearalenone (ZEN), patulin (PAT), trichothecenes (TRC), and their derivatives (Agriopoulou et al. 2020; Bulgaru et al. 2021). These mycotoxigenic fungi infect various staple crops (Paris et al. 2015), including corn, rice, and maize, as well as coffee, dried fruits, nuts, grains, spices, feed crops, vegetables, and fruits (Agriopoulou et al. 2020; Balendres et al. 2019). Table 4.2 summarizes the major genera of mycotoxins-producing fungi, the main infected cereal crops, and the associated health effects.

Aflatoxins (AFs)

AFs are the most common mycotoxins worldwide that contaminate food commodities that are synthesized by *Aspergillus* species, especially *A. flavus, A. parasiticus,* and *A. nominus* (Guan et al. 2021; Mutegi et al. 2018; Nazhand et al. 2020). There are about 20 types of AFs that were identified; however, only AFB_1, AFB_2, AFG_1, AFG_2, AFM_1, and AFM_2 are the common types (Luo et al. 2018). The most standout and toxic is AFB_1, which was classified by the IARC (International Agency for Research on Cancer) as a group A carcinogen (Ostry et al. 2017). The target organ of AFB_1 is the liver, which may, after chronic exposure, lead to hepatocellular carcinoma (Rushing and Selim 2019). AFB_1 is also mutagenic, immunotoxic (Abdel-Wahhab et al. 2016, 2017; Rushing and Selim 2019), and teratogenic (Abdel-Wahhab et al. 1998; Watson et al. 2018). It generates free radicals that cause cellular damage and several health problems to animals and humans (El-Nekeety et al. 2017a; Abdel-Wahhab et al. 2018; Eftekhari et al. 2018). Its carcinogenicity is mainly due to the metabolic activation of the reactive AFB_1 8, 9-exoepoxide by the cytochrome P450 monooxygenases (Chu et al. 2017). This reactive epoxide interacts with DNA molecules to form highly mutagenic DNA lesions such as aflatoxin-N7-guanine adducts (Wild and Turner 2002; Coskun et al. 2019).

Ochratoxin A (OTA)

OTA is elaborated by *Aspergillus ochraceus* and *Penicillium verrucosum* in temperate climates, tropical, and warm countries (Iqbal et al. 2013). Its hazardous nature is due to its high resistance to high temperatures and acidic environments as well as great stability (Schrenk et al. 2020). OTA was classified by the IARC (2002) as a group 2B carcinogen, which means probable carcinogen to humans. OTA is a hepatonephrotoxic, teratogenic, and immunosuppressive agent that has been implicated as an endocrine disruptor as well as a contributor to increased cancer risk (Abdel-Wahhab et al. 2005, 2008; El-Nekeety et al. 2017b; García-Pérez et al. 2021). The kidney is the principal target organ of OTA toxicity, and human exposure to OTA is connected to the development of Balkan endemic nephropathy and urinary bladder tumors (Doricakova and Vrzal 2015; Longobardi et al. 2022). OTA stimulates ascorbate-dependent and NADH-dependent lipid peroxidation in microsomes, using iron (Fe^{3+}) as a cofactor for forming. OTA–Fe^{3+} complex, which reduces Fe^{3+} in the presence of an NADH–cytochrome P450 reductase system. The resulting OTA-Fe2+ complex then generates OH-, leading to the peroxidation of membrane lipids (Rahimtula et al. 1988). The mechanism of the OTA effect participates in phenylalanine metabolism and inhibits protein synthesis by competing with phenylalanine in the reaction catalyzed by phenylalanyl-t-RNA synthetase (Abdel-Wahhab et al. 2005, 2008; Creppy et al. 1985). Moreover, it induces FPG-sensitive sites in kidney cell DNA (Corcuera et al. 2015).

Table 4.2 Major genera of mycotoxins-producing fungi, the main infected cereal crops, and the associated health effects

Fungus	Mycotoxins	Associated health effects	Food commonly contaminated
Aspergillus flavus	Aflatoxins	Vomiting, hepatitis	All grains, dried fruit
Aspergillus parasiticus		Liver cancer	
Fusarium verticillioides	Fumonisins	Vomiting	Maize
Fusarium proliferatum		Neural tube defects	
Aspergillus ochraceus		Esophageal cancer	
	Trichothecenes, Nonmacrocyclic		
Fusarium culmorum	Deoxynivalenol	Vomiting	Cereals (wheat, barley, maize, rice)
Fusarium graminearum			
Fusarium cerealis			
Fusarium sporotrichiodes	T-2 toxin	Alimentary toxic aleukia	
		Vomiting, hemorrhage	
	Trichothecenes, Macrocyclic		
Stachybotrys chartarum	Satratoxins	Protein synthesis inhibition	Cereals (wheat, barley, maize, rice)
Trichoderma viride		Hemorrhage	
Trichothecium roseum			
Aspergillus ochraceus	Ochratoxins	Balkan nephropathy	Coffee, cocoa
Aspergillus niger		Renal cancer	
Aspergillus alliaceus			
Penicillium verrucosum			
Penicillium expansum	Patulin	Vomiting, cancer (suspect)	Apple
Fusarium graminearum	Zearalenone	Estrogenic effects, cervical cancer (suspect)	Maize
Claviceps purpurea	Ergot alkaloids	Ergotism	

Fumonisins (FBs)

FBs are mycotoxins synthesized by *Fusarium* spp., mostly Fusarium proliferatum and F. verticillioides (Gelderblom and Marasas 2012). FBs contaminate foods such as corn and forage (Bulder et al. 2012; Szabó-Fodor et al. 2015). Twenty-eight variants of FBs were isolated and characterized; however, FB_1 was the most prevalent toxin and was classified as a group 2B carcinogenic potential for humans by the IARC (2002). This mycotoxin is resistant to the conditions that are used in the processing of food; consequently, it poses a great health hazard to humans and animals (Scott 2012). The chemical structure of FB is identical to sphinganine, which is the backbone of sphingolipids. FB_1 interrupts sphingolipid synthesis by inhibiting sphingosine-N-acetyltransferase, inducing oxidative stress, DNA replication, and cell proliferation (Voss and Riley 2013; Liu et al. 2019). FBs cause leukoencephalomalacia in horses and edema of the lungs in swine (Marasas et al. 1988; Harrison et al. 1990). FB_1 has also been reported to induce liver and kidney toxicity in rats (Voss and Riley 2013). In humans, epidemiological studies showed an increase in the incidence of cancer of the esophagus in areas widespread with FB_1, which signifies that FB_1 may have a vital role in esophageal neoplastic development (Come et al. 2019).

Zearalenone

Zearalenone (ZEN) is a nonsteroidal estrogenic mycotoxin elaborated mainly by *Fusarium* spp. (Caglayan and Üstündağ 2020; Rai et al. 2020). ZEN is commonly produced in warm climate regions and contaminates several cereals, such as barley, oats, and rice, as well as vegetables and fruits (Li et al. 2020; Wang and Xie 2020), and it was classified by the IARC as a group 3B. Due to its structural similarity with estrogen, ZEN competes with 17 β-estradiol in the binding of cytosolic estrogen receptors (Feng et al. 2022), leading to reproductive disorders in males (Bielas et al. 2017) and females (Liu et al. 2019; Samik and Safitri 2017).

Deoxynivalenol

DON is a mycotoxin synthesized by *Fusarium* spp. mainly *F. culmorum, F. crookwellense*, and *F. graminearum* (Yu et al. 2017). It contaminates several grains such as corn, wheat, and barley (Sirot et al. 2013). The ingestion of DON-contaminated food induces severe health problems, including anorexia, vomiting, disturbances in the differentiation of cell signaling, disturbances in the body's immunity and the function of the neuroendocrine system, and gastrointestinal homeostasis because of the impaired synthesis of the macromolecular (Abdel-Wahhab et al. 2018); Chen et al. 2023).

Climate change has started to have a significant role in the distribution of mycotoxins. Recently, the impact of climate change on AFs exposure and the associated health concerns in Portugal have been paid more attention (Martins et al. 2018). Global warming is suggested to increase the occurrence of hepatocellular carcinoma (HCC) resulting from aflatoxins, with forecasts suggesting a rise in AFs contamination in European crops over the next 100 years. Other reports showed that cereals and maize are expected to become more vulnerable to contamination due to climate change. High levels of mycotoxins were detected in breakfast cereals, confirming the effect of climate change on cereal-based foods (Battilani et al. 2016). These findings indicate that global warming directly affects fungal metabolism and support the claim that unfavorable conditions for fungi, such as warmer temperatures, drought, and stress, lead to increased mycotoxin production.

4.3.2 Chemical Contamination

Chemical contamination occurs at several stages, including food storage, packaging, processing, and transportation (Rather et al. 2017). Chemical contamination can also be the consequence of agrochemicals, including pesticides, veterinary medicinal residues, and/or food contact materials, or environmental contamination (toxins in industrial chemical processing), such as hydrocarbons, phthalates, perchlorate, polychlorinated biphenyls, and dioxins (Faille et al. 2013; Rather et al. 2017), or packaging-derived chemicals (bisphenol A). Furthermore, chemical contaminants may be acute with a single dose or chronic with repeated long-term doses (such as chronic exposure to AFB_1 leads to hepatocarcinoma) (Kimanya et al. 2021). Recently food contamination has become more serious because of the development of industry and the consequent environmental pollution (Song et al. 2017).

4.3.2.1 Heavy Metals (HMs)

HMs are one of the major problems in food contamination that pose a major threat to human health and marine ecosystems (Javed and Usmani 2011). They are stable, not metabolized, remain in the food, and then irreversibly bind to various organs in the body, such as Cd in the kidneys and Pb in the bones. HMs that are major toxicants even at low exposure levels are lead (Pb), chromium (Cr), cadmium (Cd), arsenic (As), and mercury (Hg). Climate change factors affecting the fate of HMs are changes in temperature and rainfall. Increasing intense and frequent rainfall patterns due to climate change are expected to increase the leaching of HMs from soils into water systems (Wijngaard et al. 2017). HMs are absorbed by the crops and enter the food chain after ingestion of these contaminated crops. Moreover, 10.2% of arable land in China was found to be contaminated with HMs, which affects about 14% of grain production (Zhang et al. 2015). Cereal and rice plants mainly bioaccumulate from the soil, threatening the health of millions of people (FAO 2020). The uptake of AS by plants was found to be increased by increasing soil temperature, which indicates that the level of AS in rice may double with the expected increase in soil temperatures (Neumann et al. 2017). Higher temperatures also cause permafrost to thaw, which enhances the release of HMs trapped in bodies of water, such as the oceans (FAO 2020). Additionally, heavy rains can also cause HMs to wash into water bodies, and the risk of their subsequent absorption by marine species should not be ignored (FAO 2020).

4.3.2.2 Pesticide and Fungicide Residues: Impact on Food Safety

Pesticides are organic synthetic compounds utilized in agriculture to protect crops from weeds, insects, fungi, and other pests. Between 20 and 40% of global crop production was damaged by pests, and according to the FAO Corporate Statistical Database, about 4.2 million tons of pesticide-active ingredients were utilized on the world's croplands in 2017 (FAO 2020). The misuse and overuse of pesticides, primarily those that degrade slowly, have had a detrimental effect on water quality and the environment with toxic impacts on human and animal health (Delcour et al. 2015; FAO 2020). As climate change is expected to alter the life cycle, regional distribution, and populations of agricultural pests and pathogens, there is expected to be an increase in the use, dosage, and frequency of application of insecticides and fungicides (Delcour et al. 2015). High temperatures can lead to a rapid increase in the number of plant pests, insects, and pathogens. For example, in viticulture, humidity and high temperatures encourage the growth of fungi and infections, such as powdery mildew on grapes (Chakraborty et al. 2000). This suggests that although the numbers of insects, pests, and pathogens may increase, the intensity of a particular insect pest may increase or decrease with increasing temperature. The combination of high concentrations of carbon dioxide and direct exposure to sunlight and high temperatures can cause dilution and increased volatility of insecticides and fungicides in plants, reducing their concentration and thus reducing their efficacy against pests and pathogens (FAO 2020). This may drive the overuse of pesticides in areas with high temperatures to minimize insect damage and fungal infection in storage areas. Although they are used in controlling the vectors of tropical diseases, such as mosquitoes, they are potentially toxic to humans and induce health risks such as cancer, fertility, and nervous or immune systems; furthermore, they cause acute and chronic effects depending on the dose and route of exposure. They must be tested for all possible health effects, and the results should be examined by experts to evaluate any risks to humans before being authorized for use.

4.4 Food Safety Regulations and Standards

The intertwining challenges of climatic changes and food toxins present a pressing landscape for global and provincial food safety regulations. As our climate undergoes transformational shifts, impacting ecosystems and agricultural practices, the regulatory frameworks governing food safety must adapt in tandem. Herein, we will delve into the intricate relationship between food safety regulations and climatic changes, exploring how global entities and regional bodies navigate the complexities to ensure the flexibility and adaptability of standards in safeguarding the safety of the global food supply.

4.4.1 Global and Regional Regulatory Bodies

Global and regional regulatory bodies play a critical role in setting standards and guidelines to ensure food safety, especially in the context of emerging challenges like climatic change and its influence on food toxins. Internationally, organizations such as the WHO and FAO collaborate to establish global standards that member countries can adopt and implement. These standards encompass various aspects of food safety, including permissible levels of contaminants and toxins.

At the regional level, regulatory bodies like the EFSA in Europe, the USFDA, the Food Safety and Standards Authority of India (FSSAI) in South Asia, the Food Safety Commission in Japan, or the National Food Safety Authority (NFSA) in Egypt work to harmonize standards within their jurisdictions. They conduct risk assessments, establish permissible limits for contaminants, and oversee compliance with these standards. In the face of climatic changes, regional regulatory bodies are tasked with collaborating with scientific communities to evaluate the effect on local food production and the possible changes in toxin profiles.

Collaboration between global and regional regulatory bodies is crucial for fostering a cohesive approach to mitigate the impact of climatic changes on food toxins. This involves information sharing, joint research initiatives, and the development of unified guidelines that address both global and region-specific challenges. As the climate continues to evolve, the vigilance and adaptability of regulatory bodies become paramount in ensuring the safety and security of the global food supply.

4.4.2 Key Food Safety Standards and Guidelines

There are globally recognized guidelines and standards that provide a framework and requirements for food safety, public health protections, and responsible food trade. Implementation across the food supply chain is critical to reduce foodborne illness risks and adverse health impacts.

4.4.2.1 FDA Food Code

The FDA Food Code, a pinnacle in the realm of safe food handling and preparation, stands as a beacon for retail food establishments across the United States. Published by the USFDA, this comprehensive guide not only sets the gold standard for food safety but also acknowledges the evolving challenges posed by climate change. In response to rising ambient temperatures, accelerated by climate change, the code meticulously outlines temperature requirements crucial for controlling pathogen growth. Anticipating the impacts of harsh weather events, it provides provisions for alternate water sources during droughts or floods while also setting restrictions on food establishments in flood-prone or coastal areas. Recognizing the increased contamination risk during flooding events, the code mandates the exclusion of food exposed to floodwaters. In a proactive stance against water safety challenges induced by climate change, it establishes rigorous requirements for water quality and testing irrigation water. As climate change intensifies, the code emphasizes employee health policies, restricting ill workers to curb the prevalence of pathogens that may proliferate further in the wake of climate impacts. Addressing the intricate dance between warming temperatures and pathogen growth, the code tightens cooling time and temperature requirements, setting critical control points to reduce the time foods spend in the danger zone. In a nod to the changing climate landscape, it requires close monitoring of cooling methods in hotter conditions and specifies tighter temperature thresholds to prevent bacteria multiplication. The FDA Food Code, a dynamic and responsive document, serves not only as a guardian of food safety but also as a resilient guide adapting to the challenges of our ever-changing climate (FDA 2006).

In adherence to the FDA Food Code's robust framework for preventive controls, food establishments are not merely urged but mandated to craft comprehensive written procedures aimed at managing the risk factors associated with foodborne illnesses. These procedures extend beyond the theoretical realm, necessitating clear instructions for vigilant monitoring, decisive corrective actions, and meticulous verification processes. Think of it as a culinary compass, guiding establishments through the intricacies of temperature control, cross-contamination prevention, employee health policies, sanitation practices, and a host of other critical considerations. The code places a premium on accountability, requiring meticulous documentation and records that chronicle the journey of monitoring, the swift response of corrective actions, and the meticulous verification activities undertaken. It is not just a code; it is a commitment to a culture of culinary excellence anchored in proactive measures and a relentless pursuit of the highest standards in food safety.

4.4.2.2 Hazard Analysis and Critical Control Points (HACCP)

HACCP is a comprehensive and systematic approach designed to ensure food safety by identifying, evaluating, and controlling potential hazards at all stages of the food production process. Originally developed by NASA and later adopted by the FDA and USDA, HACCP has become a globally recognized standard for mitigating risks connected with foodborne illnesses. While its primary objective is to ensure the safety of food products, the HACCP framework indirectly contributes to broader environmental sustainability goals.

One of the key ways HACCP intersects with environmental impact is through waste reduction. By meticulously analyzing the stages of the food production process, HACCP identifies points where waste is generated and develops strategies to minimize or eliminate it. This not only improves overall operational efficiency but also reduces the resources required for food production. The ripple effect includes a decrease in greenhouse gas emissions and other environmental impacts associated with resource-intensive food production. Sustainable agriculture practices are another area where HACCP plays a crucial role. The system can identify potential hazards linked to specific farming practices, such as the use of pesticides or fertilizers. By recognizing and strategizing to minimize these impacts, HACCP promotes sustainable agriculture. This, in turn, helps reduce the overall carbon footprint associated with food production, aligning with the broader goal of environmental sustainability (HACCP 2023).

Energy consumption in food production processes can also be addressed through HACCP's rigorous analysis. By scrutinizing energy inputs at each stage of the process, HACCP identifies opportunities for conservation and optimization. This may include reducing energy usage in food processing, transportation, and storage. The outcome is a reduction in the overall carbon footprint of food production, contributing to energy efficiency and environmental sustainability. Furthermore, HACCP can advocate for the integration of renewable energy sources in food production. Through careful analysis, the system can identify areas where renewable energy, such as solar or wind power, can be effectively incorporated into the production process. This not only reduces dependence on non-renewable energy sources but also helps mitigate greenhouse gas emissions associated with energy production (HACCP 2023).

In response to the warming seasons and shifting migration patterns impacting pest control, prerequisite programs have become pivotal in addressing these challenges. The HACCP hazard analysis extends its gaze to emerging climate-driven threats like mycotoxin formation, a consequence of climate change stressing crops and fostering mycotoxin-producing mold growth. The critical limits, recognizing the accelerated pathogen growth rates in higher temperatures, may necessitate tightening to counteract these climate-driven challenges (Awuchi 2023). Anchored in adaptability, the HACCP system requires regular reviews and updating of the food safety plan, allowing for the adjustment of controls in the face of new climate-related hazards. The hazard analysis encompasses emerging chemical, biological, and physical threats amplified by climate change, from toxins in crops to contaminants in flooded fields and new microbiological menaces. Critical control points, addressing temperature control and water quality, are on the frontline against the impacts of warming and extreme weather (Tirado et al. 2010).

As risks evolve with climate-driven food safety data and intelligence, HACCP plans must follow suit, continually adapting. This necessitates a meticulous analysis of the latest scientific literature and data on emerging pathogens, toxin-producing molds, and other microbiological hazards intensified by climate change. Monitoring and testing programs evolve to meet the challenges posed by new pathogen risks identified in hazard analyses related to climatic effects. Water activity, an often-overlooked aspect, emerges as a critical control point in certain products based on the latest climate-linked microbiological risk assessments (Tirado et al. 2010). In essence, HACCP is not just a set of guidelines; it is an adaptive, detailed guardian of our food safety, navigating the complex terrain of climate-driven challenges to ensure our food remains not only delectable but, more importantly, safe.

4.4.2.3 ISO 22000

ISO 22000, an internationally recognized standard for food safety management systems developed by the International Organization for Standardization (ISO), serves as a cornerstone for ensuring the safety of food products and meeting the evolving challenges of the twenty-first century. In today's complex and interconnected world, food safety remains a paramount concern, with an estimated 600 million people falling ill and 420,000 losing their lives annually due to unsafe food. This standard recognizes the multifaceted nature of modern challenges, including the intricacies of globalization, complex supply chains, and the direct and indirect impacts of climate change on food safety. WHO estimates underscore the urgency to address these challenges, and ISO 22000 emerges as a crucial tool to empower food producers in fulfilling their responsibility to safeguard products and prioritize consumer well-being.

ISO 22000 integrates HACCP principles with prerequisite programs, covering an array of crucial factors. Prerequisites, such as facility construction, layout, and workflow; cleaning and sanitation; pest control; personnel training; allergen management; and supplier control, are foundational operational and sanitation conditions that precede the implementation of the HACCP food safety plan. The standard emphasizes environmental monitoring, utility control, equipment suitability, preventive maintenance, calibrations, personnel competency, health status, and high-risk food control (Varzakas 2011). The standard also establishes critical limits for critical control points, considering the impact of environmental conditions and changes. For instance, critical limits for holding cold food must factor in warming warehouse temperatures due to climate change. The evolving nature of climate-related challenges necessitates ongoing adjustments based on statistical data showing shifts in local temperature averages and extremes. Furthermore, ISO 22000 tightens supplier approval and verification requirements, ensuring resilience in the face of changing transport routes, product origins, and growing conditions influenced by climate factors. Emphasizing supplier food safety capabilities and contingency plans, the standard addresses the acute or gradual climatic changes affecting their operations (ISO 2024). Therefore, ISO 22000 stands as a comprehensive and adaptive standard, providing a robust framework to navigate contemporary challenges in the food industry. Combining HACCP principles with prerequisite programs and addressing climate-related impacts not only ensures the safety of food products but also contributes to the resilience and sustainability of food systems globally.

4.4.2.4 Food Safety Modernization Act (FSMA)

The FSMA, a pivotal piece of U.S. legislation, significantly bolstered the FDA's oversight of food safety, shifting its focus from responding to outbreaks to preventing contamination. Climate change and environmental targets were required to be considered under FSMA. So, while FSMA itself is not a law directly relating to the environment, the regulatory principle requiring regard to climate targets nonetheless should have satisfied the description of an environmental law. Therefore, addressing the dynamic challenges posed by climatic changes, the FSMA should mandate a proactive approach, requiring companies to monitor and mitigate hazards influenced by climate fluctuations, including changes in irrigation water, increased flooding, temperature variations, and the presence of pest animals. With a keen emphasis on prevention, the legislation enforces supply chain transparency, necessitating contingency plans to maintain the integrity of the food supply despite disruptions caused by climate change.

4.4.2.5 European Food Safety Authority (EFSA)

The intricate relationship between climatic changes and food safety emphasizes the significant challenges posed by shifts in humidity, temperature, rainfall patterns, and extreme weather events. The sensitivity of microorganisms and pests to these climate factors suggests potential impacts on foodborne diseases, toxin-producing organisms, and invasive species. For instance, warming surface seawater contributes to the proliferation of toxin-producing algae, leading to seafood contamination outbreaks. EFSA plays a crucial role in addressing these challenges. It recognizes climate change as a driver of emerging risks to plant and animal health, food and feed safety, and nutritional quality. EFSA's activities encompass various sectors that are influenced by climate change, including animal health and plants, vector-borne diseases, and contaminants. The authority conducts assessments, develops models, and initiates projects to anticipate and address risks connected with climatic changes. Notable examples include predicting the spread of pests due to climate-induced factors and assessing the influence of temperature and humidity on the occurrence of mycotoxins in crops (EFSA 2020).

Furthermore, the Climate Change as a Driver of Emerging Risks for Food and Feed Safety, Plant and Animal Health, and Nutritional Quality (CLEFSA) project, led by EFSA, focuses on identifying and defining emerging risks related to climate change. This initiative involves anticipating multiple risks using climate change scenarios, conducting horizon scanning, crowdsourcing information, and applying multi-criteria decision analysis tools. The project underscores EFSA's commitment to understanding and mitigating the evolving risks associated with climatic changes, thereby contributing to global efforts to assure the safety of food systems (EFSA 2020).

4.4.3 Challenges and Limitations in Food Safety Regulation

The intersection of climatic changes and the presence of food toxins introduces a complex set of challenges and limitations to food safety regulation. As the climate evolves, influencing agricultural practices and the occurrence of contaminants, regulatory frameworks face heightened complexities in safeguarding the food supply chain from both environmental and biological threats.

The confluence of climatic changes and the existence of food toxins present a multifaceted challenge to the regulation of food safety. As the climate undergoes shifts, affecting agricultural practices and the prevalence of contaminants, regulatory frameworks grapple with heightened complexities in safeguarding the food supply chain from environmental and biological threats. Alterations in agricultural practices, driven by changes in temperature and precipitation, necessitate adaptive guidelines that consider the influence of climatic factors on toxin occurrence in food crops. The formulation of climate-resilient agricultural practices and early warning systems becomes imperative to address emerging toxin risks in crops.

The impacts of climatic changes extend into the processing and packaging stages, potentially compromising the safety of the food supply. Harsh weather events can disrupt processing facilities, posing challenges to maintaining hygiene standards and preventing toxin formation. Regulatory measures need to evolve to include climate-adaptive strategies for processing and packaging, encompassing resilient infrastructure, climate-resilient technologies, and protocols to address increased contamination risks associated with climate-induced disruptions. Additionally, vulnerabilities in transportation and storage conditions, influenced by climatic variations, necessitate collaboration between regulatory bodies and the transportation sector. This collaboration should focus on developing climate-resilient practices, incorporating advanced storage technologies and monitoring systems to mitigate the risk of toxin formation during the transit of perishable goods (EPA 2024; Hatfield et al. 2014).

Consumer awareness and safe handling practices represent another critical aspect impacted by changing climate patterns. Higher temperatures may elevate the risk of bacterial contamination and the formation of toxins in improperly stored or handled food products. Regulatory agencies play a pivotal role in disseminating consumer awareness programs and educating the public on safe food handling practices under diverse climatic conditions. Emphasizing proper storage, hygiene, and cooking techniques is essential to minimize the risk of exposure to food toxins, aligning consumer behavior with evolving climate scenarios. In essence, the regulatory landscape must dynamically adapt to the interplay between climatic changes and food toxins, ensuring a comprehensive approach to global food supply chain safety (Bouri et al. 2023; European Commission et al. 2020; Menini et al. 2022; NHMRC 2019).

4.5 Farm to Fork: Identifying Points of Contamination

The farm-to-fork journey of food is intricately linked to climatic changes and the presence of food toxins, both of which pose significant challenges to food safety. Changes in climate patterns can impact various stages of the food supply chain, from primary production to transportation and storage, influencing the occurrence of contaminants and the development of toxins in food products (Godde et al. 2021). Additionally, harsh weather events associated with climate change can disrupt processing facilities and transportation systems, potentially leading to contamination and the formation of toxins during these stages (Chandorkar 2023).

Identifying points of contamination in the farm-to-fork continuum requires a comprehensive understanding of how climatic changes interact with the presence of food toxins. It involves recognizing vulnerabilities in agricultural practices, processing methods, and transportation systems that may be exacerbated by changing climate conditions (Duchenne-Moutien and Neetoo 2021; James and James 2010). Furthermore, the impact of climatic changes on consumer behavior and safe food handling practices should not be overlooked, as higher temperatures may increase the risk of bacterial contamination and the formation of toxins in improperly stored or handled food products (Duchenne-Moutien and Neetoo 2021; James and James 2010). As regulatory frameworks evolve to address these challenges, there is a growing need for adaptive strategies that consider the dynamic interplay between climate-related factors and the presence of food toxins at different points along the farm-to-fork continuum.

4.5.1 Contamination Risks in Primary Production

Primary production, the initial stage of the farm-to-fork continuum, is particularly vulnerable to contamination risks influenced by climatic changes and the presence of food toxins. The effects of climate change on agricultural practices can affect the spread of contaminants and the formation of toxins in crops, introducing challenges to the safety of the food supply. Climatic factors, such as changes in temperature and precipitation patterns, can create conditions conducive to the growth of mycotoxin-producing fungi in crops (Godde et al. 2021). For example, increased temperatures and altered humidity levels may lead to elevated contamination risks from mycotoxins, such as aflatoxins and ochratoxin A, in crops such as maize and peanuts (Perrone et al. 2020). These toxins pose serious health threats to both consumers and livestock, emphasizing the need for comprehensive strategies to mitigate the impact of climatic changes on primary production.

Moreover, research suggests that the presence of mycotoxins can be exacerbated during extreme weather events and natural disasters, which are becoming more frequent and intense due to climate change (Nicholls et al. 2012; Perrone et al. 2020). These events can lead to waterlogging, mold growth, and contamination of crops with mycotoxins, further complicating the challenges faced in primary production. Regulatory frameworks must adapt to these dynamic conditions, incorporating climate-resilient agricultural practices and monitoring systems to detect and manage emerging risks (Duchenne-Moutien and Neetoo 2021; Godde et al. 2021; Nicholls et al. 2012). In conclusion, primary production stands as a critical juncture in the farm-to-fork continuum, and understanding the contamination risks influenced by climatic changes and food toxins is essential for implementing effective measures to ensure the safety and security of the global food supply.

4.5.2 Processing and Packaging: Potential Contaminants

The processing and packaging stages of the farm-to-fork continuum play a vital role in maintaining food safety. However, they are not immune to potential contaminants, and the evolving impacts of climatic changes introduce additional challenges. Research suggests that extreme weather events associated with climate change can disrupt processing facilities, leading to potential contamination risks (Bock et al. 2004; Duchenne-Moutien and Neetoo 2021; FAO 2022). Floods or power outages can compromise the hygiene standards of processing environments, contributing to the introduction of contaminants that may pose risks to the safety of food products.

Moreover, changes in the temperature and humidity patterns, influenced by climate change, can impact the efficiency of processing equipment and storage conditions, potentially leading to the formation of toxins (Ortega-Beltran and Bandyopadhyay 2023). For instance, temperature variations may affect the preservation of perishable goods, increasing the risk of bacterial contamination or the development of mycotoxins during processing and storage (Bock et al. 2004; Ortega-Beltran and Bandyopadhyay 2023). These challenges underscore the importance of adaptive strategies in processing and packaging to address the increased risks associated with climate-induced disruptions. As we navigate the challenges of processing and packaging in the context of climate change, it becomes imperative to enhance the resilience of infrastructure and materials, employing climate-adaptive technologies and protocols to ensure the continued safety of processed foods.

4.5.3 Transportation and Storage: Impact on Food Safety

The transportation and storage stages of the chain from farm to fork are critical junctures where the impact of climatic changes on food safety and the presence of food toxins becomes pronounced. Climate-related factors, including temperature fluctuations and extreme weather events, significantly influence these stages, posing challenges that demand attention to ensure the integrity of the food supply chain (James and James 2010). Research indicates that variations in temperature during transportation and storage can have profound effects on food safety. Temperature changes may accelerate the degradation of perishable goods, potentially leading to the growth of pathogens or the formation of mycotoxins (Chandorkar 2023). For example, inadequate temperature control during transportation can compromise the cold chain, resulting in the proliferation of harmful. Extreme weather events associated with climate change, such as hurricanes, floods, or wildfires, can disrupt transportation routes and storage facilities, introducing vulnerabilities to the safety of stored food (Duchenne-Moutien and Neetoo 2021; El Samra 2017). These disruptions may compromise the conditions necessary for preserving food quality, potentially leading to contamination risks and the development of toxins during transit and storage (Chandorkar 2023; FAO 2020).

Additionally, the influence of climate changes on the behavior of pests further complicates the transportation and storage dynamics. As pest behavior adapts to changing climates, there is an increased risk of contamination by pesticides or the transmission of different foodborne pathogens during storage and transportation (El Samra 2017). Thus, regulatory frameworks should address the evolving challenges associated with transportation and storage, emphasizing the integration of advanced technologies, real-time monitoring systems, and climate-adaptive measures to ensure the continued safety and quality of food products.

4.5.4 Safe Handling Practices for Consumers

The safe handling practices adopted by consumers during the consumption stage play a pivotal role in ensuring food safety and mitigating the risks associated with food toxins. This aspect is influenced by both climatic changes and the presence of food toxins, necessitating a nuanced understanding and targeted efforts to enhance these practices in response to emerging challenges.

Research indicates that increasing temperatures due to climate change can elevate the risk of bacterial proliferation and toxin formation in food products stored or consumed at home (Duchenne-Moutien and Neetoo 2021; James and James 2010). This underscores the significance of directing efforts toward educating consumers on safe storage and preparation practices, emphasizing personal hygiene and proper storage techniques to prevent bacterial contamination or toxin formation. Furthermore, the changing climate can influence consumer behavior, potentially increasing the risk of foodborne illnesses. Higher temperatures may expedite the multiplication of harmful bacteria in food (Duchenne-Moutien and Neetoo 2021; James and James 2010). Therefore, consumer awareness and health education play a vital role in reducing the risks of foodborne illnesses and the potential negative impacts of climatic changes.

Regulatory agencies play a vital role in promoting consumer awareness programs that educate the public on safe food handling practices in the context of changing climatic conditions. This includes emphasizing proper storage, hygiene, and cooking techniques to reduce the risk of foodborne illnesses and exposure to food toxins.

4.6 Minimizing Food Toxins: Prevention and Control Measures

Food safety is paramount, and minimizing food toxins requires a comprehensive approach involving prevention and control measures throughout the continuum from farm to fork. Several key strategies have been established to achieve this goal, encompassing good agricultural practices (GAP), HACCP, testing and screening methods for food contaminants, and the integration of emerging technologies and innovations.

4.6.1 Good Agricultural Practices (GAP)

Good Agricultural Practices involve adopting practices at the primary production level that minimize contamination risks and ensure the safety of crops. This includes implementing proper hygiene measures, using quality seeds and inputs, monitoring water quality, and adhering to sustainable agricultural methods. GAP serves as a foundation for minimizing food toxins by addressing potential contaminants in the early stages of food production (FAO 2013).

4.6.2 Hazard Analysis and Critical Control Points (HACCP)

HACCP is a systematic approach to identifying, evaluating, and controlling hazards throughout the food production process. This preventive system involves the analysis of critical control points in the production process where hazards can be eliminated, prevented, or reduced to acceptable levels. Applied across the entire food supply chain, HACCP is crucial in minimizing food toxins and ensuring the safety of the final food products (Cinar and Onbaş 2020; FDA 2018).

4.6.3 Testing and Screening Methods for Food Contaminants

Testing and screening methods are essential tools for identifying and quantifying the presence of contaminants in food products. Various analytical techniques, such as chromatography, mass spectrometry, and DNA-based methods, are employed to detect contaminants ranging from pesticides and heavy metals to bacterial pathogens (FAO and WHO 2011; Poltronieri et al. 2013). Regular testing and screening contribute to early identification and intervention, preventing contaminated products from reaching consumers.

4.6.4 Emerging Technologies and Innovations

Advancements in technology offer innovative solutions to minimize food toxins. From blockchain for traceability to sensor technologies for real-time monitoring, these innovations enhance transparency and control in the food supply chain. Nanotechnology, for instance, is explored for its potential in detecting and neutralizing contaminants. Integrating these emerging technologies into food safety protocols contributes to a proactive and efficient approach to minimizing food toxins (Bianchi et al. 2018; Wang et al. 2023).

4.7 Conclusions

The influence of climate change on food safety has occurred in multiple ways, including the increase of foodborne illness risks, and may affect access to safe, adequate food for millions of people around the world. Food contamination is also threatening the sustainable development goals, in particular zero hunger, good health and well-being, clean water and sanitation, reducing inequality, decent work and economic growth, responsible consumption and production, and reducing inequality. Timely global actions toward mitigation and adaptation to climate change are essential to promoting human and planetary health.

References

Abdel-Wahhab MA, Abdel-Aziem SH, El-Nekeety AA (2008) *Inula crithmoides* extract protect against ochratoxin A-induced oxidative stress, clastogenic and mutagenic alterations in male rats. Toxicon 52(4):566–573

Abdel-Wahhab MA, Abdel-Galil MM, Ellithey MM (2005) Melatonin counteracts oxidative stress in rats fed ochratoxin A-contaminated diet. J Pineal Res 38:130–135
Abdel-Wahhab MA, Aljawish A, El-Nekeety AA et al (2017) Chitosan nano particles and quercetin attenuate the oxidative stress and genotoxicity in the kidney of rats fed ochratoxin A-contaminated diet. Food Chem Toxicol 99:209–221
Abdel-Wahhab MA, El-Nekeety AA, Hassan NS et al (2018) Encapsulation of cinnamon essential oil in whey protein enhances the protective effect against single or combined sub-chronic toxicity of fumonisin B_1 and/or aflatoxin B_1 in rats. Environ Sci Pollut Res Int 25(29):29144–29161
Abdel-Wahhab MA, El-Nekeety AA, Hathout SA et al (2020) Bioactive compounds from *Aspergillus Niger* extract enhance the antioxidant activity and prevent the genotoxicity in aflatoxin B_1-treated rats. Toxicon 181:57–68
Abdel-Wahhab MA, Nada SA, Farag IM et al (1998) Potential protective effect of HSCAS and bentonite against dietary aflatoxicosis in rat: with special reference to chromosomal aberrations. Nat Toxins 6:211–218
Abdel-Wahhab MA, Salman AS, Ibrahim MIM et al (2016) Curcumin nanoparticles loaded hydrogels protects against aflatoxin B1-induced genotoxicity in rat liver. Food Chem Toxicol 94:159–171
Agriopoulou S, Stamatelopoulou E, Varzakas T (2020) Advances in occurrence, importance, and mycotoxin control strategies: prevention and detoxification in foods. Food Secur 9:137
Awuchi CG (2023) HACCP, quality, and food safety management in food and agricultural systems. Cogent Food Agric 9(1):2176280. https://doi.org/10.1080/23311932.2023.2176280
Balendres MA, Karlovsky OP, Cumagun CJR (2019) Mycotoxigenic fungi and mycotoxins in agricultural crop commodities in The Philippines: review. Food Secur 8:249
Barea JM (2015) Interactions among plants, arbuscular mycorrhizal and mycotoxigenic fungi related to food crop health in a scenario of climate change. In: Botana LM, Sainz MJ (eds) Climate change and mycotoxins. Walter de Gruyter GmbH, Berlin, pp 53–64
Barkai-Golan R, Paster N (2008) Mouldy fruits and vegetables as a source of mycotoxins: part 1. World Mycotoxin J 1:147–159
Battilani P, Toscano PHJ, Van der Fels-Klerx HJ et al (2016) Aflatoxin B1 contamination in maize in Europe increases due to climate change. Sci Rep 6:24328
Bianchi F, Giannetto M, Careri M (2018) Analytical systems and metrological traceability of measurement data in food control assessment. TrAC Trends Anal Chem 107:142–150
Bielas W, Niżański W, Nicpoń J et al (2017) Effect of zearalenone on circulating testosterone concentration, testicular and epididymal morphology and epididymal sperm characteristics in wild boars. Theriogenology 102:59–66
Boa E (2003) An illustrated guide to the state of health of trees Recognition and interpretation of symptoms and damage a FOA meeting presentation. CABI Bioscience
Bock CH, Mackey B, Cotty PJ (2004) Population dynamics of *Aspergillus flavus* in the air of an intensively cultivated region of southwest Arizona. Plant Pathol 53:422–433
Bouri M, Arslan KS, Şahin F (2023) Climate-smart pest management in sustainable agriculture: promises and challenges. Sustain For 15(5):4592. https://doi.org/10.3390/su15054592
Bulder AS, Arcella D, Bolger M et al (2012) Fumonisins. In: Safety evaluation of certain food additives and contaminants. WHO Food Additives Series 65. In: Prepared by the 74th Meeting of The Joint FAO/WHO Expert Committee on Food Additives (JECFA). WHO, Geneva, Switzerland, pp 325–527
Bulgaru CV, Marin DE, Pistol et al (2021) Zearalenone and the immune response. Toxins 13:248. https://doi.org/10.3390/toxins13040248
Caglayan MO, Üstündağ Z (2020) Detection of zearalenone in an aptamer assay using attenuated internal reflection ellipsometry and its cereal sample applications. Food Chem Toxicol 136:111081. https://doi.org/10.1016/j.fct.2019.111081
Calder L, Simmons G, Thornley C et al (2003) An outbreak of hepatitis A associated with consumption of raw blueberries. Epidemiol Infect 131:745–751

Callejón RM, Rodríguez-Naranjo MI, Ubeda C et al (2015) Reported foodborne outbreaks due to fresh produce in the United States and European Union: trends and causes. Foodborne Pathog Dis 1:32–38

Cevallos-Cevallos JM, Gu G, Danyluk MD et al (2012) *Salmonella* can reach tomato fruits on plants exposed to aerosols formed by rain. Int J Food Microbiol 158:140–146

Chakraborty S, Tiedemann AV, Teng PS (2000) Climate change: potential impact on plant diseases. Environ Pollut 108:317–326

Chandorkar S (2023) Climate change and food safety. Intech Open. https://doi.org/10.5772/intechopen.112575

Chen X, Abdallah MF, Landschoot Ş et al (2023) *Aspergillus flavus* and *Fusarium verticillioides* and their main mycotoxins: global distribution and scenarios of interactions in maize. Toxins (Basel) 15(9):577. https://doi.org/10.3390/toxins15090577

Chu YJ, Yang HI, Wu HC et al (2017) Aflatoxin B_1 exposure increases the risk of cirrhosis and hepatocellular carcinoma in chronic hepatitis B virus carriers. Int J Cancer 141:711–720

Cinar A, Onbaş E (2020) Mycotoxins: the hidden danger in foods. Intech Open. https://doi.org/10.5772/intechopen.89001

Come J, Cambaza E, Ferreira R et al (2019) Esophageal cancer in Mozambique: should mycotoxins be a concern? Pan Afr Med J 33:187. https://doi.org/10.11604/pamj.2019.33.187.18295. PMid:31565147

Corcuera LA, Vettorazzi A, Arbillaga L et al (2015) Genotoxicity of Aflatoxin B_1 and ochratoxin A after simultaneous application of the *in vivo* micronucleus and comet assay. Food Chem Toxicol 76:116–124

Coskun E, Jaruga P, Vartanian V et al (2019) Aflatoxin-guanine DNA adducts and oxidatively induced DNA damage in aflatoxin-treated mice *in vivo* as measured by Liquid Chromatography-Tandem Mass Spectrometry with isotope dilution. Chem Res Toxicol 32(1):80–89

Creppy EE, Kane A, Dirheimer G et al (1985) Genotoxicity of ochratoxin A in mice: DNA single-strand breaks evaluation in spleen, liver and kidney. Toxicol Lett 1:29–35

Delcour I, Spanoghe P, Uyttendaele M (2015) Literature review: impact of climate change on pesticide use. Food Res Int 68:7–15

Doricakova A, Vrzal R (2015) A food contaminant ochratoxin A suppresses pregnane X receptor (PXR)-mediated CYP3A4 induction in primary cultures of human hepatocytes. Toxicology 337:72–78. https://doi.org/10.1016/j.tox.2015.08.012. Epub 2015 Sep 2

Duchenne R, Ranghoo-Sanmukhiya VM, Neetoo H (2021) Impact of climate change and climate variability on food safety and occurrence of foodborne diseases, chap 25. In: Babalola OO (ed) Food security and safety: African perspectives. Springer, Cham, Switzerland

Duchenne-Moutien RA, Neetoo H (2021) Climate change and emerging food safety issues: a review. J Food Prot 84(11):1884–1897

EFSA (2020) Climate change and food safety. European Food Safety Authority. https://www.efsa.europa.eu/en/topics/topic/climate-change-and-food-safety. Accessed 1 Feb 2024

EFSA Panel on Contaminants in the Food Chain (CONTAM), Schrenk D, Bodin L, Chipman J et al (2020) Risk assessment of ochratoxin A in food. EFSA J 18(3):e06040. https://doi.org/10.2903/j.efsa.2020.6040

Eftekhari A, Ahmadian E, Panahi-Azar V et al (2018) Hepatoprotective and free radical scavenging actions of quercetin nanoparticles on aflatoxin B_1-induced liver damage: *in vitro/in vivo* studies. Artif Cells Nanomed Biotechnol 46:411–420

El Samra GH (2017) Climate change, food security, food safety and nutrition. Egypt J Occup Med 41(2):217–236

El-Nekeety AA, Abdel-Wahhab KG, Abdel-Aziem SH et al (2017b) Papaya fruits extracts enhance the antioxidant capacity and modulate the genotoxicity and oxidative stress in the kidney of rats fed ochratoxin A-contaminated diet. J Appl Pharm Sci 7(7):111–121

El-Nekeety AA, Salman AS, Hathout AS et al (2017a) Evaluation of the bioactive extract of actinomyces isolated from the Egyptian environment against aflatoxin B_1-induce cytotoxicity, genotoxicity and oxidative stress in the liver of rats. Food Chem Toxicol 105:241–255

EPA (2024) Climate impacts: agriculture and food supply. Climate change: Chicago's response to a changing climate. Environmental Protection Agency. https://climatechange.chicago.gov/climate-impacts/climate-impacts-agriculture-and-food-supply#ref1. Accessed 1 Feb 2024

European Commission, Directorate General for Health and Food Safety, Feijao C, d'Angelo C, Flanagan I et al (2020) Development of future scenarios for sustainable pesticide use and achievement of pesticide-use and risk reduction targets announced in the Farm to Fork and Biodiversity Strategies by 2030. Publications Office of the European Union

European Food Safety Authority (EFSA) (2014) Panel on Biological Hazards. Scientific Opinion on the risk posed by pathogens in food of non-animal origin. Part 2 (Salmonella and Norovirus in berries). EFSA J 12:95

Faille C, Cunault C, Dubois T et al (2013) Hygienic design of food processing lines to mitigate the risk of bacterial food contamination with respect to environmental concerns. Innov Food Sci Emerg Technol 46:65–73. https://doi.org/10.1016/j.ifset.2017.10.002

FAO (2013) Good agricultural practices for greenhouse vegetable crops. Food and Agriculture Organization of the United Nations, Rome

FAO (2020) Climate change: unpacking the burden on food safety. In: FAO (ed) Food Safety and Quality Series No 8. FAO, Rome Available at: https://doi.org/10.4060/ca8185en. Accessed 16 Nov 2020

FAO (2022. Rome, Italy) Thinking about the future of food safety – a foresight report, p 158

FAO, WHO (2011) Safety evaluation of certain contaminants in food. World Health Organization, Geneva, and Food and Agriculture Organization of the United Nations, India

FDA (2006) Analysis of microbial hazards related to time/temperature control of foods for safety. Compr Rev Food Sci Food Saf 2(s2):33–41

FDA (2018) Hazard Analysis Critical Control Point (HACCP). In: U.S. Food and Drug Administration. https://www.fda.gov/food/guidanceregulation/haccp. Accessed 6 Feb 2024

Feng YQ, Zhao AH, Wang JJ et al (2022) Oxidative stress as a plausible mechanism for zearalenone to induce genome toxicity. Gene 829:146511. https://doi.org/10.1016/j.gene.2022.146511

García-Pérez E, Ryu D, Lee C et al (2021) Ochratoxin A induces oxidative stress in HepG2 Cells by impairing the gene expression of antioxidant enzymes. Toxins 13:271

Ge C, Lee C, Lee J (2012) The impact of extreme weather events on Salmonella internalization in lettuce and green onion. Food Res Int 45:1118–1122

Gelderblom WC, Marasas WF (2012) Controversies in fumonisin mycotoxicology and risk assessment. Hum Exp Toxicol 31(3):215–235

Godde CM, Mason-D'Croz D, Mayberry DE et al (2021) Impacts of climate change on the livestock food supply chain; a review of the evidence. Glob Food Secur 28:100488. https://doi.org/10.1016/j.gfs.2020.100488. PMID: 33738188; PMCID: PMC7938222

Gracias KS, McKillip JL (2004) A review of conventional detection and enumeration methods for pathogenic bacteria in food. Can J Microbiol 50(11):883–890

Guan Y, Chen J, Nepovimova E et al (2021) Aflatoxin detoxification using microorganisms and enzymes. Toxins 13:46. https://doi.org/10.3390/toxins13010046

HACCP (2023) HACCP is climate friendly. In: Green HACCP. HACCP Food Safety. https://ehaccp.org/haccp-is-good-for-the-environment/02/05/2023/09/07/. Accessed 23 Jan 2024

Harrison LR, Colvin BM, Greene JT et al (1990) Pulmonary edema and hydrothorax in swine produced by fumonisin B_1, a toxic metabolite of *Fusarium moniliforme*. J Vet Diagn Invest 2(3):217–221

Hatfield J, Takle G, Grotjahn R et al (2014) Agriculture. Climate change impacts in the United States: The Third National Climate Assessment. In: Melillo JM, Richmond TC, Yohe GW (eds) US Global Change Research Program. Washington, DC, pp 150–174

Herrera M, Anadón R, Iqbal SZ et al (2016) Climate change and food safety. In: Selamat J, Iqbal SZ (eds) Food safety-basic concepts, recent issues, and future challenges. Springer, Cham, Switzerland, pp 113–123

IARC (2002) Some traditional herbal medicines, some mycotoxins, naphthalene and styrene. In: Monograph on the evaluation of carcinogenic risk to humans, vol 82. IARC, Lyon

Iqbal SZ, Asi MR, Zuber M et al (2013) Natural occurrence of aflatoxins and ochratoxin A in commercial chilli and chilli sauce samples. Food Control 30(2):621–625

ISO (2024) ISO 22000 Food Safety Management System – Training Courses. https://pecb.com/en/education-and-certification-for-individuals/iso-22000. Accessed 23 Jan 2024

James SJ, James C (2010) The food cold-chain and climate change. Food Res Int 43(7):1944–1956

Javed M, Usmani N (2011) Accumulation of heavy metals in fishes: a human health concern. Int J Environ Sci 2:659

Kimanya ME, Routledge MN, Mpolya E et al (2021) Estimating the risk of aflatoxin-induced liver cancer in Tanzania based on biomarker data. PLoS One 16(3):e0247281. https://doi.org/10.1371/journal.pone.0247281

Lacetera N (2019) Impact of climate change on animal health and welfare. Anim Front 9:26–31

Li N, Liu X-L, Zhang F-L et al (2020) Whole-transcriptome analysis of the toxic effects of zearalenone exposure on ceRNA networks in porcine granulosa cells. Environ Pollut 261:114007

Liu C, Hofstra N, Franz C (2013) Impacts of climate change on the microbial safety of pre-harvest leafy green vegetables as indicated by *Escherichia coli* O157 and *Salmonella* spp. Int J Food Microbiol 163:119–128

Liu X, Fan L, Yin S et al (2019) Molecular mechanisms of fumonisin B_1-induced toxicities and its applications in the mechanism-based interventions. Toxicon 167:175. https://doi.org/10.1016/j.toxicon.2019.06.009. PMid:31173793

Longobardi C, Ferrara G, Andretta E, Montagnaro S et al (2022) Ochratoxin A and kidney oxidative stress: the role of nutraceuticals in veterinary medicine-a review. Toxins (Basel) 14(6):398. https://doi.org/10.3390/toxins14060398. PMID: 35737059; PMCID: PMC9231272

Luo Y, Liu X, Li J (2018) Updating techniques on controlling mycotoxins A review. Food Control 89:123–132

Maggiore A, Afonso A, Barrucci F et al (2020) Climate change as a driver of emerging risks for food and feed safety, plant, animal health and nutritional quality. EN- 1881. European Food Safety Authority, Parma, Italy

Marasas WF, Kellerman TS, Gelderblom WC et al (1988) Leukoencephalomalacia in a horse induced by fumonisin B_1 isolated from *Fusarium moniliforme*. Onderstepoort J Vet Res 55(4):197–203

Marin S, Ramos AJ, Cano-Sancho G et al (2013) Mycotoxins: occurrence, toxicology, and exposure assessment. Food Chem Toxicol 60:218–237

Marques A, Nunes ML, Moore SK et al (2010) Climate change and seafood safety: human health implications. Food Res Int 43:1766–1779

Martins C, Assunção R, Cunha SC et al (2018) Assessment of multiple mycotoxins in breakfast cereals available in the portuguese market. Food Chem 239:132–140

Menini A, Mascarello G, Giaretta M et al (2022) The critical role of consumers in the prevention of foodborne diseases: an ethnographic study of Italian families. Food Secur 11(7). https://doi.org/10.3390/foods11071006

Milićević D, Petronijevic R, Petrovic Z et al (2019) Impact of climate change on aflatoxin M_1 contamination of raw milk with a special focus on climate conditions in Serbia. J Sci Food Agric 99:5202–5210. https://doi.org/10.1002/jsfa.9768

Moretti A, Pascale M, Logrieco AF (2018) Mycotoxin risks under a climate change scenario in Europe. Trends Food Sci Technol 84:38–40

Multari RA, Multari RA, Cremers DA et al (2023) Detection of biological contaminants on foods and food surfaces using laser-induced breakdown spectroscopy (LIBS). J Agric Food Chem 61(36):8687–8694

Mutegi C, Cotty P, Bandyopadhyay R (2018) Prevalence and mitigation of aflatoxins in Kenya (1960-to date). World Mycotoxin J 11:341–357

Nazhand A, Durazzo A, Lucarini M et al (2020) Characteristics, occurrence, detection and detoxification of aflatoxins in foods and feeds. Food Secur 9:644. https://doi.org/10.3390/foods9050644

Neumann RB, Seyfferth AL, Teshera-Levye J et al (2017) Soil warming increases arsenic availability in the rice rhizosphere. Agric Environ Lett 2:170006

NHMRC (2019) Australian guidelines for the prevention and control of infection in healthcare. National Health and Medical Research Council, Canberra

Nicholls N, Easterling D, Goodess CM et al (2012) Changes in climate extremes and their impacts on the natural physical environment. In: Field CB, Barros V, Stocker TF et al (eds) Managing the risks of extreme events and disasters to advance climate change adaptation. A special report of working groups I and II of the intergovernmental panel on climate change (IPCC). Cambridge University Press, Cambridge, UK, and New York, NY, USA, pp 109–230

Nleya N, Adetunji M, Mwanza M (2018) Current status of mycotoxin contamination of food commodities in Zimbabwe. Toxins 10:89. https://doi.org/10.3390/toxins10050089

Ortega-Beltran A, Bandyopadhyay R (2023) Aflatoxin biocontrol in practice requires a multidisciplinary, long-term approach. Front Sustain Food Syst 7:1110964. https://doi.org/10.3389/fsufs.2023.1110964

Ostry V, Malir F, Toman J et al (2017) Mycotoxins as human carcinogens-the IARC Monographs classification. Mycotoxin Res 33(1):65–73

Paris MPK, Liu YJ, Nahrer K et al (2015) Climate change impacts on mycotoxin production. In: Botana LM, Sainz MJ (eds) Climate change and mycotoxins. Walter de Gruyter GmbH, Berlin, pp 133–149

Perrone G, Ferrara M, Medina A et al (2020) Toxigenic fungi and mycotoxins in a climate change scenario: ecology, genomics, distribution, prediction and prevention of the risk. Microorganisms 8(10). https://doi.org/10.3390/microorganisms8101496

Polo Q (2023) Food contamination: understanding the risks and prevention. J Food Microbiol Saf Hyg 8:192

Poltronieri P, Mezzolla V, Primiceri E et al (2013) Biosensors for the detection of food pathogens. Food Secur 3(3):511–526

Rahimtula AD, Bérézíat JC, Bussacchini-Griot V et al (1988) Lipid peroxidation as a possible cause of ochratoxin A toxicity. Biochem Pharmacol 37:4469–4477

Rai A, Das M, Tripathi A (2020) Occurrence and toxicity of a fusarium mycotoxin, zearalenone. Crit Rev Food Sci Nutr 60(16):2710–2729

Rather IA, Koh WY, Paek WK et al (2017) The sources of chemical contaminants in food and their health implications. Front Pharmacol 8:830. https://doi.org/10.3389/fphar.2017.00830. PMID: 29204118; PMCID: PMC5699236

Rushing BR, Selim MI (2019) Aflatoxin B_1: a review on metabolism, toxicity, occurrence in food, occupational exposure, and detoxification methods. Food Chem Toxicol 124:81–100

Safavieh M, Nahar S, Zourob M et al (2014) Microfluidic biosensors for high throughput screening of pathogens in food. In: Bhunia AK, Kim MS, Taitt CR (eds) High throughput screening for food safety assessment, 1st edn. Woodhead Publishing, Cambridge, pp 327–357

Samik A, Safitri E (2017) Mycotoxin binders potential on histological of ovary mice exposed by zearalenone. Vet World 10(3):353–357

Scott PM (2012) Recent research on fumonisins: a review. Food Addit Contam Part A Chem Anal Control Expo Risk Assess 29:242–248

Sirot V, Fremy JM, Leblanc JC (2013) Dietary exposure to mycotoxins and health risk assessment in the second French total diet study. Food Chem Toxicol 52:1–11

Song Q, Zheng YJ, Xue Y et al (2017) An evolutionary deep neural network for predicting morbidity of gastrointestinal infections by food contamination. Neurocomputing 226:16–22

Szabó-Fodor J, Kachlek M, Cseh S et al (2015) Individual and combined effects of subchronic exposure of three *Fusarium* toxins (Fumonisin B, Deoxynivalenol and Zearalenone) in rabbit bucks. Clin Toxicol 5:4–15

Tan ETT, Materne CM, Silcock RG et al (2016) Seasonal and species variation of the hepatotoxin indospicine in Australian Indigofera legumes as measured by UPLC-MS/MS. J Agric Food Chem 64:6613–6621

Tirado MC, Clarke R, Jaykus LA et al (2010) Climate change and food safety: a review. Food Res Int 43(7):1745–1765. https://doi.org/10.1016/j.foodres.2010.07.003

Uyttendaele M, Liu C, Hofstra N (2015) Special issue on the impacts of climate change on food safety. Food Res Int 68:1–6

Van der Spiegel M, Van der Fels-Klerx HJ, Marvin HJP (2012) Effects of climate change on food safety hazards in the dairy production chain. Food Res Int 46:201–208
Varzakas TH (2011) Application of ISO22000, failure mode, and effect analysis (FMEA) cause and effect diagrams and pareto in conjunction with HACCP and risk assessment for processing of pastry products. Crit Rev Food Sci Nutr 8:762–782
Voss KA, Riley RT (2013) Fumonisin toxicity and mechanism of action: overview and current perspectives. Food Safety 1(1):2013006. https://doi.org/10.14252/foodsafetyfscj.2013006
Wang B, Wang H, Lu X et al (2023) Recent advances in electrochemical biosensors for the detection of foodborne pathogens: current perspective and challenges. Food Secur 12(14):2795. https://doi.org/10.3390/foods12142795
Wang J, Xie Y (2020) Review on microbial degradation of zearalenone and aflatoxins. Grain Oil Sci Technol 3:117–125
Watson S, Moore SE, Darboe MK et al (2018) Impaired growth in rural Gambian infants exposed to aflatoxin: a prospective cohort study. BMC Public Health 18:1247
WHO (2018) Natural Toxins in food (Factsheet). Available online: https://www.who.int/newsroom/factsheets/detail/natural-toxins-in-food. Accessed on 10 Feb 2020
Wijngaard RR, van der Perk M, van der Grift B et al (2017) The impact of climate change on metal transport in a lowland catchment. Water Air Soil Pollut 228:107. https://doi.org/10.1007/s11270-017-3261-4
Wild CP, Turner PC (2002) The toxicology of aflatoxins as a basis for public health decisions. Mutagenesis 17:471–481
Wu X, Lu Y, Zhou S et al (2016) Impact of climate change on human infectious diseases: empirical evidence and human adaptation. Environ Int 2016(86):14–23
Yu M, Chen L, Peng Z et al (2017) Mechanism of deoxynivalenol effects on the reproductive system and fetus malformation: current status and future challenges. Toxicol In Vitro 41:150–158
Zhang X, AuZhong T, Liu L, Ouyang X (2015) Impact of soil heavy metal pollution on food safety in China. PLoS One 10:e0135182. https://doi.org/10.1371/journal.pone.0135182

Climate Change Impacts on Pillars of Food Security and Human Health

5

Wafaa A. Kandeel, Nayra Sh. Mehanna, Lamia S. Ellaithy, and Doaa Y. Hammad

Abstract

Climate change threatens the accessibility, availability, well-being, and food systems stability. Changes in rainfall patterns, temperature, and the frequency of extreme weather events significantly impact food security. Climate change affects food safety and leads to higher foodborne illness and poisoning rates. Changing rainfall patterns, increasing temperatures, and increased rates of extreme weather events put children, pregnant women, and elders at high risk of health problems. Climate change leads to nutrition-related diseases such as malnutrition, overweight, and obesity due to decreased water quality, altered food production, and inadequate food consumption. Stillbirth is a common consequence of climate change, which has the greatest impact on pregnant women and newborns through heatwaves and extreme events. Furthermore, low- and middle-income countries may face severe socioeconomic impacts from climate change due to food insecurity. This chapter will discuss all these health impacts and how to prevent the impacts of unhealthy food among vulnerable communities.

W. A. Kandeel (✉) · D. Y. Hammad
Biological Anthropology Department, Institute of Medical Research and Clinical Studies, National Research Centre, Giza, Egypt

N. S. Mehanna
Dairy Science Department, Institute of Food Science and Nutrition, National Research Centre, Giza, Egypt

L. S. Ellaithy
Environmental and Occupational Medicine Department, Environment and Climate Change Research Institute, National Research Centre, Giza, Egypt

A. Saad-Hussein et al. (eds.), *Climate Change Impacts on Toxins and Health Effects*,
https://doi.org/10.1007/978-981-96-3416-3_5

Keywords

Climate change · Food security · Food safety · Malnutrition · Obesity · Extreme weather events

Abbreviations

C. Botulinum	Clostridium Botulinum
CO²	Carbon Dioxide
DNA	Deoxyribonucleic Acid
ECDC	European Center for Disease Prevention and Control
EFSA	European Food Safety Authority
EMME	The Eastern Mediterranean and Middle East
FAO	Food and Agriculture Organization
FBI	Foodborne Illness
HIV	Human Immunodeficiency Virus
IFAD	International Fund for Agricultural Development
IPCC	Intergovernmental Panel on Climate Change
NASBA	Nucleic Acid Sequence-Based Amplification
NGS	Next-Generation Sequencing
NCV	North Sea Whereas Toxigenic
PCR	Polymerase Chain Reaction
PTSD	Posttraumatic Stress Disorder
RT-PCR	Real-Time Polymerase Chain Reaction
SARS	Severe Acute Respiratory Syndrome
SDGS	Sustainable Development Goals
SFS	Sustainable Food System
SSA	Sub-Saharan Africa
SSPS	Shared Socioeconomic Pathways
UK	United Kingdom
UNFSS	United Nations Convened the Food System Summit
VBD	Vector-Borne Diseases
WFP	World Food Program
WHO	World Health Organization

5.1 A Focus on Climate Change and its Impact on Food Security

Food security, according to the 1996 World Food Summit, is defined as: "*When all people, at all times, have physical and economic access to enough safe and nutritious food to meet their nutritional needs and food preferences for an active and healthy life*" (World Food Summit 1996). The four main features of food security are:

- *Food Availability:* It is estimated by stock levels, food production levels, and net trade and addresses the "supply side" of food security.
- *Economic and Physical Access to Food*: It is the sufficient national or international food supply level, but this does not guarantee food security at the household level. Food insecurity concerns have led to a heightened focus on income, spending, markets, and pricing as key factors in achieving food security objectives.
- *Food Utilization:* It is commonly understood as the process through which the body extracts the most nutrients from food. Individuals' adequate energy and nutrient intake results from proper feeding practices, diet diversity, food preparation, and intra-household food distribution. The individual's nutritional status is influenced by both the quantity and quality of food consumed, as well as how efficiently the body absorbs and uses nutrients.
- *Long-Term Stability of the Other Three Dimensions*: Despite adequate food intake today, a lack of consistent access to food can lead to food insecurity, potentially compromising nutritional status. Bad weather, political unrest, or socioeconomic factors (gender diversity, low income, or unemployment, besides the high food prices) can affect food security.

To achieve food security goals, all four dimensions must be met at the same time (Clay 2002).

5.1.1 Effect of Climate Change on Food Production and Access

5.1.1.1 Effects on Crops

Climate change has substantial impacts on crops, with higher concentrations of CO_2 and temperature having alternative effects on crop yields. However, temperature extremes can also have negative effects. CO_2 affects some crops more (wheat, cotton, rice, potato, soy) than some others (sorghum, corn, sugar). Temperature variability and extreme weather episodes can also negatively affect crops, with shorter maturation periods and lower grain production. Ozone, another anthropogenic pollutant, has direct impacts on reproductive function and physiological processes. Increased temperatures can spread vegetable diseases and pests in higher latitudes, while the use of pesticides and climate changes contribute to decreased populations of pollinating insects. This sensitivity to climate and air pollution has led to a marked decrease in crop yields globally and greater variability in food provision. The sixth report of the IPCC estimates a 2.3% decrease in corn crop yields, a 3.3% decrease in soy, a 0.7% decrease in rice, and a 1.3% decrease in wheat (Rakhmankulova et al. 2021; Wang et al. 2022; Bezner Kerr et al. 2022).

5.1.1.2 Effects on the Composition and Quality of Food Products

Climate change can negatively impact food quality by altering carbon, nutrient absorption, and biochemical processes. This can affect human and animal health, as well as economic value. Increases in CO_2 levels can lead to lower protein concentrations and decreased trace elements in grains and green parts of wheat (Myers et al. 2014).

5.1.2 Effects of Climate Change on Animal-Origin Food Products

5.1.2.1 Livestock

The production of fodder impacts animal feeding and food products, with decreased vegetable crop yields leading to higher production costs. Increased average temperatures in cold zones may extend the growing season but decrease the quality, causing variability in production. Extreme temperatures also influence public health, with increased morbidity-mortality and animal stress. Genetically selected animals have higher sensitivity to environmental changes, leading to decreased dairy production, heat-related mortality, lower growth index, fertility changes, and increased mortality. Variable precipitation and extreme drought also cause water provision problems and increased production costs. Vector-borne diseases like West Nile Virus and Crimean-Congo Fever also pose risks (Godde et al. 2021; Thornton et al. 2021).

5.1.2.2 Fishing

It is a crucial sector for the global population and is at risk because of climate change, particularly in developing countries. Elevated water temperature and ocean acidification are causing a redistribution of fish catch potential toward higher latitudes, affecting lower latitudes closer to the tropics. This has led to decreased captures of calcareous fish and coral reef death, negatively impacting the sector in tropical developing countries. Research indicates that climate change might contribute to declining fish catches in lakes and rivers, although it is challenging to separate this impact from human activities (Magel et al. 2020).

5.1.2.3 Effects on Food Prices

Global food prices generally declined throughout the twentieth century, following the principles of supply and demand. However, since 2007, there have been intermittent periods of sharp price increases. This trend is attributed to a combination of factors, including rising global demand for food and increased production of biodiesel. The latter is driven by energy policies and fluctuations in biofuel prices (Mueller et al. 2011).

5.1.2.4 Effects on Food Safety

There is an influence on food safety and nutritional quality. For example, the seasonal behavior of salmonellosis and campylobacteriosis, the most common foodborne diseases in Europe, shows a summer peak that coincides with rising temperatures (EFSA and ECDC 2021). Due to control measures put in place by public authorities, indicating that the potential risks posed by rising temperatures associated with climate change can be addressed through concerted public health action (Kyncl et al. 2021). Predictive models simulating temperature increases of 2 or 5 °C in Europe have generated risk maps highlighting a potential rise in aflatoxin-contaminated corn, particularly in southern regions such as Spain (Battilani et al. 2016). Ocean rise levels, continental glaciers, the coalescence of polar ice, changes in precipitation patterns, and contributions of ocean and lake precipitation, as well as other extreme meteorological phenomena, such as increased frequency of major storms, affect the marine system in terms of reducing salinity and transporting pollutants. This is associated with species modification, the emergence of fecal pathogens (failure of the sewage system due to flooding), flood-induced sewage overflows, and marine pathogens such as *Vibrio* spp. (due to changes in water temperature and salinity) (Froelich and Daines 2020), and increases in the size and frequency of dinoflagellate toxin blooms (Estevez et al. 2019).

The global community has committed to eradicating poverty, ending hunger, and combating climate change by 2030 through sustainable development goals (FAO 2015). With the expected population growth worldwide to one-third more, agricultural production must be increased to 60% to meet increased demands for food and feed. About a 60% increase in global food demand is expected to occur by 2050 due to the increased population growth rate and the changes in food patterns (Gunaratne et al. 2021). Hunger has been on the rise globally since 2014, with 25.9% of the global population experiencing hunger or not having regular access to sufficient and nutritious food (FAO 2020). Malnutrition affects a staggering two billion people worldwide. Estimates suggest that a 60% increase in food availability is required to feed the global population by the middle of this century (Fig. 5.1).

Impact of Climate Change on Diet and Nutrition-Related Diseases

Climate change affects food utilization by altering food safety conditions and changing disease distribution and patterns, for example, changes in water, vector, and foodborne diseases. The primary concern about food security and climate change is that changing climatic conditions can initiate a vicious circle where infectious diseases usually are caused or compounded by poor nutritional status, affecting populations` immunological status, and making them more susceptible to vigorous infectious diseases. This vicious cycle will be completed by the decline in labor productivity of the infected population, leading to increased poverty and even mortality. This will increase the number of individuals exposed to water- and foodborne diseases (e.g., cholera), which lowers their capacity to effectively utilize the consumed food. Therefore, the main pathways in which climate change affects diet and nutrition-related diseases are shown in Fig. 5.2.

Reduction in food quality, which can be a result of high CO_2 concentration in the atmosphere, could impact the crops' trace elements, which consequently increases the risk of mineral deficiency in populations consuming these crops, such as zinc or iron deficiency. Also, there may be an alteration in healthy food consumption, which leads to an elevation of the risk of non-communicable diseases, including diabetes mellitus, cancer, cardiovascular diseases, or even death (Fanzo and Downs 2021).

Climate change is altering various environmental conditions, which can increase the transmission of infectious diseases. Pathogens spread through water, air, food, and vectors are particularly vulnerable to these changes. Additionally, climate

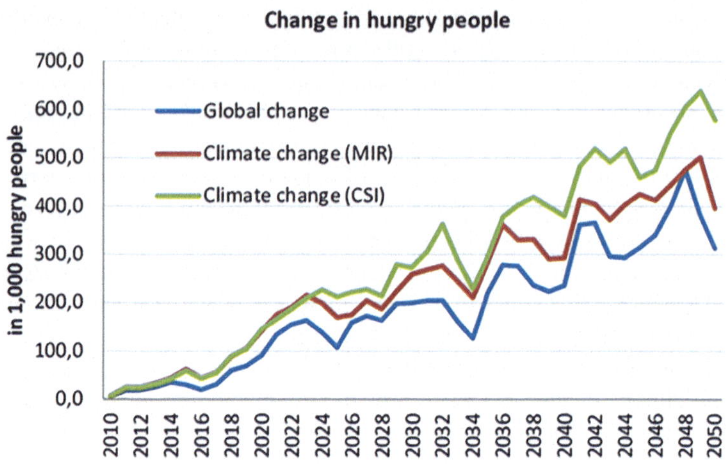

Fig. 5.1 Effect of climate change on food security (Breisinger et al. 2011)

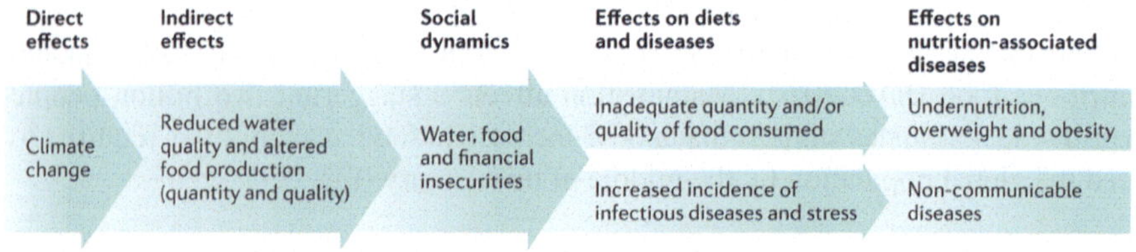

Fig. 5.2 Impacts of climate change on nutrition-associated diseases (Fanzo and Downs 2021)

change indirectly affects nutritional status by influencing the prevalence of infectious diseases. Malnutrition can worsen the impact of infections, while infections can also contribute to malnutrition due to factors like reduced water availability and nutrient deficiencies. Low- and middle-income countries are particularly susceptible to the global rise in food prices. Furthermore, these countries are more prone to contagious illnesses, placing an additional burden on their public health infrastructure (Romanello et al. 2021). Factors causing malnutrition include the high cost of a healthy diet as well as wide income inequality. In 2019, the cumulative impact of previous factors contributed to a global food affordability crisis, affecting nearly three billion people (FAO [Food and Agriculture Organization] 2020).

5.2 Effect of Climate Change on Food Availability and Human Well-Being

The World Health Organization states that "*well-being is a state of complete physical, mental, and social well-being, not merely the absence of disease or infirmity.*" Well-being refers to a state of well-being, health, and happiness, and although happiness is an integral part of personal health, well-being includes other things, including a sense of purpose, achieving long-term goals, and how to take control of life. Maintaining a healthy body and mind requires a combination of exercise and proper nutrition. Physical health is not just about being free of diseases; it means that the body is functioning well enough to prevent lifestyle diseases. Mental health is the ability to adapt in everyday life and to be resilient and recover quickly from life's setbacks. Emotional well-being is the ability to control feelings, thoughts, and behaviors. Being emotionally well allows you to express feelings, feel connected to the world around you, and have a positive subjective experience. Appreciation Social well-being is the extent of the feeling of belonging, inclusion, and stability (Kühn and Rieger 2017).

Globally, climate change is jeopardizing food availability, a critical factor in human well-being and mental health. Climate-induced food insecurity disproportionately impacts specific populations. Understanding the intricate connections between climate change, food insecurity, and mental health challenges requires considering diverse factors. Food systems and mental health determinants vary across cultures, regions, and historical contexts. The relationships between climate change, food insecurity, and mental health are multifaceted. Furthermore, climate change's effects often intersect with other societal and environmental pressures, including pollution, political instability, and economic hardship (WHO 2018; Xu et al. 2020).

Climate change negatively affects food availability by reducing crop yields, fish stocks, and livestock productivity. Additionally, extreme weather events, deteriorating soil health, altered crop nutrient content, and food contamination contribute to post- and pre-harvest losses, further affecting the environment. As illustrated in Fig. 5.3, the effects of these consequences threaten the stability of global food supplies (FAO 2016).

Climate change presents a substantial risk to human health. Its impacts, including air pollution, diseases, extreme weather events, displacement, mental health stressors, and food insecurity, are already affecting populations worldwide. As climate change intensifies, the risks of food and water scarcity will likely increase (Zhao et al. 2022).

The health of our planet is inextricably connected to human health and well-being. A healthy environment safeguards us from the escalating risks of extreme weather events, such as droughts, heatwaves, storms, flooding, and unpredictable climate patterns. Our access to oceans, rivers, forests, and bushlands provides vital recreational opportunities that contribute to our physical and mental health. Vulnerable populations, including those living in poverty, rural areas, or isolated communities, as well as individuals with disabilities, children, the elderly, pregnant women, and unborn children, are disproportionately influenced by the impacts of climate change (Al-Delaimy 2020: Linares et al. 2020).

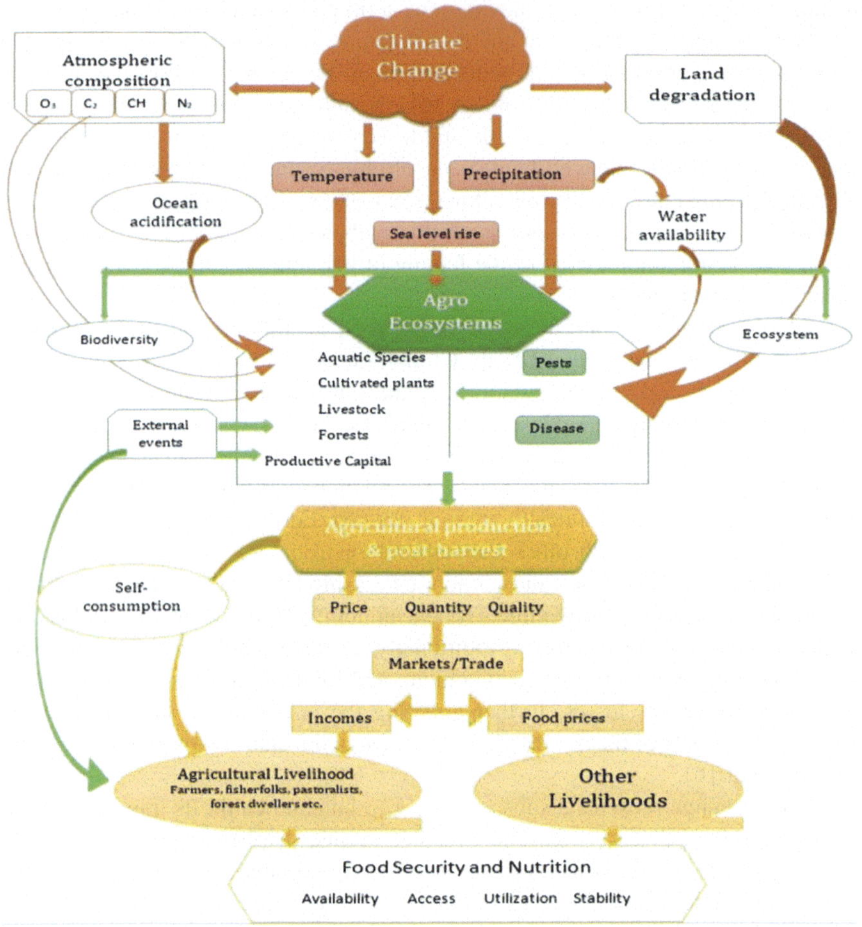

Fig. 5.3 The complex relationship between climate change and food security, focusing on the ways in which climate change affects nutritional outcomes. (Adapted from FAO 2016)

5.3 The Food Systems Approach to Food Security

Food systems play a crucial role in achieving the sustainable development goals by 2030, but the progress has been insufficient until now. The United Nations convened the Food System Summit (UNFSS) in September 2021 and identified bold new actions to enable progress. Extensive research over the past decade has explored the workings of food systems and strategies for enhancing their sustainability. A recurring theme in this literature is the inability of current food systems to ensure food security while minimizing environmental impact. However, perspectives on the root causes of these challenges vary significantly (Béné et al. 2019a) and various focus and disciplinary orientations lead to various types of solutions (Brouwer et al. 2020). Several authors highlight the necessity of understanding the interconnectedness of system components (Allen and Prosperi 2016; Béné et al. 2019b; Dury et al. 2019; Melesse et al. 2019). Many agree that food systems analysis should go beyond traditional production-focused approaches to food security. This involves recognizing the diverse roles food systems play, such as environmental sustainability and social equity (Chaudhary et al. 2018; Dury et al. 2019; Gustafson et al. 2016; Westhoek et al. 2016). The increasing uncertainty and complexity in the food system's overtime demands improved resilience. Despite the consensus on the need for sustainable food systems, there is a lack of clarity on operationalizing the conceptual approach, understanding sustainability determinants, and designing pathways to improve them (Tendall et al. 2015; Allen and Prosperi 2016; Dury et al. 2019; Fanzo et al. 2021).

Food systems encompass the interconnected network of actors, activities, and influencing factors within the food supply chain. These elements interact to shape long-term sustainability outcomes and impacts, which, in turn, can influence other components through feedback loops (David-Benz et al. 2022).

Food security is defined as: "*Food security is when all people, at all times, have physical and economic access to enough safe and nutritious food to meet their nutritional needs and food preferences for an active and healthy life*" (World Food Summit 1996).

Food systems components. This concept encompasses the various components of food systems, such as the factors influencing production, the individuals and organizations involved in the food supply chain, and the resulting outcomes.

Drivers are "internal or external processes that intentionally or unintentionally affect or impact a food system over a sufficiently long period such that their effects permanently alter the activities of that system, and thus its outcomes" (Béné et al. 2019b).

Outcomes are the short- and mid-term consequences of food system performance on various aspects of sustainability and are known as outcomes. These outcomes can be intended or unintended, positive or negative.

Feedback loops are "circular effects between the effects of a food system and the drivers that influence it. This is particularly the case with respect to the dimensions of environmental, social, economic, and regional balance, which influence and are affected by food systems" (David-Benz et al. 2022).

Food System Approach

The food system is evolving due to factors such as population growth, wealth, urbanization, globalization, changing consumption patterns, climate change, and resource depletion. While these developments have led to increased off-farm employment opportunities and a wider range of food choices, they also present challenges for food security and nutrition. These include the availability of highly processed, low-nutritional-value, and high-calorie food items; limited access to small-scale producers and businesses; high food loss and waste; increased food safety and health issues; and an increased energy intensity and ecological footprint. So, the functions of diverse food systems are crucial to minimize their negative impacts and maximize their positive contributions.

Structure of the Food System

The food system is structured around FAO's main goals of poverty reduction, proper nutrition, and food security. These goals are embedded in the system's performance in economic, environmental, and social dimensions. System performance is influenced by the behavior of different actors or stakeholders in the food system. The system consists of a core system and societal and natural elements. The core system includes activities like processing, production, aggregation, distribution, consumption, and services supporting the flow of food products. These activities are established in a societal context and natural environment, including policies, regulations, laws, infrastructures, socio-cultural norms, and organizations as shown in Fig. 5.4.

The global food system is unsustainable, and urgent action is needed to address this imbalance. Transitioning food systems for sustainable, healthy diets with minimal environmental impact is a major goal of local and international organizations. The food system includes activities involved in the production, processes, packaging, distribution, retailing, and food consumption, which form the basis of the four food security pillars: accessibility, availability, stability, and utilization (Marambe 2022).

Food security is a complex issue that involves various outcomes that contribute to environmental outcomes. While sufficient global food production is not guaranteed, it is essential to focus on food quantity over quality. The concept of food security has evolved to include nutritional adequacy. Webb et al. (2020) propose four policy actions to transition food systems toward sustainable diets: availability, accessibility, affordability, and desirability. The "missing middle" (Veldhuizen et al. 2020) refers to the paradox of persistent hunger regardless of growing global food availability. Overcoming food insecurity requires a deep comprehension of the multifaceted interactions between food and nutrition security strategies and global environmental alterations. A range of stakeholders have utilized food system paradigms to promote interdisciplinary investigation of the two-way linkages between these elements (Ericksen et al. 2010).

Dietary patterns and food systems vary widely across cultures and regions, making a one-size-fits-all approach impractical (Webb et al. 2020). Food security and sustainability cannot be attained without simultaneously addressing agricultural practices and the social and ecological challenges that perpetuate inequality (Cadieux and Slocum 2015; Paganini and Lemke 2020). This perspective employs a descriptive methodology that interrogates a range of food system transformation measures that collectively underpin socio-environmental local contexts, livelihoods, nutritional habits, and ecosystems. Intersectionality factors determine the adaptation of people to different socio-ecological changes (Erwin et al. 2021), which consequently differentially shape their food and nutrition security experiences (Williams-Forson and Wilkerson 2011).

A key factor to achieve sustainable health today and for future generations is the use of "sustainable diets," which have low environmental impacts that contribute to nutritional security and healthy living.

Sustainable diets are safe, healthy, protective, nutritionally adequate, acceptable socially, accessible, economically adequate, and respectful of ecosystems (Fig. 5.5).

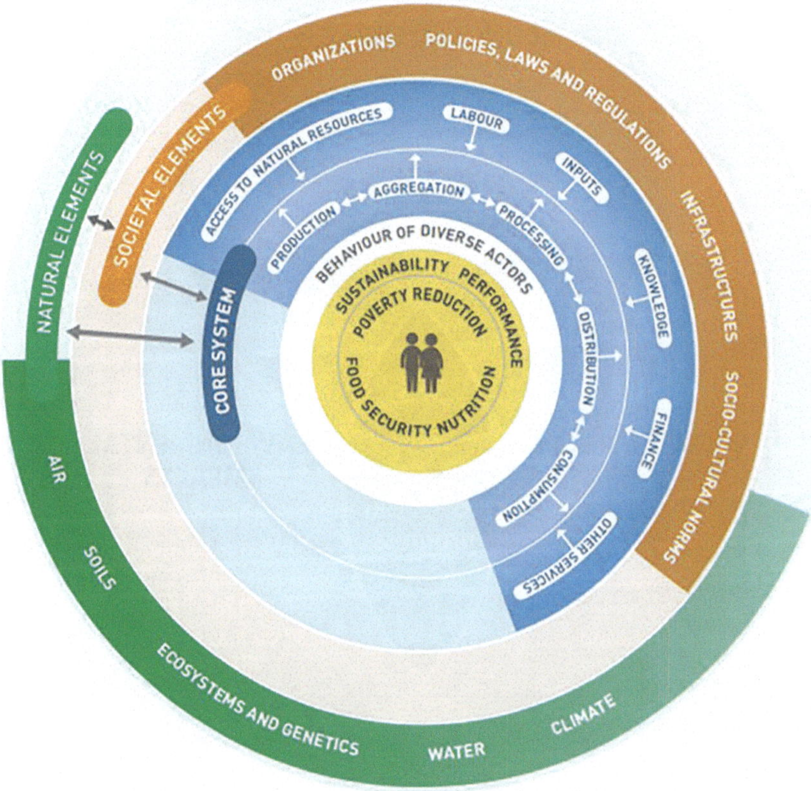

Fig. 5.4 Food system wheel framework (FAO 2018)

There is a relationship between nutritional value and environmental impact. Nutritionists suggest prioritizing plant-based foods like fruits and vegetables for their health benefits and lower environmental footprint while limiting the consumption of resource-intensive foods. There are multiple potential ways to slow, and perhaps even reverse, increases in chronic non-communicable diseases and diet-related environmental stressors.

One such strategy is to move to diets that contain fewer calories from animal-source foods and diets in which caloric intake is sufficient to meet metabolic requirements. Several studies have shown that reduction of meat consumption can decrease greenhouse gas emissions while maintaining adequate nutrition.

Diet-related greenhouse emissions and premature mortality could be substantially reduced by adopting a low-meat diet that meets nutritional standards for vegetables, fruits, and caloric needs.

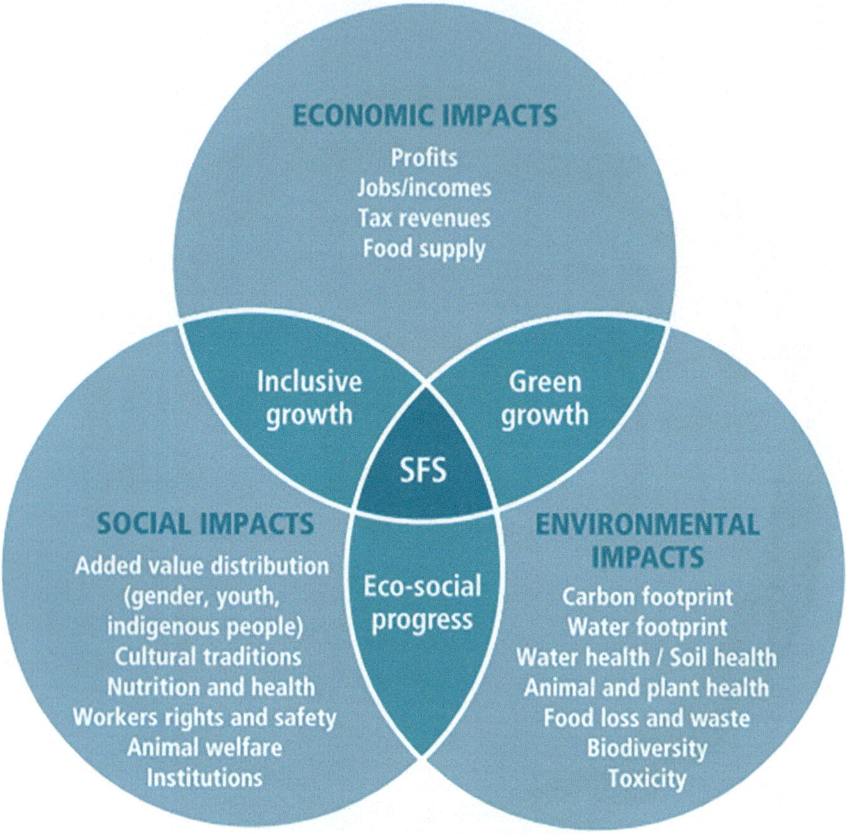

Fig. 5.5 Sustainable food system (FAO 2018)

A sustainable food system (SFS) provides nutrition security for all customers without compromising the socioeconomic and environmental foundations of ensuring food security for future generations. SFS are profitable (economic sustainability), have benefits for society (social sustainability), and have positive or neutral effects on the environment (environmental sustainability) (Fig. 5.5).

5.4 Effect of Climate Change on Foodborne Diseases

Foodborne Illness: Understanding the Symptoms and Outbreaks
Getting sick from contaminated food or drinks is called foodborne illness (FBI), or food poisoning. This happens due to the pathogenic micro-organisms in the food, with most cases caused by over 250 different ones. The common signs include feeling sick, vomiting, stomach pain, and diarrhea. According to the WHO estimation, nearly 1 in 10 people worldwide get sick from foodborne diseases every year. That is a staggering 600 million cases/year. This leads to about 420,000 deaths, with

young children under 5 being the most tragically affected (125,000 deaths). Diarrheal diseases are a major global health threat, causing hundreds of millions of illnesses and hundreds of thousands of deaths each year. They are caused by bacteria, viruses, and parasites. The most common causes of FBI include norovirus, *Campylobacter, Taenia solium* (tapeworm), *Salmonella typhi*, and hepatitis A virus. Aflatoxin and viruses, on the other hand, are the most common causes of death (Holland et al. 2023).

The main symptoms of foodborne diseases are based on the source of food. The most commonly observed symptoms are abdominal pain, vomiting, bloody or simple diarrhea, chills, fever, and other pathogen-specific signs. The disease can lead from acute to chronic cases depending upon the range of symptoms, which can last from a few days to months. Certain bacteria, such as C. botulinum, can cause alarming situations by damaging the nervous system, resulting in symptoms such as paralysis, blurred vision, dizziness, headaches, weakness, and numbness or tingling of the skin (Shakeel et al. 2023).

Foodborne Illness and Their Causes

The foodborne infection entails different pathways; thus, these criteria of convergence are undefined and complex. Food production from animal sources is thought to be the primary cause of the spread of foodborne infections. The food products from animals, such as meat, milk, etc., may have the highest rate of contaminants, pathogen contents, adulterants, and natural toxins. It has been observed that with the high consumption of food products derived from animals, the rate of foodborne diseases is amplifying day by day, leading to an increase in the death rate.

The contamination of food products may occur in any division of the food chain or maybe at the food production level, where different processes are taking place in different food units to make purified food products. The toxification of the food depends on the hygienic condition of the place for food production and the health condition of workers. With proper knowledge and practice to keep up the hygienic condition, the risk of food contamination can be reduced to a minimum level. (Shakeel et al. 2023).

Effect of Climate Change on Foodborne Illness

Environmental factors, especially climate change, such as variations in temperature, humidity, precipitation, and soil properties, are playing a crucial role in influencing the spread and survival of zoonotic pathogens. Changes in these factors, such as unpredictable rainfall, rising temperatures, and droughts, can make it easier for these pathogens to thrive. Climate change could worsen more than half of all known infectious diseases. It does not just affect current diseases but also increases the risk of new diseases emerging or old, controlled diseases re-emerging. For example, water scarcity caused by climate change could make treated wastewater a common irrigation source. This can potentially introduce dangerous pathogens into our food supply (Mora et al. 2022).

Climate change creates unstable weather patterns that disrupt habitats and affect our food supply. This can increase the risk of FBIs. Additionally, climate-related water shortages lead to the use of treated wastewater for irrigation, potentially exposing crops to harmful pathogens (Dietrich et al. 2023).

Diseases Diminished by Climatic Hazards

Most studied diseases worsen due to climate change, but a small number (63 out of 286) actually see a decrease. This decrease might be linked to warmer temperatures making it harder for certain viruses to survive or even strengthening our immune systems against them, as seen with influenza, SARS, and some intestinal illnesses. However, it is important to note that this positive effect is not consistent. Even diseases that benefit from one climate hazard can be worsened by another, or even the same one in different circumstances. Extreme weather events can have complex and sometimes contradictory effects on the spread of diseases. While floods might sometimes limit the habitat of disease-carrying creatures like snails, in other situations, they can actually increase human contact with these creatures and even help them spread further. Similarly, droughts can either decrease the breeding grounds of disease-carrying mosquitoes or concentrate them in smaller water sources, leading to a rise in their density (Mora et al. 2022).

Pathogens Strengthened by Climatic Hazards

Climatic hazards not only make it easier for people to come into contact with disease-causing organisms, but they also help these pathogens thrive. They become better suited to the climate, reproduce faster, are active for longer periods, spread more easily (for example, through shorter incubation times), and become more harmful. Warmer temperatures, for instance, have aided mosquitoes in growing, surviving, biting, and spreading viruses like West Nile.

Rising ocean temperatures promote the rapid growth of harmful algae and bacteria, including those that cause diseases like Vibriosis. Warmer water, especially when combined with heavy rain that dilutes saltwater, creates ideal conditions for these organisms to thrive. This also sometimes leads to increased animal populations, which can spread diseases like plague and hantaviruses. Storms and flooding create perfect conditions for mosquitoes to breed and spread diseases like malaria, yellow fever, dengue, and more. In warmer weather, some of these diseases can also become more dangerous. Heat can make certain bacteria, like *Vibrio*, express genes that help them spread, attach, enter cells, survive longer, and cause more harm to the infected organism. Scientists think heatwaves may cause some viruses to become "heat-resistant." If these viruses then infect humans, they could cause more severe illness because they can better handle the body's natural fever defense mechanism. When there are droughts and less food, bats' immune systems may become weaker. This can cause them to shed more viruses, making outbreaks like those caused by the Hendra virus more likely (Mora et al. 2022).

Climate Influence on Campylobacter Infections

Campylobacter infections are more common in the summer. As climate change leads to warmer weather and longer warm seasons, we can expect to see more of these infections. More cases of *Campylobacter* infections and outbreaks of diarrheal illness caused by these bacteria have been seen following heavy rain and floods. This is likely because people are exposed to more contaminated water from overflowing sewage or unclean drinking sources. In contrast, *Campylobacter* cases are expected to decrease after periods of drought. As climate change makes water increasingly scarce, we may have to rely more on treated wastewater for crop irrigation, even though it could contain disease-causing organisms from animal or human waste. Studies suggest that using treated wastewater for irrigation could increase the presence of harmful bacteria like *Campylobacter* in plant-based foods, which could pose a threat to food safety. Additionally, research predicts a potential doubling of *Campylobacter* infections by 2080 in several European countries (Oberheim et al. 2020).

Climate Influence on Salmonella Infections

Temperature plays a key role in how quickly *Salmonella* bacteria can grow in food. The warmer it is, the faster they multiply, increasing the risk of contamination. This is why proper food storage temperatures are so important. Outdoor temperatures are closely linked to the occurrence of Salmonella outbreaks (EFSA 2021).

According to Zhang et al., a rise in weekly maximum temperature by 1 °C is associated with an 8.8% increase in weekly cases. Similarly, a 1 °C rise in weekly minimum temperature is associated with a 5.8% increase in weekly cases (Zhang et al. 2010).

Climate Influence on Vibrio Infections

Vibrio bacteria are more likely to be found in water that is warmer than 12 °C with a low or moderate amount of salt (1–25 g/L). Global warming promotes the growth of *Vibrio* bacteria, potentially leading to the wider spread of disease. Rising water temperatures will likely increase *Vibrio* contamination in European seafood and make this contamination possible year-round, rather than just in the warmer months. While pathogenic Vibrio species are less common in bodies of water with fluctuating temperatures like the Baltic and North Seas, toxigenic non-cholera vibrio (NCV) are increasingly detected in warmer waters, from which some seafood is imported. The warmer the water, the more likely it is to contain seafood that can make people sick (FAO 2020).

Climate Influence on Parasites

Cryptosporidium and Giardia are highly resilient parasites, making them a potential health threat as climate change worsens. Extreme weather washes them into water supplies and contaminates crops. To protect yourself, practice strict kitchen hygiene (Rumsey and Waseem 2021).

Preventive Measures

The majority of foodborne diseases are avoidable. When developing control techniques for foodborne diseases, technical information on pathogen transmission channels is essential. A One Health strategy is required to decrease food waste and maximize nutrient utilization by adopting closed circular food systems. However, for these methods to be effective, consumer behavior, food trends, economic incentives, trade, and politics must all be considered (Koutsoumanis et al. 2019). Because of this, effective One Health programs are built on an awareness of the socioeconomic settings of farmers, food industry professionals, and consumers (Osterhaus et al. 2020). Innovations in refrigeration, transportation (by air, sea, and land), and instant communication enable food distribution systems to deliver perishable foods quickly, offer just-in-time resupply of non-perishable foodstuffs, and benefit from economies of scale (Koutsoumanis et al. 2019). Control measures for major reservoirs will minimize continued host exposure, regardless of the route of transmission and carrier (Yenealem et al. 2020).

Notable Foodborne Illness Outbreaks

While many foodborne outbreaks happen closer to the point of sale, contamination originating at the farm or processing facility can lead to widespread contamination throughout the food supply chain, affecting a large number of people in various locations.

There are three standard components to the regulatory response and investigation process, each contributing to the challenge of identifying the source: (i) detection of the outbreak, (ii) identification of the food vector responsible for the outbreak, and (iii) identification of the location source of the outbreak at a farm or processing center.

The availability of new analytical tools and the increasing availability of data sources improve the ability for (i) outbreak detection and (ii) determination of the food type or specific product carrying the disease.

It is challenging to trace the location of the outbreak source due to the complexity of the food supply network and the absence of integrated labeling and distribution records. The established procedure typically involves convergence analysis, which entails tracing the distinct dissemination channels of goods from various sites to pinpoint a common source within the distribution chain, such as a shared date and location of production or place of manufacture (Horn and Friedrich 2019).

Detecting and Diagnosing Foodborne Illness

Even though food production and safety have improved greatly, foodborne illnesses remain a common problem. This is because harmful germs (pathogens) in food can still make people sick. To protect public health, scientists have developed various methods, both traditional and new, to find these pathogens in food. Researchers have extensively explored a wide range of methods to detect foodborne pathogens. These include traditional culture-based techniques, as well as advanced molecular methods like PCR variations (multiplex PCR, real-time PCR, RT-PCR), DNA microarrays, nucleic acid sequence-based amplification (NASBA), next-generation sequencing (NGS), along with immunological and nanotechnology-based approaches (Aladhadh 2023).

5.5 Health Risks of Climate Variability and Change

Our health depends heavily on the climate, and any changes in it can significantly impact people's well-being, both directly and indirectly. How much each person is affected depends on various factors, like their age, where they live, their job, and how much money they have. Even their health and home situation play a role (Neira et al. 2023).

Our planet's changing climate is wreaking havoc on our health and healthcare systems. From more frequent diseases to crumbling hospitals, the damage is widespread. Just look at the numbers: between 2005 and 2019, climate disasters like floods, droughts, and heatwaves destroyed an average of 412 health facilities each year, and things are only getting worse. This means more people getting sick from climate-sensitive diseases and even fewer places to get treatment when they need it most. It is clear: climate change is a health crisis, and we need to act now.

Coming decades will likely see climate change dramatically reshape where and how many people suffer from health problems linked to the environment. It is expected to strain public health systems too. Unless we take action, many more people will get sick and die from various health issues in the years ahead (Neira et al. 2023).

People who are already struggling and places less able to cope will be hit harder by climate change, widening the gap between rich and poor. Up until the year 2050, other factors like access to healthcare, income, education, and community support will likely have a bigger impact on health than climate change itself. These factors can directly affect health issues or change how climate change impacts people.

The Eastern Mediterranean and Middle East (EMME) is a vast geographic region that brings several countries together, including Bahrain, Egypt, Cyprus, Greece, Iran, Turkey, Iraq, Jordan, Palestine, Lebanon, Qatar, Syria, Kuwait, Oman, Saudi Arabia, Yemen, and the United Arab Emirates (Neira et al. 2023).

The rise of environmental temperatures, as well as more frequent, intense heat waves, are major dangers to our health due to climate change. Heat exposure can cause stroke, heat stress, kidney and heart problems in adults, electrolyte imbalances, fever, and respiratory and kidney problems in children. It can even lead to mental and behavioral problems like sleep issues, thinking problems, aggressive behavior, and even violence and suicide (Palinkas and Wong 2020).

Older adults (individuals over 65 years old) are especially at risk when the heat gets intense, as research shows (Watts et al. 2021). Other folks who have a hard time dealing with the heat include those who are struggling financially, living with disabilities or chronic health illnesses, working outside or in hot places without temperature control, and people living in areas where it is already super-hot, pushing the limits of what our bodies can handle (Watts et al. 2021).

Certain health problems make people more vulnerable to heat, like diabetes, endocrine disorders, neurological and mental problems, such as schizophrenia, dementia, Alzheimer's, Parkinson's, or chronic health problems, such as obesity, cardiovascular disease, respiratory diseases, renal diseases, and cystic fibrosis (Smith 2019). This is because these conditions already affect how the body regulates temperature in hot environments, increasing your risk of heat-related illness. Also, physiological, behavioral, and cognitive problems are normally linked with these conditions, which directly disrupt the ability of the affected individual to successfully cope with rising temperatures. Medications taken to manage their conditions can actually make it harder for their bodies to cool down, making them even more vulnerable to heatstroke and other heat-related problems (Li et al. 2015).

While heatwaves are becoming more common globally, their impact on human lives is not equal everywhere. A study by Vicedo-Cabrera et al. (2018) found that overall heat-related deaths have actually declined in many countries over recent decades. However, this good news does not apply everywhere. Even within regions like continental Europe, with similar social and economic structures, the picture is mixed. De'Donato et al. (2015) found that heat-related deaths have decreased in warmer southern Europe, particularly around the Mediterranean. However, northern Europe has not seen the same benefit, with heat-related deaths actually increasing (de'Donato et al. 2015).

In Egypt, it has been predicted that by 2100, everyday temperatures could be hotter than 80% of the hottest days we experienced in the past decade (2006–2015) (Mostafa et al. 2019). This intense heat, coupled with rising humidity, could fuel the spread of some infectious diseases that are linked to climate changes like fungal keratitis, an eye infection. Researchers predict that cases could keep increasing until at least 2030, mirroring the rise in greenhouse gas emissions in the region (Saad Hussein et al. 2011).

The exact details are unclear, but uncovering the link between climate change and dust storms is an ongoing research challenge. Researchers like Klingmüller et al. (2016) suggested that hotter temperatures and drier air, linked to climate change over the past decade, have led to more dust storms in the Middle East. Naderi Beni et al. (2021) found a strong connection between dust storms in the Persian Gulf region and the increasing dryness and water shortages caused by climate change. Even though the pathogenicity and viability of bio-agents in dust is unclear, breathing in lots of dust over time has definitely been linked to various health problems like allergies, lung scarring (silicosis), and infections like granulomatosis, meningococcal meningitis, and sarcoidosis. It has been reported that injuries observed in lung cells following exposure to dust extracts can be due to inhalation of dust-containing bacterial material (such as endotoxins) that causes lung injury. Just like with extreme heat, dust storms pose a serious threat to certain groups. Young children (whose immune systems are still immature) and older adults (whose immunity weakens with age) are especially at risk. People with chronic heart and lung problems are also vulnerable. In fact, studies in Iran found a connection between hot, humid weather with dust storms and spikes in shigellosis cases during summer (Aminharati et al. 2018).

There is a complex interaction between temperature and the transmission of vector-borne diseases (VBD), often non-linear, and variable among various vector and pathogen combinations. It has been suggested that global warming will enhance the potential for transmission of VBDs through changing regional distribution and seasonality in the spread of different vector species, which will allow them to survive in unsuitable areas or seasons due to temperature variations. In contrast, it has also been suggested that global warming might push temperatures beyond what certain vectors and pathogens can tolerate in specific areas. This could shrink the VBD range, change when they spread during the year, or even make them less of a problem in some places (Mordecai et al. 2020).

To assess the immediate health impacts of extreme weather events in the future, it is essential to understand how people will deal with gradual, long-term environmental changes in the future, such as changes in attitudes and habits, including the development of early warning systems and response strategies (Madrigano et al. 2021).

In a worldwide ecological study, researchers discovered a close connection between fluctuations in temperature and the number of meningitis cases reported. In addition, they reported that Central Sub-Saharan Africa, Australasia, and High-income North America are most vulnerable to increases in meningitis cases due to increased temperature variability. Moreover, different shared socioeconomic pathways (SSPs) could lead to severe consequences for both temperature variability and meningitis incidence rate. Therefore, to meet the WHO's goal of eradicating meningitis by 2030, we need to double down on efforts by not only fighting the disease directly but also addressing the growing threat posed by climate change, which exacerbates meningitis outbreaks (Chen et al. 2023).

Studies have shown that meningitis cases might change in the future, depending on how much greenhouse gas emission and various climate policies are in place, which can be represented by SSPs. These pathways imagine how society and the economy might develop by 2100. They are ranked from least to most severe, with SSP126 representing a "best-case" scenario where we choose a sustainable lifestyle and keep global warming low. On the contrary, SSP585 shows a gray picture with rising emissions of greenhouse gases and intense global warming. In contrast, SSP245 and SSP370 offer a more moderate path, like taking the middle lane on the highway toward a greener future.

To protect our health from climate change, there are two main approaches: mitigation (fighting the problem at its source by reducing our impact on the climate) and adaptation (adjusting to the changes already happening to minimize harm). Both mitigation and adaptation policies are not equally exclusive. Reducing greenhouse gases not only helps the planet, but it also improves our health. For example, hospitals using cleaner energy sources to avoid power outages during disasters are both good for the environment and prepare communities for extreme weather. Even though we cannot predict climate change perfectly, ignoring it leaves us vulnerable to its worst effects. By investing in solutions now, we can protect ourselves and future generations (Chen et al. 2023).

5.6 Climate Change and Females in Child-Bearing Period

Climate change has harmful effects on pregnant women and their babies. These risks can come directly from heat waves, storms, and dirty air, or indirectly through other effects. Climate change can hurt the health of both moms and newborns right away and in the long run, and it can do this in many different ways (Roos 2021).

Pregnancy and childbirth make women and newborns more susceptible to harm from high temperatures due to physical changes, body functions, and social and cultural factors. Extreme heat can be too much for a pregnant woman's body to regulate its temperature, especially during labor. This can lead to dehydration, hormone problems, and issues with the placenta. It can also increase the risk of high blood pressure, gestational diabetes, preterm birth, and stillbirth. Hot weather can also lead to more infections, make it harder for healthcare workers to do their jobs, and discourage people from seeking medical care. Climate change with its rising temperatures and limited resources is making this a growing concern (Chersich et al. 2023).

African women face a double threat from extreme heat during pregnancy. In urban areas, many pregnant women in growing African cities live in poorly built homes or crowded informal settlements. These areas trap heat even more than the surroundings, making them dangerously hot. In rural areas, pregnant women often have limited access to things that could help them cool down, like clean drinking water or cool spaces. Many people do not realize the dangers of high heat, and some cultural practices might even make things worse by keeping pregnant women outdoors for long periods (Chersich et al. 2023).

Pathophysiologic Processes and Indirect Effects of High Ambient Temperatures on Maternal and Newborn Health

Heat waves pose a particular threat to pregnant women and newborns. This is because pregnancy changes how a woman's body regulates temperature. Pregnant women have a higher metabolism, increased body fat, and a growing baby, all of which generate more heat. Normally, sweating and blood flow to the skin help the body cool down. But during extreme heat, this system can become overwhelming. Studies suggest that heat stress can harm both mothers and babies. It may increase the risk of premature birth, low birth weight, and stillbirth. Dehydration from sweating can also lead to early or prolonged labor. Heat stress can raise stress hormones, potentially reducing blood flow to the placenta and affecting fetal growth. Additionally, extreme heat may contribute to high blood pressure and pre-eclampsia, a serious pregnancy complication (Chersich et al. 2020).

The link between heat waves and hypertension is not entirely clear. Heat can actually vasodilate blood vessels, which might lower blood pressure. There is some suggestion that heat exposure early in pregnancy could raise the risk of birth defects, but more studies are needed to confirm this. It is also uncertain if heat waves increase pregnancy-related deaths, even though they do for other groups who struggle to regulate body temperature. We do know, however, that extreme weather events can lead to more cases of postpartum depression, PTSD, preterm birth, and low birth weight babies.

Studies around the world have shown that hot weather can harm the health of both pregnant women and their newborns. However, the exact temperature that becomes risky can vary depending on several factors. These include things like the usual climate in a place, the woman's health before pregnancy, and her individual risk factors. Interestingly, the relationship between hot weather and health problems is not always straightforward. Often it follows a U- or J-shaped curve. This means that mild temperatures around 20 °C are generally safe. But as it gets hotter, or if there is a heatwave, the risks for health problems increase much more quickly.

How to Protect Pregnant Females from Climate Change Effects

- To beat the heat, stay cool and hydrated by drinking plenty of fluids, finding air-conditioned spaces or shade to rest in, using inexpensive misters or water sprays to cool down, and wearing loose-fitting, breathable clothing like cotton. Be aware of your body's signals and watch for signs of heat illness like dizziness, sweating, tiredness, and chills with goosebumps.
- Stay hydrated by drinking plenty of fluids. Find ways to cool down, like getting in air conditioning, sitting in the shade, or using a mister. Wear loose-fitting, breathable clothing, preferably cotton. Be aware of your body and watch for signs of heat illness like dizziness, sweating more than usual, tiredness, and chills with cold, sweaty skin.
- Many areas in Africa have widespread antenatal care and readily available community health services. This allows pregnant women to have regular check-ups with healthcare providers, creating numerous chances to encourage and solidify positive changes in their behavior.
- Birth preparedness includes helping women to plan in advance for their transport at labor time, which may be particularly important during periods of high ambient temperatures.
- The program offers pregnant women support during hot weather. Community health workers can visit them at home to check on their well-being. Additionally, "cooling areas" are available in the community or at maternity facilities to provide relief from the heat. During childbirth, healthcare providers should be extra vigilant for complications that could arise from heat exposure. This includes monitoring a woman's hydration status, how long labor lasts, blood loss, and any signs of infection, particularly group B strep. In hot climates, even mild dehydration can be serious, so making sure pregnant women stay hydrated is crucial. Healthcare systems can also help by tracking heat-related illnesses in pregnant women and newborns and by ensuring that medications like oxytocin are kept cool during extreme heat events.
- "Heat proofing" includes enhancing the heat resistance of buildings by increasing the reflectivity of surfaces and providing localized solutions such as mist sprays and evaporative materials. White reflective paint for rooftops to reflect solar radiation.

- While air conditioning offers critical health benefits for vulnerable groups during heat waves, it is also seen as a contributor to climate change. Despite drawbacks like high energy consumption and pollution, air conditioning can be a valuable tool for protecting mothers and newborns. Strategically placed "cool spaces" in maternity units, similar to maternity waiting homes, could be a cost-effective solution during extreme heat. This approach echoes the concept of providing safe havens for pregnant women.

5.7 Climate Crisis Will Affect the Future Children's Lives

Childhood health faces a triple threat: obesity, undernourishment, and climate change. These problems appear together, worsen each other's effects (both physically and mentally), and stem from the same root causes.

Despite efforts to reduce child malnutrition, it remains a major public health problem. Most countries are failing to achieve goals for reducing stunting, wasting, and overweight in children. Children also face a "double burden" of malnutrition, where lack of proper nutrition coexists with obesity and other chronic diseases linked to diet (Agostoni et al. 2023).

Climate Change and Malnutrition

A large amount of research shows climate change is harming people's health, especially children's health. Our environment and our well-being are deeply connected. As extreme weather events become more common due to climate change, there is growing evidence that malnutrition will become a bigger problem. In Africa alone, climate change-linked malnutrition caused an estimated 17,000 deaths and millions of years of healthy life lost. Rising temperatures and lower crop yields could lead to 12 million more people suffering from malnutrition.

Researchers in Mali discovered a link between climate change and chronic malnutrition, specifically stunting. Their study predicted that climate change could lead to an additional six million malnourished people, with children making up 75% of that number (Noushadali 2021).

Soaring food prices squeeze household budgets, forcing people, especially those already struggling, to eat less and choose simpler, cheaper foods. This lack of variety and quality in their diets puts them at greater risk of malnutrition.

Environmental challenges make things worse. Polluted water causes diarrhea, making it harder for people to absorb nutrients from the food they eat. Additionally, climate change brings more floods and droughts, disrupting food production and pushing prices even higher. This cycle of hunger and malnutrition has serious, long-term health consequences (Noushadali 2021).

Climate Change Interactions with the Agro-Ecosystem and Effects on Child Health

The effects of climate change are expected to amplify health challenges for humans.

Our well-being, the world around us, and the planet itself all depend on a healthy food system. Climate change threatens this system by disrupting the delicate balance of plants, animals, and the environment they share (the agro-ecosystem). This disruption could lead to problems growing enough food and ensuring everyone has access to the nutrients they need.

Problems with the amount, quality, availability, and cost of food grown on farms, caught in fisheries, and raised as livestock make it harder for people who already struggle to get enough nutrients to eat a healthy diet. This lack of good food makes them more vulnerable to malnutrition and other health problems. Additionally, the way we currently produce and consume food contributes to climate change (Agostoni et al. 2023).

Climate change is creating a double whammy for health, causing both obesity and undernutrition to rise together. This is like a global syndemic, where these problems worsen each other and share similar root causes in society. Undernutrition, especially in children, is the biggest health threat linked to climate change. The changing climate might also make it harder for food to have the nutrients we need, and it could increase the spread of diseases that make it harder for people to absorb nutrients (Lieber et al. 2022).

Warmer temperatures might be making us fatter in two ways. First, people may exercise less as the weather gets hotter. Second, rising temperatures can hurt the production of fruits and vegetables, making them more expensive. This could lead people to choose cheaper, processed foods and drinks, which are often higher in calories and lower in nutrients.

Additionally, global warming could disrupt entire ecosystems, leading to less food being produced overall. This includes lower crop yields, decreased livestock production, and smaller fish catches. Forests, which play a vital role in our environment, may also suffer, impacting the resources they provide (Lieber et al. 2022).

Climate change is already hurting the yields of essential crops like wheat and maize, and this trend is expected to continue even with modest warming. By 2100, under high-emission scenarios, climate change is projected to reduce maize yields by 20–45%, wheat yields by 5–50%, rice yields by 20–30%, and soybean yields by 30–60% (Agostoni et al. 2023).

Crops like wheat, barley, rice, and potatoes are packing less protein due to rising CO2 levels. Climate change is making things worse by helping pests spread faster and reach new areas. On top of that, it is expected to cause more foodborne and waterborne illnesses by disrupting the balance of microbes in the environment.

A World Health Organization (WHO) report warns of a troubling rise in deaths among children under 15 caused by climate change. Diarrheal diseases are expected to be responsible for an additional 48,000 deaths by 2030, with that number sharply increasing to 330,000 by 2050 (Lieber et al. 2022).

Heat stress significantly harms livestock production, animal health, and biodiversity. Studies reveal that dairy cows in southern Europe experiencing heat stress can lose up to an average of 5.5 kg of milk per day. Furthermore, severe droughts in Sub-Saharan Africa over the past 20 years have caused livestock population losses ranging from 20% to 60%.

Just like warming alters fish habitats by changing the water's chemistry (saltiness, oxygen levels, and acidity), it can also stunt fish growth and reduce the amount of fish that can be caught. Moreover, warming likely reduced long-chain polyunsaturated fatty acid and iron contents in seafood (Agostoni et al. 2023).

A major concern is the increasing threat climate change poses to forests and trees. These vital ecosystems play crucial roles in our environment, such as providing clean water, preventing landslides and soil erosion, supporting diverse wildlife, and supplying valuable resources. Research indicates that rising temperatures and altered precipitation patterns are contributing to higher tree mortality rates. The loss of tree species can significantly diminish the ecosystem's ability to perform these essential functions (Agostoni et al. 2023).

These Are Actions That Pediatricians Can Promote in Their Homes, Clinics, and Hospitals

While many believe climate change solutions are solely for governments and large organizations, and others focus on individual efforts like reducing energy use, there is a gap that pediatricians and pediatric associations can fill. They can play a crucial role in leading by example and advocating for more action.

Because pediatricians focus on the well-being of children, they have a unique obligation to fight climate change. This includes promoting environmentally friendly practices in healthcare and ensuring communities can adapt to a changing environment to keep people healthy. To effectively address this issue, medical professionals and children's health organizations need to work together to expand their ongoing education programs to cover climate change and its impact.

To better prepare future doctors, climate change and its health impacts should be integrated into various learning resources for child health. This includes teaching modules, online courses, and residency training programs. Additionally, pediatric associations at all levels (national, regional, and global) should incorporate climate change education into their conferences and training programs for doctors. It is important to organize these events in an eco-friendly way to minimize their environmental impact (Bhutta et al. 2019).

Since climate change harms children the most, pediatricians and pediatric groups need to take a strong stance. They should push governments, businesses, and other organizations to take action to slow climate change. In addition, more research is needed to help children and their families deal with the challenges of a changing climate.

Future research needs to explore how climate change impacts children differently. This includes understanding if there are variations based on gender, age, and socioeconomic background. Are some groups facing a heavier burden than others?

Healthcare professionals can play a key role in raising awareness. They are trusted sources of information and can educate their patients and communities about the link between climate change and health. These efforts can start locally and spread to broader areas, influencing regional and national conversations.

The impact of climate change on children's health should not be confined to scientific journals. We need to translate this research into action by making it a central topic in public discussions and policy changes (Bhutta et al. 2019).

5.8 Climate Change and Healthy Aging

As climate change intensified, public health faced a growing challenge. Record-breaking national temperatures and more frequent extreme weather events posed significant health risks. Notably, older adults (over 65) are more susceptible to climate-related health problems and rely more heavily on a well-functioning healthcare system compared to younger demographics (Cai et al. 2022).

Older adults (over 65) are especially at risk from climate change's health effects. This is clear from the high number of heatwave deaths in this age group (76% in 2021). As our population ages, these risks are growing faster for older adults than for the general population. For example, wildfires exposed older people to more danger between 2017 and 2021 compared to the historical average, and this increase was sharper than for the whole population. The same pattern holds for extreme weather events like heavy rainfall and droughts. Finally, air pollution indoors is much more deadly for older adults, with death rates 4.7 times higher than the average population (Neira et al. 2023).

As older adults are more vulnerable to extreme heat, an important factor to consider is the world's changing population. In fact, for the first time ever, in 2018, there were more people over 65 than children under 5 worldwide for the first time in history (Chang et al. 2022), and it is expected to reach 16% of the world's population by 2050 (United Nations 2019). Older adults and pregnant women are especially vulnerable to heat waves. As people age, their bodies become less able to regulate temperature (thermoregulatory capacity), and they may move around less (reduced mobility). Additionally, older adults often have chronic health problems and weaker social support networks, making them more likely to suffer during heat events. Research also suggests that heat waves can be more dangerous for older women than men. Pregnant women and their unborn babies are another high-risk group for heatstroke and other heat-related illnesses (Zhang et al. 2019).

Aging populations, chronic illnesses, and rapidly growing cities in Europe and the Eastern Mediterranean make these areas especially at risk from heat waves. This danger has been clear in recent years with record-breaking heatwaves causing tens of thousands of deaths across Europe. Just in Spain and Portugal, over 1700 heat-related deaths occurred in the first half of 2022, and more are expected before summer ends (WHO 2022).

Even though older adults are more susceptible to health problems caused by climate change and face greater risks, they are largely ignored in media discussions about health and climate change. This lack of attention extends to government resources—for example, climate change was not mentioned in the latest report on healthy aging for older people. Additionally, compared to other age groups, older adults seem to be less involved in advocating for climate action related to health.

Focusing on the specific needs of older adults can significantly improve their health and well-being. In China, for instance, many older people rely on solid fuels for cooking and heating, which can cause indoor air pollution. By providing clear and targeted information on safer fuel options, millions of deaths from air pollution could be prevented. However, if China does not take action to address these issues as its population ages, the progress made in health outcomes could be lost (Neira et al. 2023).

Prevention

To safeguard the health of older adults facing climate change, we need a two-pronged approach. First, climate action plans must consider the health risks they pose to this age group. Second, existing healthcare strategies need to be adapted to address the new threats posed by a changing climate.

Healthcare leaders need to sound the alarm about the health risks of climate change. These risks will especially impact people who are already more susceptible to health problems, like the elderly, people with disabilities, those with chronic illnesses, and those who rely on community support.

In response to the escalating threat of heat-related illnesses, leaders must take decisive action. This includes: (1) providing vulnerable populations with tailored meteorological and early warning services; (2) issuing timely guidelines for safe outdoor activities during heat stress for different demographics; (3) conducting community-based health education initiatives for older adults; and (4) creating a regional dataset on health and aging to inform effective interventions (Cai et al. 2022).

5.9 Conclusion

While older adults across Canada (or specify provinces if desired) are more vulnerable to climate change's health impacts, they are often overlooked in media reports and policy discussions. Additionally, compared to younger generations, they seem to be less involved in climate and health issues (Cai et al. 2022).

5.10 How Will the Future of Food Security Impact Health?

Food security is crucial for health and well-being, as per the World Health Organization (Giller 2020; WHO 2002). Poor health is a key component of poverty, and governments should actively reduce mortality and preventable illnesses. Food insecurity, a social determinant of health, is linked to severe and costly health problems, as highlighted by the Food Research and Action Center (Food Research and Action Center 2017).

The FAO, International Fund for Agricultural Development (IFAD), and World Food Program (WFP) define food insecurity as "*a situation that exists when people lack secure access to sufficient amounts of safe and nutritious food for normal growth and development and an active and healthy life*" (FAO, IFAD, WFP 2013). It is widely accepted that food security and nutrition are critical to human health and development. According to studies, millions of people suffer from food insecurity, which is one of the most serious risks to human health. In 2017, approximately one in every four people worldwide (1.9 billion) was moderately or severely food insecure, with the highest numbers in Sub-Saharan Africa (SSA) and South Asia (FAO et al. 2018). In 2018, 9.2% of the global population faced severe food insecurity, with 11% undernourished (820 million). This number has increased significantly since 2015, making it a significant challenge to achieve the SDGs of zero hunger by 2030 (Word Count 2020).

Recent global climate changes have exacerbated the food crisis, as extreme weather is a major contributor to world hunger. Heatwaves, droughts, floods, storms, and wildfires are all expected to increase as global temperatures and sea levels rise. These conditions make it difficult for farmers to produce food and needy people to receive it. According to scientific studies, extreme weather events will become more common or severe as a result of human-caused climate change. The climate crisis is characterized by natural disasters such as floods, storms, and heat waves, and it directly contributes to a food security crisis. This increases the difficulty of feeding people, as shown in Fig. 5.6 (Baars et al. 2023).

Despite the fact that the hunger crisis affects individuals of all genders and ages, it has a particularly devastating impact on children. According to statistics from the United Nations, there are approximately 165 million children globally who suffer from chronic malnutrition, resulting in stunted growth or being underweight. It is distressing to note that malnutrition is responsible for claiming the life of almost one child under the age of five every 2 min across the globe. Food insecurity not only directly impacts health outcomes but also indirectly contributes to disordered eating patterns, as well as fluctuations in blood cholesterol levels, serum albumin, hemoglobin, vitamin A levels, and overall physical and mental health (Gulliford et al. 2003; Stuff et al. 2005; Parker et al. 2010). Micronutrient deficiencies in iodine, iron, and zinc are commonly identified across all age groups. Vitamin A deficiency affects approximately 190 million preschoolers and 19 million pregnant women (WHO 2009). While hidden hunger is often associated with pregnant women, children, and teenagers, it can also have implications for individuals' health at all stages of life (Saltzman et al. 2014).

Magnitude of the Problem

Various structural factors, such as climate, socioeconomic status, social dynamics, and local food availability, play a significant role in determining people's food security. Food insecurity can have a profound impact on individuals' health through nutritional, mental health, and behavioral pathways (Weiser et al. 2015). In terms of nutrition, food insecurity can affect overall caloric intake, dietary quality, and nutritional well-being (Kirkpatrick and Tarasuk 2008). When food supplies are limited, hunger and undernutrition may arise, potentially leading to issues such as wasting, stunting, and immunological deficiencies (Schaible and Kaufmann 2007).

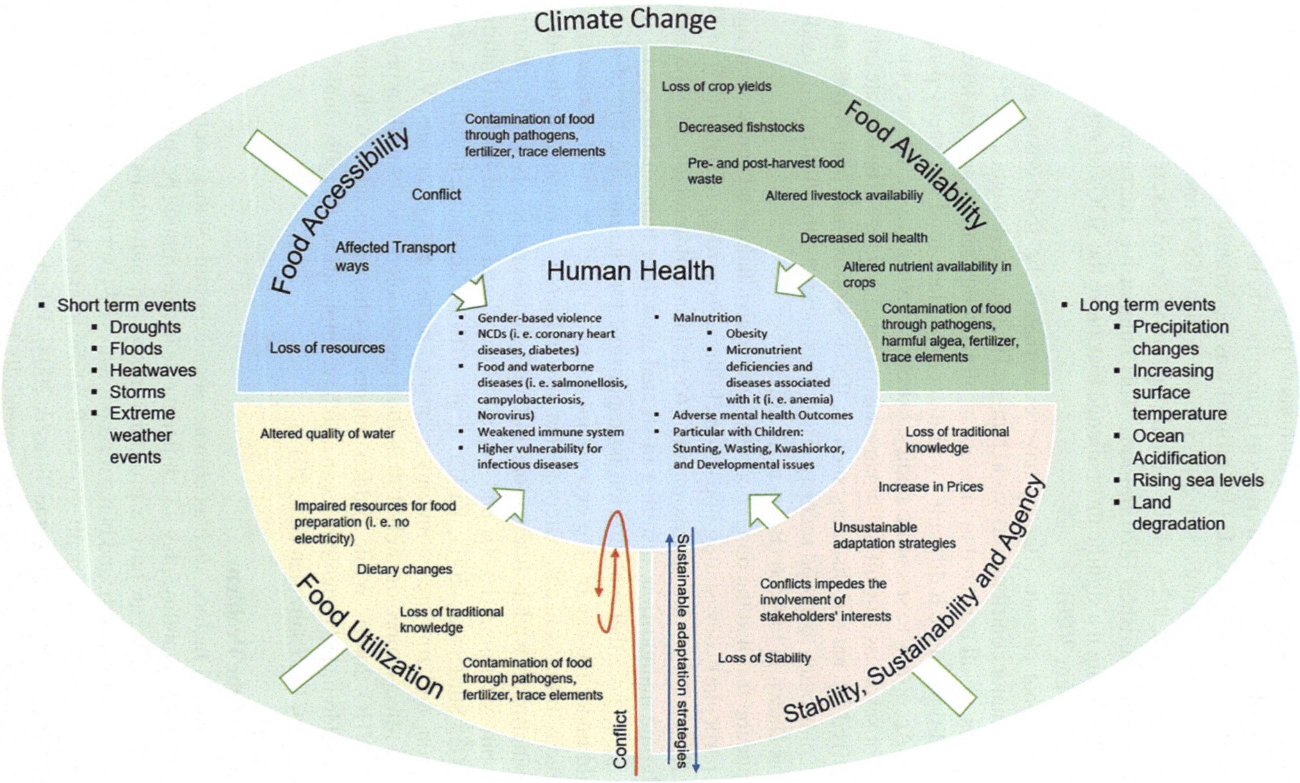

Fig. 5.6 How can climate change impact human health via food security (Baars et al. 2023)

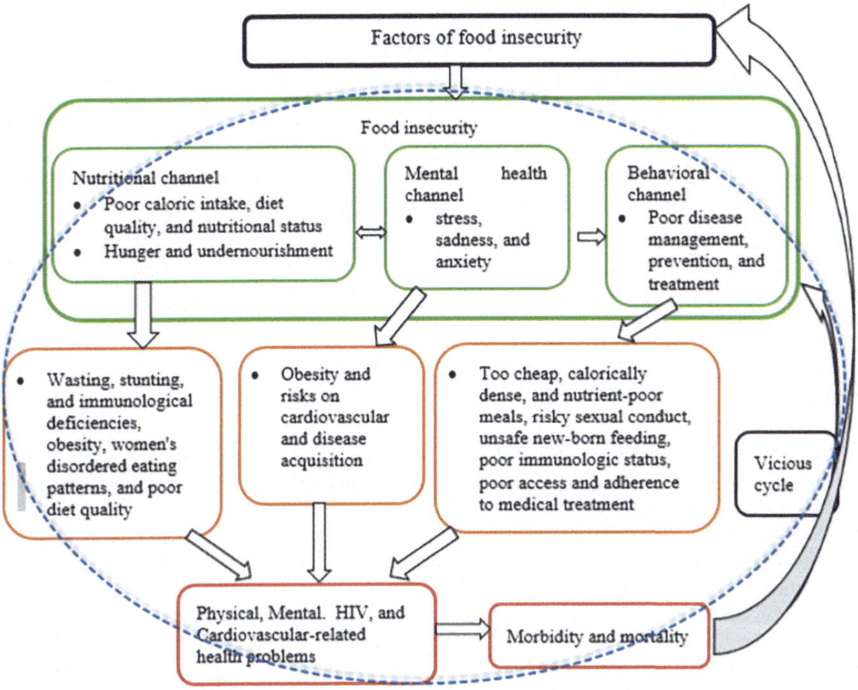

Fig. 5.7 Conceptual framework of food insecurity and health (Weiser et al. 2015)

Food insecurity negatively impacts health by contributing to obesity, diabetes, hypertension, disordered eating patterns, and poor dietary habits (Laraia et al. 2013). Mental health conditions such as stress and anxiety are linked to obesity and cardiovascular risk. Food insecurity also affects health behaviors, such as unhealthy choices and risky sexual behaviors (Weaver et al. 2021). It can lead to decreased access to medical care and increased susceptibility to disease. Food insecurity also increases the risk of exposure to HIV (Vogenthaler et al. 2013), increasing the likelihood of unsafe sexual practices and transmission of infection through unsafe infant feeding practices. It also leads to decreased adherence to antiretroviral therapy, deteriorating physical health, poor immune status, decreased viral suppression, and increased mortality among HIV-infected individuals.

Food insecurity can impact physical, emotional, and behavioral health as shown in Fig. 5.7 (Weiser et al. 2015).

The Future of Food Security

The worldwide food future until 2050 will include periods of intense heat and drought in one season and flooding rains in another. A tidal surge of change brought on by population growth, global warming, digital futures, and profound structural problems with food futures through 2050 and beyond is also being observed. Food security, scarcity, and industrial advances like precision agriculture and biotechnology are all impacted by these problems.

Food security influences our health, especially children. Research reveals that food insecurity can lead to persistent health problems in the short and long term. The first 1000 days (from conception to age 2) of a baby's life are crucial for his or her healthy growth and development. For children, acute food insecurity is linked to chronic health conditions such as wasting, stunting, and anemia.

A diet that lacks enough protein, calories, minerals, and vitamins will hinder the growth and development of children from birth through adolescence. Malnutrition during pregnancy can have negative consequences for the baby, including low birth weight, increased risk of infant mortality, preterm delivery, and impaired cognitive development. So, it is clear that hunger and health are deeply connected.

References

Agostoni C, Baglioni M, La Vecchia A, Molari G, Berti C (2023) Interlinkages between climate change and food systems: the impact on child malnutrition—narrative review. Nutrients 15:416. https://doi.org/10.3390/nu15020416

Aladhadh M (2023) A review of modern methods for the detection of foodborne pathogens. Microorganisms 11:1111. https://doi.org/10.3390/microorganisms11051111

Al-Delaimy W (2020) Health of people, health of planet and our responsibility: climate change, air pollution and health. In: Al-Delaimy W, Ramanathan V, Sanchez SM (eds) Vulnerable populations and regions: Middle East as a case study. Springer Open, Cham, Switzerland, pp 121–133

Allen T, Prosperi P (2016) Modeling sustainable food systems. Environ Manag 57(5):956–975. https://doi.org/10.1007/s00267-016-0664-8

Aminharati F et al (2018) The effect of environmental parameters on the incidence of Shigella outbreaks in Yazd province. Iran Water Supply 18(4):1388–1395. https://doi.org/10.2166/ws.2017.205

Baars C, Barbir J, Paulino JH (2023) How can climate change impact human health via food security? A bibliometric analysis. Environments 10:196. https://doi.org/10.3390/environments10110196

Battilani P, Toscano P, Van Der Fels-Klerx HJ, Moretti A, Camardo Leggieri M, Brera C, Rortais A, Goumperis T, Robinson T (2016) Aflatoxin B1 contamination in maize in Europe increases due to climate change. Sci Rep 6:24328. https://doi.org/10.1038/srep24328

Béné C, Oosterveer P, Lamotte L, Brouwer ID, Haan SD, Prager SD, Talsma EF, Khoury CK (2019a) When food systems meet sustainability—current narratives and implications for actions. World Dev 113:116–130. https://doi.org/10.1016/j.worlddev.2018.08.011

Béné C, Prager SD, Achicanoy HAE, Toro PA, Lamotte L, Cedrez CB, Mapes BR (2019b) Understanding food systems drivers: a critical review of the literature. Glob Food Sec:149–159. https://doi.org/10.1016/j.gfs.2019.04.009

Bezner Kerr R, Hasegawa T, Lasco R, Bhatt I, Deryng D, Farrell A, Gurney-Smith H, Ju H, Lluch-Cota S, Meza F, Nelson G, Neufeldt H, Thornton P (2022) Food, fibre, and other ecosystem products supplementary material. In: Pörtner H-O, Roberts DC, Tignor M, Poloczanska ES, Mintenbeck K, Alegría A, Craig M, Langsdorf S, Löschke S, Möller V, Okem A, Rama B (eds) Climate change 2022: impacts, adaptation, and vulnerability. Contribution of Working Group II to the Sixth Assessment Report of the Intergovernmental Panel on Climate Change. Available from https://www.ipcc.ch/report/ar6/wg2/

Bhutta ZA, Aimone A, Akhtar S (2019) Climate change and global child health: what can paediatricians do? Arch Dis Child 104(5):417–418. https://doi.org/10.1136/archdischild-2018-316694. Epub 2019 Mar 16

Breisinger C, Ecker O, Al-Riffai P, Robertson R, Thiele R, Wiebelt M (2011. Kiel Working Paper, No. 1747) Climate change, agricultural production and food security: evidence from Yemen. Kiel Institute for the World Economy (IfW)

Brouwer Inge D, McDermott John, Ruben Ruerd (2020) Food systems everywhere: improving relevance in practice. Glob Food Sec 26:6100398. ISSN 2211-9124, https://doi.org/10.1016/j.gfs.2020.100398. (https://www.sciencedirect.com/science/article/pii/S2211912420300523)

Cadieux K, Slocum R (2015) What does it mean to do food justice? J Polit Ecol 22:1–26. https://doi.org/10.2458/v22i1.21076

Cai W, Zhang C, Zhang S, Bai Y, Callaghan M, Chang N, Chen B, Chen H, Cheng L, Cui X, Dai H, Danna B, Dong W, Fan W, Fang X, Gao T, Geng Y, Guan D, Hu Y, Hua J, Huang C, Huang H, Huang J, Jiang L, Jiang Q, Jiang X, Jin H, Kiesewetter G, Liang L, Lin B, Lin H, Liu H, Liu Q, Liu T, Liu X, Liu X, Liu Z, Liu Z, Lou S, Lu C, Luo Z, Meng W, Miao H, Ren C, Romanello M, Schöpp W, Su J, Tang X, Wang C, Wang Q, Warnecke L, Wen S, Winiwarter W, Xie Y, Xu B, Yan Y, Yang X, Yao F, Yu L, Yuan J, Zeng Y, Zhang J, Zhang L, Zhang R, Zhang S, Zhang S, Zhao Q, Zheng D, Zhou H, Zhou J, Fung MFC, Luo Y, Gong P (2022) The 2022 China report of the Lancet Countdown on health and climate change: leveraging climate actions for healthy ageing. Lancet Publ Health 7(12):e1073–e1090. https://doi.org/10.1016/S2468-2667(22)00224-9. Epub 2022 Oct 29. PMID: 36354045; PMCID: PMC9617661

Chang AY, Tan AX, Nadeau KC, Odden MC (2022) Aging hearts in a hotter, more turbulent world: the impacts of climate change on the cardiovascular health of older adults. Curr Cardiol Rep 24(6):749–760. https://doi.org/10.1007/s11886-022-01693-6

Chaudhary A, Gustafson D, Mathys A (2018) Multi-indicator sustainability assessment of global food systems. Nat Commun 9(1):848. https://doi.org/10.1038/s41467-018-03308-7

Chen J, Jiao Z, Liang Z, Ma J, Xu M, Biswal S, Ramanathan M, Sun S, Zhang Z (2023) Association between temperature variability and global meningitis incidence, 107649. Environ Int 171. https://doi.org/10.1016/j.envint.2022.107649. ISSN 0160-4120. https://www.sciencedirect.com/science/article/pii/S0160412022005761

Chersich MF, Pham MD, Areal A et al (2020) Associations between high temperatures in pregnancy and risk of preterm birth, low birth weight, and stillbirths: systematic review and meta-analysis. BMJ 371:m3811

Chersich MF, Scorgie F, Filippi V, Luchters S, Climate Change and Heat-Health Study Group (2023) Increasing global temperatures threaten gains in maternal and newborn health in Africa: a review of impacts and an adaptation framework. Int J Gynecol Obstet 160:421–429. https://doi.org/10.1002/ijgo.14381

Clay E (2002) Food security: concepts and measurement., Paper for FAO Expert Consultation on Trade and Food Security: Conceptualising the Linkages Rome, 11-12 July 2002. Published as Chapter 2 of Trade Reforms and Food Security: conceptualising the linkages. FAO, Rome

David-Benz H, Sirdey N, Deshons A, Orbell C, Herlant P (2022) Conceptual framework and method for national and territorial assessments–catalysing the sustainable and inclusive transformation of food systems. FAO, CIRAD and European Union, Rome, Montpellier, Bruxelles. https://doi.org/10.4060/cb8603en

de'Donato F et al (2015) Changes in the effect of heat on mortality in the last 20 years in nine European cities. Results from the PHASE project. Int J Environ Res Publ Health 12(12):15567–15583. https://doi.org/10.3390/ijerph121215006

Dietrich J, Hammerl JA, Johne A, Kappenstein O, Loeffler C, Nöckler K, Rosner B, Spielmeyer A, Szabo I, Richter MH (2023) Impact of climate change on foodborne infections and intoxications. J Health Monit 8(Suppl 3):78–92. https://doi.org/10.25646/11403. PMID: 37342431; PMCID: PMC10278375

Dury S, Bendjebbar P, Etienne, Hainzelin, Giordano T, Bricas N (2019) Food systems at risk. In: New trends and challenges. FAO, CIRAD et Commission europ´eenne, p 132. http://www.fao.org/3/ca5724en/CA5724EN.pdf

EFSA and ECDC (European Food Safety Authority and European Centre for Disease Prevention and Control) (2021) The European Union One Health 2020 Zoonoses report. EFSA J 19(12):6971. https://doi.org/10.2903/j.efsa.2021.6971

Ericksen P, Stewart B, Dixon J, David B, Loring P, Anderson M et al (2010) The value of a food system approach. In: Ingram J, Ericksen P, Liverman D (eds) Food security and global environmental change. Earthscan, London, pp 25–45

Erwin A, Ma Z, Popovici R, Salas O'Brien EP, Zanotti L, Zeballos Zeballos E et al (2021) Intersectionality shapes adaptation to social-ecological change. World Dev 138:105282. https://doi.org/10.1016/j.worlddev.2020.105282

Estevez P, Castro D, Pequeño-Valtierra A, Giraldez J, Gago-Martinez A (2019) Emerging marine biotoxins in seafood from European coasts: incidence and analytical ahallenges. Foods 8(5):149. https://doi.org/10.3390/foods8050149

Fanzo JC, Downs SM (2021) Climate change and nutrition- associated diseases. Nat Rev Dis Primers 7:90. https://doi.org/10.1038/s41572-021-00329-3

Fanzo J, Rudie C, Sigman I, Grinspoon S, Benton TG, Brown ME, Covic N, Fitch K, Golden CD, Grace D, Hivert M-F, Huybers P, Jaacks LM, Masters WA, Nisbett N, Richardson RA, Singleton CR, Webb P, Willett WC (2021) Sustainable food systems and nutrition in the 21st century: a report from the 22nd Annual Harvard Nutrition Obesity Symposium. Am J Clin Nutr 115:18. https://doi.org/10.1093/ajcn/nqab315

FAO (2015) Climate change and food systems: global assessments and implications for food security and trade

FAO (2018) Sustainable food systems – concept and framework. Food and Agriculture Organization of the United Nations, Rome

FAO (2020) The state of food security and nutrition in the world. Available: https://www.fao.org/3/ca9692en/ca9692en.pdf

FAO IFAD, UNICEF, WFP, and WHO (2018) The state of food security and nutrition in the world. Building climate resilience for food security and nutrition. Technical report. FAO. Available from: https://www.fao.org/3/I9553EN/i9553en.pdf

FAO, IFAD, WFP (2013) The State of Food Insecurity in the World 2013. The multiple dimensions of food security. Food and Agriculture Organization of the United Nations, Rome. Available from: http://www.fao.org/3/i3434e/i3434e00.htm

Food and Agriculture Organization (FAO) (2016) Climate change and food security: risks and responses. In: Food and Agriculture Organization of the United Nations, Rome

Food and Agriculture Organization of the United Nations (FAO), WHO (2020) Risk assessment tools for *Vibrio parahaemolyticus* and *Vibrio vulnificus* associated with seafood, Microbiological risk assessment series, vol 20. Rome. https://apps.who.int/iris/handle/10665/330867

Food Research and Action Center (2017) Hunger and health: the impact of poverty, food insecurity, and poor nutrition on health and well-being. Food Research & Action Center, Washington, DC. https://frac.org/research/resource-library/hunger-health-impact-poverty-food-insecurity-poor-nutrition-health-well

Froelich BA, Daines DA (2020) In hot water: effects of climate change on vibrio-human interactions. Environ Microbiol 22(10):4101–4111. https://doi.org/10.1111/1462-2920.14967

Giller KE (2020) The food security conundrum of sub-Saharan Africa. Glob Food Sec 2020(26):100431

Godde C, Mason-D'Croz D, Mayberry D, Thornton PK, Herrero M (2021) Risk of climate-related impacts on the livestock sector: a review of the evidence. Glob Food Sec 28(100488):10.1016/j.gfs.2020.100488

Gulliford MC, Mahabir D, Rocke B (2003) Food insecurity, food choices, and body mass index in adults: nutrition transition in Trinidad and Tobago. Int J Epidemiol 32:508–516

Gunaratne SM, Firdaus RBN, Rathnasooriya IS (2021) Climate change and food security in Sri Lanka: towards food sovereignty. Humanit Soc Sci Commun 8(29):1–13. https://doi.org/10.1057/s41599-021-00917-4

Gustafson D, Gutman A, Leet W, Drewnowski A, Fanzo J, Ingram J (2016) Seven food system metrics of sustainable nutrition security. Sustain For 8(3). https://doi.org/10.3390/su8030196

Holland D, Clifford R, Mahmoud Zadeh N et al (2023) Can foodborne illness estimate from different countries be legitimately compared case study of rates in the UK compared with Australia, Canada and USA. BMJ Open Gastro 10:e001009. https://doi.org/10.1136/bmjgast-2022-001009

Horn AL, Friedrich H (2019) Locating the source of large-scale outbreaks of foodborne disease. J R Soc Interface 16:20180624. https://doi.org/10.1098/rsif.2018.0624

Kirkpatrick SI, Tarasuk V (2008) Food insecurity is associated with nutrient Inadequacies among Canadian adults and adolescents. J Nutr 138(3):604–612

Klingmüller K, Pozzer A, Metzger S, Stenchikov GL, Lelieveld J (2016) Aerosol optical depth trend over the Middle East. Atmos Chem Phys 16(8):5063–5073. https://doi.org/10.5194/acp-16-5063-

Koutsoumanis K et al (2019) Salmonella control in poultry flocks and its public health impact. EFSA J 17:e05596

Kühn S, Rieger UM (2017) Health is a state of complete physical, mental and social well-being and not merely absence of disease or infirmity. Surg Obes Relat Dis 13(5):887. https://doi.org/10.1016/j.soard.2017.01.046

Kyncl J, Spackova M, Fialova A, Kyseli J, Maly M (2021) Influence of air temperature and implemented veterinary measures on the incidence of human salmonellosis in The Czech Republic during 1998–2017. BMC Public Health 21:55. https://doi.org/10.1186/s12889-020-10122-8

Laraia B, Epel E, Siega-Riz AM (2013) Food insecurity with the experience of restrained eating is a recipe for increased gestational weight gain. Appetite 65:178–184

Li M, Gu S, Bi P, Yang J, Liu Q (2015) Heat waves and morbidity: current knowledge and further direction-a comprehensive literature review. Int J Environ Res Publ Health 12(5):5256–5283. https://doi.org/10.3390/ijerph120505256. MDPI AG

Lieber M, Chin-Hong P, Kelly K, Dandu M, Weiser SD (2022) A systematic review and meta-analysis assessing the impact of droughts, flooding, and climate variability on malnutrition. Glob Public Health 17:68–82. [CrossRef] [PubMed]

Linares C, Díaz J, Negev M, Martínez GS, Debono R, Paz S (2020) Impacts of climate change on the public health of the Mediterranean Basin population – current situation, projections, preparedness and adaptation. Environ Res 182:109107. https://doi.org/10.1016/j.envres.2019.109107. Academic Press Inc

Madrigano J, Shih RA, Izenberg M, Fischbach JR, Preston BL (2021) Science policy to advance a climate change and health research agenda in the United States. Int J Environ Res Publ Health 18(15). https://doi.org/10.3390/ijerph18157868

Magel JMT, Dimoff SA, Baum JK (2020) Direct and indirect effects of climate change-amplified pulse heat stress events on coral reef fish communities. Bull Ecol Soc Am 101(3):e01706. https://doi.org/10.1002/bes2.1706

Marambe B (2022) Strengthening climate-resilient food systems to tackle climate crisis and to ensure food security. Ceylon J Sci 51(4):331. https://doi.org/10.4038/cjs.v51i4.8050

Melesse MB, Van den Berg M, Béné C, Brouwer ID, de Brauw A (2019. https://www.ifpri.org/publication/improving-diets-through-food-systems-low-and-middle-income-countries-metrics-analysis) Improving diets through food systems in low- and middle-income countries. Met Anal 12:1085

Mora C, McKenzie T, Gaw IM et al (2022) Over half of known human pathogenic diseases can be aggravated by climate change. Nat Clim Chang 12:869–875. https://doi.org/10.1038/s41558-022-01426-1

Mordecai EA, Ryan SJ, Caldwell JM, Shah MM, LaBeaud AD (2020) Climate change could shift disease burden from malaria to arboviruses in Africa. Lancet Planet Health 4(9):e416–e423. https://doi.org/10.1016/S2542-5196(20)30178-9. Elsevier B.V

Mostafa AN et al (2019) Past (1950–2017) and future (–2100) temperature and precipitation trends in Egypt. Weather Clim Extrem 26. https://doi.org/10.1016/j.wace.2019.100225

Mueller S, Anderson J, Wallington T (2011) Impact of biofuel production and other supply and demand factors on food price increases in 2008. Biomass Bioenergy 35:1623–1632. https://doi.org/10.1016/j.biombioe.2011.01.030

Myers SS, Zanobetti A, Kloog I, Huybers P, Leakey ADB, Bloom AJ, Carlisle E, Dietterich LH, Fitzgerald G, Hasegawa T, Holbrook NM, Nelson RL, Ottman MJ, Raboy V, Sakai H, Sartor KA, Schwartz J, Seneweera S, Tausz M, Usui Y (2014) Increasing CO2 threatens human nutrition. Nature 7503(510):139–142. https://doi.org/10.1038/nature13179

Naderi Beni A et al (2021) Climate change: a driver of future conflicts in the Persian Gulf Region? Heliyon 7(2):e06288. https://doi.org/10.1016/J.HELIYON.2021.E06288

Neira M, Erguler K, Ahmady-Birgani H, Al-Hmoud ND, Fears R, Gogos C, Hobbhahn N, Koliou M, Kostrikis LG, Lelieveld J, Majeed A, Paz S, Rudich Y, Saad-Hussein A, Shaheen M, Tobias A, Christophides G (2023) Climate change and human health in the Eastern Mediterranean and Middle East: Literature review, research priorities and policy suggestions. Environ Res 216(Pt 2):114537. https://doi.org/10.1016/j.envres.2022.114537. Epub 2022 Oct 21. PMID: 36273599; PMCID: PMC9729515

Noushadali M (2021) Climate change impact on nutritional status of preschoolers: a cross sectional study. Indian J Econ Dev 9:5. https://doi.org/10.17485/IJED/v9.2020.9

Oberheim J, Hoser C, Luchters G et al (2020) Small-scaled association between ambient temperature and campylobacteriosis incidence in Germany. Sci Rep 10(1):17191

Osterhaus AD et al (2020) Make science evolve into a One Health approach to improve health and security: a white paper. One Health Outlook 2(1):1–32

Paganini N, Lemke S (2020) "There is food we deserve, and there is food we do not deserve" Food injustice, place and power in urban agriculture in Cape Town and Maputo. Local Environ 25:1000–1020. https://doi.org/10.1080/13549839.2020.1853081

Palinkas LA, Wong M (2020) Global climate change and mental health. Curr Opin Psychol 32:12–16. https://doi.org/10.1016/j.copsyc.2019.06.023. Elsevier B.V

Parker ED, Widome R, Nettleton JA, Pereira MA (2010) Food security and metabolic syndrome in U.S. adults and adolescents: findings from the National Health and Nutrition Examination Survey 1999–2006. Ann Epidemiol 20:364–370

Rakhmankulova Z, Shuyskaya E, Toderich K, Voronin P (2021) Elevated atmospheric CO2 concentration improved C4 xero-halophyte kochia prostrata physiological performance under saline conditions. Plants 10(3):491. https://doi.org/10.3390/plants10030491

Romanello M, McGushin A, Di Napoli C, Drummond P, Hughes N, Jamart L, Kennard H, Lampard P, Solano Rodriguez B, Arnell N, Ayeb-Karlsson S, Belesova K, Cai W, Campbell-Lendrum D, Capstick S, Chambers J, Chu L, Ciampi L, Dalin C, Dasandi N, Dasgupta S, Davies M, Dominguez-Salas P, Dubrow R, Ebi KL, Eckelman M, Ekins P, Escobar LE, Georgeson L, Grace D, Graham H, Gunther SH, Hartinger S, He K, Heaviside C, Hess J, Hsu SC, Jankin S, Jimenez MP, Kelman I, Kiesewetter G, Kinney PL, Kjellstrom T, Kniveton D, Lee JKW, Lemke B, Liu Y, Liu Z, Lott M, Lowe R, Martinez-Urtaza J, Maslin M, McAllister L, McMichael C, Mi Z, Milner J, Minor K, Mohajeri N, Moradi-Lakeh M, Morrissey K, Munzert S, Murray KA, Neville T, Nilsson M, Obradovich N, Sewe MO, Oreszczyn T, Otto M, Owfi F, Pearman O, Pencheon D, Rabbaniha M, Robinson E, Rocklöv J, Salas RN, Semenza JC, Sherman J, Shi L, Springmann M, Tabatabaei M, Taylor J, Trinanes J, Shumake-Guillemot J, Vu B, Wagner F, Wilkinson P, Winning M, Yglesias M, Zhang S, Gong P, Montgomery H, Costello A, Hamilton I (2021) The 2021 report of the Lancet Countdown on health and climate change: code red for a healthy future. Lancet 398(10311):1619–1662. https://doi.org/10.1016/S0140-6736(21)01787-6. Epub 2021 Oct 20. Erratum in: Lancet. 2021 Dec 11;398(10317):2148. https://doi.org/10.1016/S0140-6736(21)02429-6. PMID: 34687662; PMCID: PMC7616807

Roos N, Kovats S, Hajat S, Filippi V, Chersich M, Luchters S, Scorgie F, Nakstad B, Stephansson O, CHAMNHA Consortium (2021) Maternal and newborn health risks of climate change: a call for awareness and global action. Acta Obstet Gynecol Scand 100(4):566–570. https://doi.org/10.1111/aogs.14124. Epub 2021 Mar 4

Rumsey P, Waseem M (2021) Giardia Lamblia Enteritis. In: StatPearls. StatPearls Publishing, Treasure Island, FL

Saad Hussein A, El Mofty HM, Hassanien MA (2011) Climate change and predicted trend of fungal keratitis in Egypt. EMHJ - East Mediterr Heal J 17(6):468–473

Saltzman A, Birol E, Wiesman D, Prasai N, Yohannes Y, Menon P, Thompson J (2014) Global hunger index: the challenge of hidden hunger. International Food Policy Research Institute, Washington, DC

Schaible UE, Kaufmann SH (2007) Malnutrition and infection: complex mechanisms and global impacts. PLoS Med 4(5):e115

Shakeel J, Khalil T, Khalil M, Shakeel N, Shukat R, Aleem MT, Shaukat I, Shaukat A, Asrar R, Sharafat H (2023) Impact of food-borne diseases in association to one health concept and efforts of their prevention. In: Aguilar-Marcelino L, Abbas RZ, Khan A, Younus M, Saeed NM (eds) One health triad, vol I. Unique Scientific Publishers, Faisalabad, Pakistan, pp 150–157. https://doi.org/10.47278/book.oht/2023.23

Smith CJ (2019) Pediatric thermoregulation: considerations in the face of global climate change. Nutrients 11(9). https://doi.org/10.3390/nu11092010

Stuff JE, Casey PH, Szeto K, Gossett J, Weber J, Simpson P et al (2005) Household food insecurity and adult chronic disease in the lower Mississippi delta. J Federation Am Soc Experiment Biol 19:A986

Tendall DM, Joerin J, Kopainsky B, Edwards P, Shreck A, Le QB, Kruetli P, Grant M, Six J (2015) Food system resilience: defining the concept. Glob Food Sec 6:17–23. https://doi.org/10.1016/j.gfs.2015.08.001

Thornton P, Nelson G, Mayberry D, Herrero M (2021) Increases in extreme heat stress in domesticated livestock species during the twenty-first century. Glob Chang Biol 27:5762–5772. https://doi.org/10.1111/gcb.15825

United Nations (2019) World population ageing 2019: highlights. UN, New York

Veldhuizen LJL, Giller KE, Oosterveer P, Brouwer ID, Janssen S, van Zanten HHE et al (2020) The missing middle: connected action on agriculture and nutrition across global, national and local levels to achieve sustainable development goal 2. Glob Food Secur 24:100336. https://doi.org/10.1016/j.gfs.2019.100336

Vicedo-Cabrera AM et al (2018) A multi-country analysis on potential adaptive mechanisms to cold and heat in a changing climate. Environ Int 111:239–246. https://doi.org/10.1016/j.envint.2017.11.006

Vogenthaler NS, Kushel MB, Hadley C, Frongillo EA, Riley ED, Bangsberg DR et al (2013) Food insecurity and risky sexual behaviors among homeless and marginally housed HIV-infected individuals in San Francisco. AIDS Behav 17(5):1688–1693

Wang Y, Wild O, Ashworth K, Chen X, Wu Q, Qi Y, Wang Z (2022) Reductions in crop yields across China from elevated ozone. Environ Pollut 292(A):118218. https://doi.org/10.1016/j.envpol.2021.118218

Watts N et al (2021) The 2020 report of the lancet countdown on health and climate change: responding to converging crises. Lancet, Lancet Publishing Group 397(10269):129–170. https://doi.org/10.1016/S0140-6736(20)32290-X

Weaver LJ, Owens C, Tessema F, Kebede A, Hadley C (2021) Unpacking the "black box" of global food insecurity and mental health. SocSci Med 282:114042

Webb P, Benton TG, Beddington J, Flynn D, Kelly NM, Thomas SM (2020) The urgency of food system transformation is now irrefutable. Nat Food 1(10):584–585

Weiser SD, Palar K, Hatcher AM, Young SL, Frongillo EA (2015) Food insecurity and health: a conceptual framework. In: Ivers L (ed) Food insecurity and public health, 1st edn. CRC Press, Boca Raton, pp 23–50

Westhoek H, Ingram J, van Berkum S, Ozay L, Hajer M, United Nations Environment Programme, International Resource Panel, Working Group on Food Systems and Natural Resources (2016) Food Systems and Natural Resources. United Nations Environment Programme. https://books.google.fr/books?id=QjM3DQEACAAJ

WHO (2022) Climate change and health vulnerability and adaptation assessment. World Health Organization, Geneva. Licence: CC BY-NC-SA 3.0 IGO

Williams-Forson P, Wilkerson A (2011) Intersectionality and food studies. Food Cult Soc 14:7–28. https://doi.org/10.2752/175174411X12810842291119

Word Count (2020) How many people die from hunger each year. The World Counts, Denmark. Available from: https://www.theworldcounts.com/challenges/people-and-poverty/hunger-and-obesity/how-manypeople-die-from-hunger-each-year/story

World Food Summit 1996 (1996) Rome declaration on World Food Security. FAO. Report of the World Food Summit, FAO, Rome. Available: https://www.un.org/en/development/devagenda/

food.shtml#:~:text=At%20the%20World%20Food%20Summit,in%20the%20world%20by%202015
Xu R, Yu P, Abramson MJ, Johnston FH, Samet JM, Bell ML et al (2020) Wildfires, global climate change, and human health. N Engl J Med 383(22):2173–2181
Yenealem DG et al (2020) Food safety practice and associated factors among meat handlers in Gondar town: a cross-sectional study. J Environ Public Health 2020:1
Zhang Y, Bi P, Hiller JE (2010) Climate variations and *Salmonella* infection in Australian subtropical and tropical regions. Sci Total Environ 408(3):524–530
Zhang S, Zhang J, Yue T, Jing X (2019) Impacts of climate change on urban rainwater harvesting systems. Sci Total Environ 665:262–274
Zhao Q, Yu P, Mahendran R, Huang W, Gao Y, Yang Z, Yeb T, Wen B, Wu Y, Li S, Guo Y (2022) Global climate change and human health: pathways and possible solutions. Eco-Environ Health 1:53–62. https://doi.org/10.1016/j.eehl.2022.04.004

Climate Change Impacts on Pesticide Use and Fate in the Environment

6

Nour El-Hoda A. Zidan, Khaled Y. Abdel-Halim, and Fawzy Eissa

Abstract

Climate change is an international issue with profound consequences for multiple components of the ecosystem, including the usage and fate of pesticides, posing significant challenges to agricultural systems and ecosystem health. Pesticides are frequently employed to manage pests in the agricultural and public health sectors, but their use, efficacy, frequency of applications, transport, movement, degradation processes, and persistence in the environment are influenced by climate-related factors, including temperature increases, elevated CO_2 concentrations, and precipitation patterns. The objective of this chapter is to investigate how pesticide use and environmental fate are impacted by climate change. The effects of climate change on pesticide usage and fate have important implications for environmental contamination, ecological effects, and human health. Understanding these effects is critical to identifying the possible risks of pesticide usage and establishing appropriate mitigation and adaptation strategies. The knowledge gathered from this brief overview can assist in addressing the complicated challenges caused by climate change and reducing the detrimental impact of pesticide use on the environment. The necessity of taking proactive steps to

N. E.-H. A. Zidan (✉)
Pesticides Chemistry and Toxicology Department, Faculty of Agriculture, Kafrelsheikh University, Kafr El-Sheikh, Egypt

National Committee of Toxicology, Academy of Scientific Research and Technology (ASRT), Ministry of Higher Education and Scientific Research, Cairo, Egypt

K. Y. Abdel-Halim
Mammalian & Aquatic Toxicology Department, Central Agricultural Pesticides Laboratory (CAPL), Agricultural Research Center (ARC), Giza, Egypt

F. Eissa
Environment and Bio-Agriculture Department, Faculty of Agriculture, Al-Azhar University, Cairo, Egypt

A. Saad-Hussein et al. (eds.), *Climate Change Impacts on Toxins and Health Effects*,
https://doi.org/10.1007/978-981-96-3416-3_6

lessen the negative effects is generally highlighted in this chapter, which also explores the significance of implementing integrated pest management (IPM) practices, developing pesticide formulations that are climate-resilient, incorporating climate change considerations into pesticide policy and regulations, and promoting organic and precision agriculture approaches.

Keywords

Climate change · Pesticide efficacy · Pest dynamics · Pesticide persistence · Adaptation and mitigation strategies

6.1 Introduction

Pesticides are used globally to boost food production and prevent vector-borne diseases. Crop losses due to pests range from 25% to 50% depending on the crop (Oerke 2006), which require pesticide application to achieve high yields of products. Between 1990 and 2020, the use of pesticides in agriculture climbed largely by 57%. Globally, the consumption of pesticides reached 2.7 million metric tons in terms of active ingredients in 2020, equating to 7.2 million metric tons of formulated pesticides, with a market value of $41.1 billion. According to the Food and Agriculture Organization (FAO 2022), herbicides accounted for 52% of the total agricultural pesticide use in 2020, while fungicides and insecticides comprised 23% and 18%, respectively.

Globally, climate change remains one of the biggest threats to human life and livelihoods (IPCC 2018). Increased ocean acidity, rising sea levels, melting ice sheets and snow cover, more frequent heatwaves, floods, and droughts are some of the changes that have been documented. Anthropogenic greenhouse gas (GHG) emissions from the combustion of fossil fuels, industrial activities, agricultural intensification, and changes in land use have been linked to climate change (IPPC Secretariat 2021). Article 1 of the United Nations Framework Convention on Climate Change defines climate change as "*a change of climate which is attributed directly or indirectly to human activity that alters the composition of the global atmosphere and which is in addition to natural climate variability observed over comparable time periods.*"

Every step of the manufacture, delivery, and application of pesticides has an effect on climate change. Carbon dioxide (CO_2), methane (CH_4), and nitrous oxide (N_2O) are the three main GHGs emitted during the manufacture of pesticides (Heimpel et al. 2013; Liang et al. 2021). Synthetic pesticide manufacture is energy intensive, emitting more GHGs per kg than synthetic fertilizer production. In Austria, between 2000 and 2019, insecticide levels climbed by 58%, fungicide amounts increased by 29%, and herbicide amounts fell by 29%; corresponding GHG emissions followed similar patterns (Cech et al. 2022). An average of 14.79 to 18.91 kg of CO_2 equivalent (CO_2e) is generated during the manufacture of insecticides, while an average of 11.94 to 29.19 kg of CO_2e is generated during the production of fungicides. The most widely used herbicide in the world, glyphosate, emits 31.29 kg of CO_2e per kg. The energy required for manufacturing the quantity of glyphosate used globally in 2014 is equivalent to the energy required to fuel 6.25 million cars for a whole year (Audsley et al. 2009).

However, climate change-related factors impact pesticide application, resulting in increased usage and pollution (Tudi et al. 2021). Zhang et al. (2018) presented a case study in China, demonstrating that climate change effects can undermine pesticide reduction efforts due to the increased pressure from pests and diseases. The findings suggest that a rise in both temperature and rainfall leads to an increased application of pesticides, projected to grow by 1.1% to 2.5% by 2040, 2.4% to 9.1% by 2070, and 2.6% to 18.3% by the century's end in 2100. It is anticipated that climate change will alter the global patterns of distribution, population behavior, life cycles, and food chain relationships among different insect pests affecting agriculture (Lehmann et al. 2018). Climate change facilitates the spread and proliferation of pests through migration and range shifts, leading to more frequent pest infestations and changing the patterns of vector transmission (Miraglia et al. 2009). Additionally, the occurrence of plant diseases is predominantly influenced by factors such as temperature, precipitation, moisture levels, sunlight exposure, and dew presence (Patterson et al. 1999). It is projected that the frequency of late blight on potatoes will rise with the advent of warmer and more humid conditions during spring and summer in the coming years (Delcour et al. 2015). Studies have also shown that a rise in atmospheric CO_2 levels can directly enhance the resistance of weeds to herbicides (Gutierrez et al. 2008). Moreover, an increase in temperatures leading to more partial closure of plant stomata can decrease the uptake and effectiveness of herbicides (Jackson et al. 2011). Changes in climate, such as higher temperatures and increased levels of CO_2, could result in pests having more generations throughout the year. This, combined with the extended use of pesticides due to longer growing seasons, could enhance the resistance of pests to these pesticides (Matzrafi 2019). Elevated temperatures are also known to accelerate the growth rates of both plants and their pests (Delcour et al. 2015). In the United States, the number of insecticide treatments for Lepidoptera pests in corn varies significantly with climate; in the warmer state of Florida, there are 15 to 30 applications annually, while in the cooler state of New York, there are none to five applications (Hatfield et al. 2011). Climate change leads to a diminished residual effect of pesticides on crops, which in turn heightens their susceptibility to pests and diseases. Consequently, farmers might be compelled to apply pesticides more frequently throughout the crop season in the future. An increase in pest or disease prevalence will likely necessitate more frequent and voluminous applications of pesticides. The potential employment of illegal or adulterated pesticides, which tend to be less effective during pest infestations, could aggravate these issues even further. For example, many pesticides are less effective under dry weather conditions, possibly requiring increased dosages or more regular applications to ensure crop protection (Tirado et al. 2010). As a result, the concentrations of residues found on crops might increase for certain chemicals, while for others, they may degrade more rapidly (Delcour et al. 2015). Moreover, the relative efficiency of some bio-pesticides may be considerably reduced by a hot climate (Gomez-Zavaglia et al. 2020). The success of pesticides is greatly influenced by the surrounding environmental conditions, which affect how pesticides are absorbed, penetrated, transported within the plant, and broken down. The combination of higher temperatures and an increase in CO_2 levels can lead to a decrease in pesticide effectiveness, either by changing how the pesticide is metabolized and translocated within the plant or by indirectly boosting the plant's ability to detoxify the pesticide, thereby reducing its availability to the intended pest. Changes in the plant's stress response signaling pathways and other physiological changes may also be connected to faster pesticide detoxification when plants are subjected to changing climate conditions (Matzrafi 2019).

Climate change significantly influences how pesticides are distributed within the soil, sediment, and water ecosystems by changing the basic processes that govern their movement between these compartments (Noyes et al. 2009). The fate and movement of pesticides in surface and groundwater systems can be affected by climate change, with key factors, including shifts in the timing and intensity of rainfall and rising temperatures (Bloomfield et al. 2006). Additionally, the absorption and transport of pesticides within plants are impacted by rainfall and will be restricted when transpiration is reduced in arid conditions (Keikotlhaile 2011). Climate change has the potential to influence various steps of human exposure to pesticides, impacting the release of these substances into the environment, their environmental movement, and their transformation (Balbus et al. 2013). It is estimated that climate change will raise the existing average external cost associated with pesticide application in US agriculture (currently US$42 per hectare), potentially reaching as much as $72 per hectare by 2100 (Koleva and Schneider 2009).

Finally, it is essential to comprehend how climate change affects pesticide efficacy, pest dynamics, and pesticide fate and persistence in the environment. This is crucial for developing effective mitigation and adaptation strategies. Hence, we will detail different approaches to mitigate and adapt to the effects of climate change on the usage and environmental fate of pesticides.

6.2 Climate Change and Pesticide Use

6.2.1 Climate-Related Factors Influencing Pesticide Use

Many classes of pests and pathogens prefer warm and humid conditions, which influence crop–pest interactions. These patterns necessitate adjustments in pesticide use and plant protection strategies (Bloomfield et al. 2006; Delorenzo et al. 2009). Some pesticides are more toxic to insects at high temperatures (Noyes and Lema 2015). Conversely, another study by de' Beeck et al. (2017) indicated that chlorpyrifos degrades at high temperatures, resulting in lower mortality. Other studies suggest that climate change could lead to reduced pesticide concentrations due to heightened rates of volatilization and accelerated decomposition, both heavily influenced by higher temperatures, increased humidity, and direct sunlight exposure (Noyes et al. 2009; Zhang et al. 2006; Otieno et al. 2013). Furthermore, hot and dry conditions may reduce the need for fungicide applications. Conversely, rising temperatures may necessitate greater pesticide usage (Jager et al. 1998). Also, changes in temperature can change plant productivity and pesticide use after harvest during storage or post-harvest treatment (Eid et al. 2007). Additionally, the balance between the absorption and release of semi-volatile pesticides in plant leaves is swiftly reached under warmer conditions (Bloomfield et al. 2006). Moreover, the timing and intensity of rainfall are critical in determining the stability and effectiveness of pesticides (Bailey 2003).

6.2.2 Impact of Climate Change on Pest Dynamics

Population dynamics are defined as a portion of an inhabitant's ecology connecting with factors upsetting change in population densities (Karuppaiah and Sujayanad 2012). Insects can be essential indicators for evaluating the effects of climate change on more complex organisms. As poikilotherms, their developmental and reproductive traits are directly affected by temperature changes (Beresford and Sutcliffe 2009). It is anticipated that insects will respond favorably to warmer temperatures, displaying increased reproductive capabilities, for example. The bark beetle, *Ips typographus*, known to prosper in such conditions, benefits from faster growth rates. This rapid development enables it to complete its life cycles more quickly, potentially allowing for the production of additional generations within a single season (Jönsson et al. 2009). Warm temperatures reduce the time required for the spruce beetle, *Dendroctonus rufipennis*, to reproduce by half (Berg et al. 2006). Pine beetle populations significantly decline at −16 °C, which is lethal, yet they exhibit enhanced survival and activity during warm winters. In the case of aphids, a temperature increase of 2 °C can result in 1–5 additional life cycles within the season (Yamamura and Kiritani 1998).

Additionally, changes in rainfall patterns can impact wireworm populations, leading to significant crop damage (Gregory et al. 2009). However, heavy rainfall can be advantageous as it helps wash away small pests such as aphids, mites, jassids, and whiteflies (Pathak et al. 2012). Severe droughts can lead to increased densities of bark beetles (Netherer et al. 2019), while stressed trees may develop greater resistance to these beetles (Netherer et al. 2015). Prolonged droughts, followed by significant rainfall, can cause a surge in populations of the oriental armyworm, *Mythimna separata* (Walker) (Lepidoptera: Noctuidae), as drought conditions adversely affect the activity and numbers of its natural predators (Sharma et al. 2010). Furthermore, water stress in plants can reduce their vital functions, rendering them more vulnerable to diseases and pests (Zayan 2019). Research also indicates that aphids on water-stressed plants exhibit decreased rates of parasitism, due to reduced host size or availability (Ahmed et al. 2017).

Climate change is affecting how pathogens evolve and interact with their hosts, which may result in the development of new disease variants. This alteration could broaden the geographical spread of pathogens, consequently heightening the occurrence of plant diseases in previously unaffected regions (Singh et al. 2023). Recent research underscores an increase in both the frequency and intensity of disease outbreaks, presenting serious risks to agricultural production, worldwide food security, and ecological diversity (Velasquez et al. 2018). Warm temperatures shorten the incubation period for pathogens, resulting in a greater abundance of pathogens throughout the growing season. Furthermore, elevated humidity is linked to increased severity of diseases, such as those caused by the potato blight pathogen, *Phytophthora infestans* (Gustafson et al. 2022). The severity of Fusarium head blight in wheat is expected to increase as the milder *Fusarium culmorum*, which thrives in cool and wet environments, is replaced by the more aggressive *Fusarium graminearum* that favors warm and wet conditions (Parikka et al. 2012). Higher CO_2 levels trigger specific reactions from both pathogens and hosts. Studies indicate that elevated CO_2 concentrations have exacerbated the severity of powdery mildew in cucurbits caused by *Sphaerotheca fuliginea* (Khan and Rizvi 2020) and have also aggravated head blight and blotch in wheat, caused by *Fusarium* spp. and *Septoria tritici*, respectively (Var et al. 2015). The occurrence of drought and similar stressful conditions leads to a higher rate of root infections by fungi such as *Armillaria* spp. and those that cause cankers, like *Botryosphaeria*, *Diplodia*, and *Phytophthora*. Additionally, drought stress has been linked to increased viral infections in plants, notably maize dwarf mosaic virus and beet yellows virus (Clover et al. 1993). Conversely, according to Roos et al. (2011), moist conditions enhance the germination of spores, the dispersion and efficacy of zoospores, and the growth of fungi and bacteria.

Research shows that pest resistance to chemicals frequently stems from genetic variations that boost metabolic processes and detoxification or from a decreased responsiveness of the target sites to pesticide impacts (Li et al. 2007). Environmental factors may also enhance the occurrence of certain resistance genes within a population or contribute to the development of resistance to pesticides. Increased temperatures might heighten the selection pressure for these resistance alleles due to more frequent pesticide use (Delcour et al. 2015). Moreover, such climatic conditions might hasten the breeding cycle of pests, thus increasing the rate at which resistance emerges (Altermatt 2010).

Research indicates that climate changes are prolonging the growing seasons, evident from the later arrival of winter and the earlier onset of spring. This shift is also accompanied by a rise in average winter temperatures and more frequent extreme weather events, such as winter heatwaves (Marshall et al. 2020). These climatic shifts are significantly affecting the reproductive patterns, life cycles, and growth trajectories of various species, with noticeable impacts during the winter months that enhance their survival through the colder season (Boggs and Inouye 2012). Diapause, a survival strategy employed by many organisms to cope with harsh conditions, significantly benefits them by boosting nutrient reserves, lowering metabolic activity, increasing resistance to cold, and enhancing their ability to withstand dry conditions. This state provides them with a substantial defensive edge against environmental stressors compared to those not in diapause (Denlinger 2022). Numerous factors suggest that the overwintering survival of *Aedes albopictus* is enhanced by winter heat waves. Such warm periods may provide the optimal conditions for *Ae. albopictus* to eliminate metabolic waste and recover from the detrimental effects of cold exposure. Furthermore, *Ae. albopictus*, which originates from the temperate climates of Southeast Asia, has extended its distribution to additional locales within both its native and invasive ranges (Porretta et al. 2012). It is also noted that some insect species possess broader geographical distributions, rendering them less susceptible to the energy deficits experienced during overwintering, a challenge predominantly faced by species in more northerly latitudes (Williams et al. 2012).

Figure 6.1 presents a network visualization generated using VOSviewer, illustrating a keyword co-occurrence network derived from 524 articles indexed in the Scopus database. This network specifically explores the interconnections between "climate change" and associated terms such as "pest," "insect," "weed," and "plant disease." Nodes in the visualization represent keywords, with larger nodes indicating more frequent occurrences. Colors group nodes into thematic clusters, suggesting related research areas, while edges (lines) between nodes represent the co-occurrence of keywords in publications, with thicker lines indicating stronger relationships. The spatial distribution of nodes shows thematic relatedness, with closer nodes likely sharing more connections. Central to the network is "climate change," highlighting its prominence in the research. Surrounding nodes like "crop pest," "insect," "food security," and "plant disease" underscore a significant interest in the impact of climate change on these areas. Additional clusters focus on specific pests, geographic regions, agricultural issues, and ecological impacts, illustrating the broad scope and interconnected nature of research in this field. This visualization serves as a useful tool for researchers to explore key focus areas, trends, and potential gaps in understanding the nexus between climate change and its ecological and agricultural effects.

6.3 Climate Change and Pesticide Fate in the Environment

The influence of climate change on the fate of pesticides is significant concerning environmental pollution, ecological effects, and human health. Comprehending these effects is vital for evaluating the potential risks of pesticide use and developing effective mitigation strategies.

6.3.1 Influence of Climate on Pesticide Transport and Movement

Pesticides may find their way into the environment through multiple pathways, such as runoff, leaching into the groundwater, evaporating into the air, and sorption. These processes can be influenced by climate change, which may result in increased pesticide contamination in different environmental compartments. Numerous research efforts have been directed at understanding how climate change affects the transport and movement of pesticides, shedding light on the following processes:

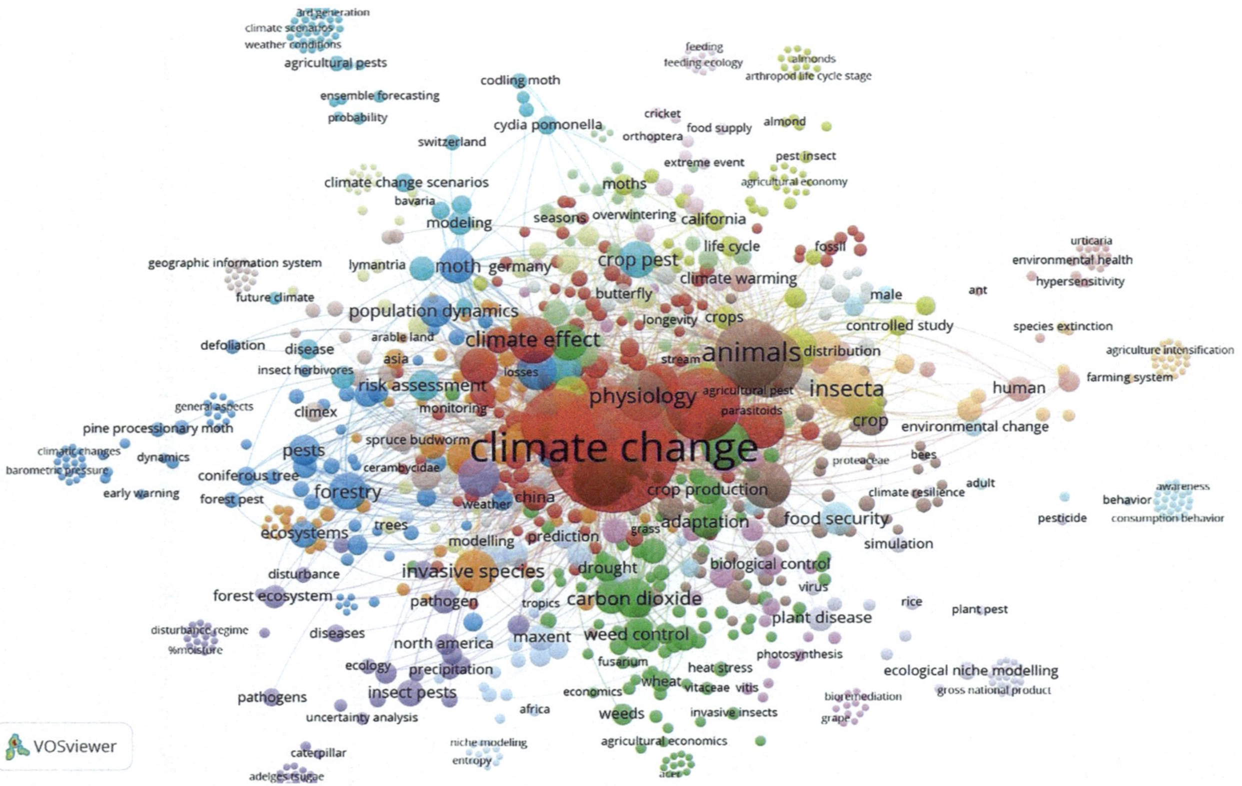

Fig. 6.1 A network visualization created with VOSviewer, depicting a keyword co-occurrence network from 524 scientific publications available on the Scopus database on “climate change” and its interactions with “pest,” “insect,” “weed,” and “plant disease”

6.3.1.1 Runoff

Runoff occurs when pesticides are transported by surface water during periods of rainfall or irrigation. Forecasted alterations in the patterns, intensity, and scheduling of precipitation due to climate change could affect how the frequency and magnitude of runoff occur. These changes in the hydrological regime can affect how pesticides are carried into adjacent water bodies, potentially raising the levels of contamination in the water (Bloomfield et al. 2006). Climate change is anticipated to significantly elevate the risk of insecticide runoff, posing a profound threat to the ecological health of freshwater habitats, particularly in central and northern Europe (Kattwinkel et al. 2011). Adil et al. (2023) highlighted the necessity of understanding how climate change influences pesticide runoff and the resulting consequences for aquatic ecosystems.

6.3.1.2 Leaching

Leaching is the process by which pesticides are transported downward within the soil layers and eventually make their way into the groundwater. Lipophilicity, the amount of organic matter in the soil, soil pH, and prevailing weather conditions influence the leaching of pesticides into groundwater (Nicholls 1988). Changes in climate that affect soil moisture levels, the rate of evapotranspiration, and the distribution and amount of rainfall can all influence how much and how deeply pesticides leach into the soil. Climate change has the potential to affect soil moisture levels and infiltration rates, which can have an effect on the leaching of pesticides. Changes in precipitation patterns and increased soil erosion can affect the movement of pesticides through the soil layers, possibly increasing the threat of groundwater pollution. Such changes could lead to greater leaching occurrences, which jeopardize water quality and pose risks to both ecological systems and human health (Delcour et al. 2015). Additionally, rising temperatures and changes in precipitation patterns may intensify leaching processes, thus increasing the likelihood of groundwater pollution (Bloomfield et al. 2006). Chen and McCarl (2001) underscored the link between climate change, the application of pesticides, and the resulting increase in pesticide leaching. The extent and nature of the projected shifts in pesticide leaching from the current situation to the future are highly dependent on the specific climate change scenario under consideration (Steffens et al. 2013a). Understanding temperature-related processes is critical for projecting future pesticide losses, as rising temperatures might lead to more leaching of chemicals that are weakly bound to soils, while reducing leaching for those that are moderately or strongly bound (Steffens et al. 2013b). Conversely, while climate change might marginally lower the concentrations of herbicides that leach, the expanse of land vulnerable to groundwater pollution could increase twofold due to indirect consequences such as alterations in land-use practices and herbicide application (Steffens et al. 2015). These studies indicate that climate change influences leaching and persistence in the environment through factors such as temperature, soil moisture levels, precipitation trends, soil characteristics, and changes in pesticide formulation and application, with potential impacts on groundwater contamination and bioavailability in the soil.

6.3.1.3 Volatilization

Volatilization refers to the process by which pesticides convert from a liquid state to a gaseous state and are then emitted into the air. It represents a primary route by which pesticides are transferred into the atmosphere, contributing between 2% and 90% of the initial amount of pesticide applied. This can take place through the pesticide evaporating from a liquid, sublimating from a solid, evaporating from an aqueous solution, or being desorbed from the soil (Bedos et al. 2002). Climate change has the potential to affect how pesticides vaporize from both soil and plant surfaces. Elevated temperatures and higher levels of moisture in the atmosphere can alter the rate at which pesticides volatilize. Changes in the climate can alter the vapor pressure and the speed at which pesticides evaporate, which may result in a greater volatilization of these chemicals into the atmosphere and their subsequent long-range transport. Climatic factors, such as temperature and wind velocity, are pivotal in determining the rate of volatilization. As climate change impacts these factors, it can influence pesticide volatilization rates and patterns, potentially leading to more pollution in the atmosphere (Noyes et al. 2009). Climate change has the potential to impact the rate at which pesticides volatilize by altering the physical and chemical properties of the pesticides, the surrounding environmental conditions, and the methods by which they are managed. Higher temperatures and increased atmospheric humidity can amplify the volatilization process, resulting in higher atmospheric concentrations and potential long-range transport (Delcour et al. 2015). Increasing the relative humidity from 60% to 85% can lead to a rate of pesticide volatilization that is up to eightfold greater under arid conditions (Schneider et al. 2013). While climate change may lead to a rise in pesticide volatilization from the soil, altering land use by just 10% has the potential to neutralize this effect (Komprda et al. 2013). The findings from these studies indicate that climate change tends to increase the volatilization of pesticides as a result of factors such as higher temperatures, shifts in precipitation, and greater humidity.

6.3.1.4 Sorption

Sorption refers to the process by which pesticides attach to soil particles or other substrates. Climate change-induced alterations in soil characteristics, including organic matter levels, temperature, and pH, can affect the ability of soil to retain pesticides. These alterations may affect how pesticides move through the environment and their readiness for uptake by organisms, thereby influencing their ultimate environmental fate and ecological impact (Adil et al. 2023). Changes in soil properties due to climate change can affect pesticide availability and persistence in the environment, which in turn could impact their transportation to aquatic systems or their uptake by plants (Schiedek et al. 2007). Key factors related to climate change, such as temperature and the level of moisture in the soil, play a significant role in determining how readily pesticides are available in the soil (Kaka et al. 2021).

Figure 6.2 depicts a network map, generated via VOSviewer, that illustrates the co-occurrence of keywords within a corpus of 926 academic publications indexed in the Scopus database. The central theme of these articles pertains to the intersection of "pesticide" with the concepts of "fate," "transport," or "movement." In this map,

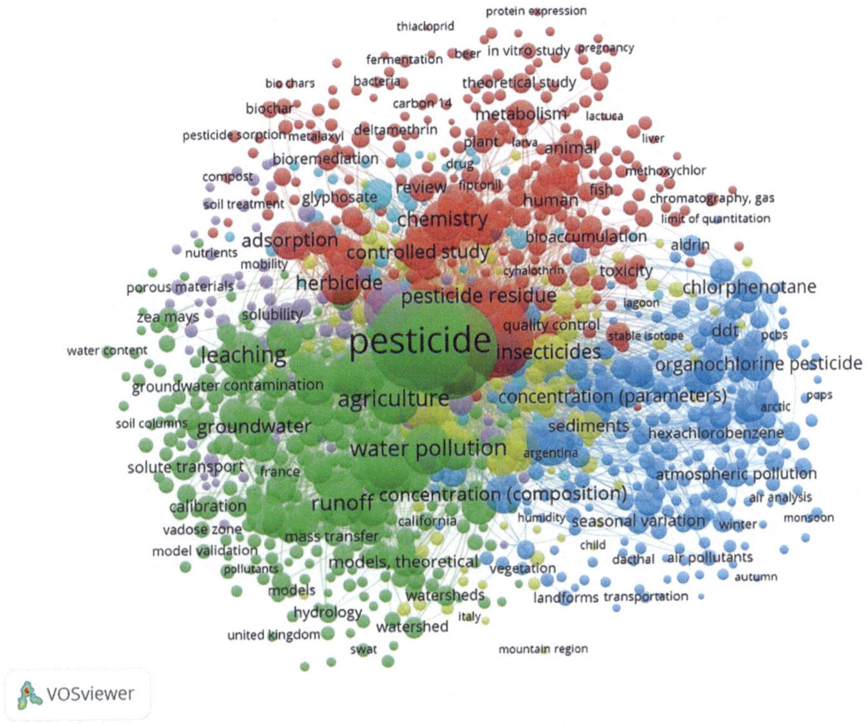

Fig. 6.2 A network map created with VOSviewer, depicting keyword co-occurrence in 926 scientific publications sourced from the Scopus database and focusing on the terms "pesticide" in conjunction with "fate," "transport," or "movement"

nodes represent keywords, with their size indicating the frequency of their occurrence in the literature. Colors differentiate thematic clusters of keywords that frequently appear together, signifying related research areas. The spatial arrangement of nodes reveals the strength of relationships, with closer nodes being more strongly related. Keywords with larger labels are more prominent within the field. The map distinctly highlights a key term, "pesticide," surrounded by related clusters: a red cluster focusing on chemical properties and toxicology (e.g., "chemistry," "herbicide," and "toxicity"), a green cluster examining environmental transport and effects (e.g., "leaching," "groundwater," and "runoff"), and a blue cluster addressing broader environmental issues and detection methods (e.g., "water pollution" and "sediments"). This network map serves as a useful tool for researchers to visualize the concentration of studies, interconnected topics, and potential gaps in the field of pesticide research.

6.3.2 Alterations in Pesticide Degradation Processes

The degradation of pesticides through processes such as photodegradation, microbial degradation, and chemical degradation is crucial for diminishing their persistence and potential ecological risks. Climate change can modify these degradation pathways, thus changing the environmental fate of pesticides. The affected mechanisms include:

6.3.2.1 Photodegradation

Photodegradation is the process by which pesticides decompose due to sunlight exposure. Variations in weather patterns, including higher temperatures and changes in cloudy cover, can affect the intensity and amount of time that solar radiation reaches the Earth's surface. Such alterations may influence the rate and extent of pesticide photodegradation, possibly leading to prolonged persistence and increased environmental exposure (Schiedek et al. 2007). Additionally, fluctuations in levels of ultraviolet (UV) light, owing to shifts in cloud cover and the thinning of the stratospheric ozone layer, can alter how rapidly pesticides degrade (Noyes et al. 2009). This may affect pesticide persistence in the environment and their toxic effects.

6.3.2.2 Microbial Degradation

Microbial degradation is the disintegration of pesticides through the action of microorganisms found in the soil, water, and other environmental compartments. Climate change can affect the microbial populations that break down pesticides in terrestrial and aquatic ecosystems by changing conditions like temperature, moisture levels, and the availability of nutrients. These changes could alter the activity, metabolic processes, and variety of pesticide-degrading microorganisms, consequently influencing pesticide degradation rates and efficiency (Kannan 2023). Moreover, variations in temperature and rainfall patterns due to climate change can have an impact on microbial activity levels and thus modify the degradation rates of pesticides in both soil and aquatic environments (Matzrafi 2019).

6.3.2.3 Chemical Degradation

Chemical degradation is the process by which pesticides decompose as a result of environmental chemical reactions, including hydrolysis, reduction, and oxidation. Changes in climate-related factors such as temperature, pH, and moisture can impact the rates and pathways of chemical degradation, thus influencing pesticide persistence in the environment and their level of toxicity (Duchenne-Moutien and Neetoo 2021). Climate change-induced alterations in these factors may lead to variations in pesticide degradation rates, which in turn might affect their persistence and potential effects on the environment (Bloomfield et al. 2006).

Figure 6.3 is a network visualization created using VOSviewer, depicting the co-occurrence of keywords within a corpus of 1250 scientific articles sourced from the Scopus database, focusing on the terms "pesticide" and "degradation." This visualization illuminates the interrelationships and thematic clusters prevalent within this

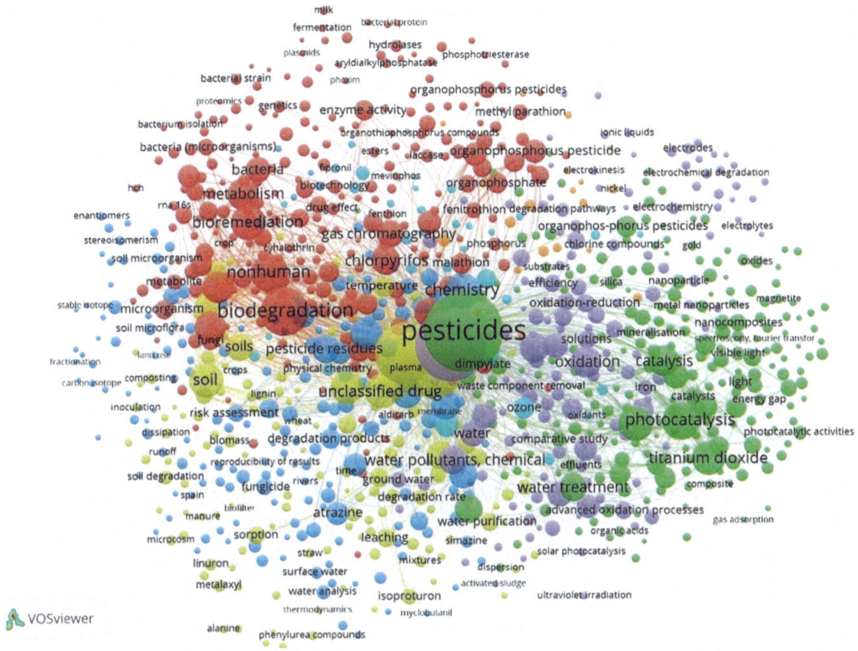

Fig. 6.3 A network visualization created using VOSviewer, depicting the co-occurrence of keywords within a corpus of 1250 scientific articles sourced from the Scopus database, focusing on the terms "pesticide" and "degradation"

research area. Each node in the network represents a keyword, with the node's size denoting the frequency of its occurrence and the thickness of the connecting lines indicating the strength of co-occurrence between keywords. The various colors delineate distinct clusters of keywords, each signifying a particular subtopic within the broader field. The central, significantly prominent cluster, colored red, centers on "biodegradation" and "soil," indicating a strong research focus within this area. Adjacently, a yellow cluster highlights issues related to "water treatment" and pertinent aquatic concerns, whereas a blue cluster correlates with agricultural practices and their environmental impacts. Additionally, a purple cluster is dedicated to "photocatalysis" and other advanced degradation techniques. This network visualization serves as a strategic tool for researchers, enabling them to identify the most extensively studied areas, comprehend the interconnectedness of various research themes, and pinpoint under-explored topics warranting further investigation. Moreover, it offers valuable insights into the dynamic evolution of research within this field over time.

To summarize, the impact of climate change on the environmental fate of pesticides is considerable. The changes in water cycles, soil characteristics, and weather patterns that accompany climate change can alter the fate of pesticides, which may result in greater pollution and ecological risks. These changes can have significant consequences for pesticide exposure, efficacy, and the broader ecological footprint of their application. It is critical to comprehend these effects to devise robust pesticide regulation tactics, adopt pest management practices that can withstand climate changes, and enhance monitoring and modeling efforts to reduce the ecological and health risks tied to pesticide application in a changing climate. Additional studies are necessary to more thoroughly understand and manage the relationship between climate change and the environmental fate of pesticides.

6.3.3 Interaction Between Climate Change and Pesticide Pre-Harvest Interval (PHI)

The effect of climate change on pesticide preharvest interval (PHI) is a significant concern for agricultural systems worldwide. PHI, which is the time span between the last pesticide application and the harvesting of the crop, stands as a critical factor in the implementation of best agricultural practices (Leong et al. 2020). The PHI plays a crucial role in ensuring product food safety, protecting the environment, and complying with regulatory requirements. Majed et al. (2021) showed that the PHI is greatly influenced by weather factors such as sunlight, humidity, and temperature, emphasizing the importance of establishing PHIs that are specific to local climates.

Climate change impacts various environmental factors, including temperature fluctuations, changes in rainfall, the frequency of extreme weather events, and pest dynamics, all of which can affect pesticide efficacy and residual effects. With rising temperatures, pesticides may degrade more quickly, which could shorten the PHI or decrease the time needed for pesticide residues to diminish to safe levels. However, increased temperatures might also result in greater pest pressure, necessitating more pesticide applications or higher doses, potentially prolonging the PHI. Additionally, when plants are under heat stress, their capacity to metabolize pesticides might be compromised, leading to the need for extended PHIs to guarantee the safety of the food supply.

Extreme weather events, including storms, droughts, and floods, can interrupt the timing of pesticide applications and make it difficult to follow PHI recommendations. For instance, heavy rainfall or floods can delay the application of pesticides, necessitating urgent pesticide use to manage sudden pest and disease outbreaks, which could lead to the need for quick revisions of PHI timescales to address immediate concerns. In regions where there is an increase in precipitation and humidity, the spread of fungal diseases may rise, leading to a higher frequency of fungicide applications. If these wet conditions also cause pesticides to be washed off the plants more rapidly, this could necessitate more frequent applications and modifications to PHIs. In contrast, if a region becomes more arid due to climate change, the resulting drought may impact pest populations and how long pesticides remain effective, potentially leading to changes in the periods of PHIs.

Climate change can significantly impact the geographic range, abundance, and phenology of pests. Increases in temperature and changes in seasonal patterns can lead to alterations in when pests emerge, their population dynamics, and their life cycle. Such shifts may necessitate changes in the timing of pesticide applications and the corresponding PHI requirements.

However, the relationship between climate change and PHI is complex and has not been extensively explored (Besil et al. 2019). It's vital to revise PHI guidelines to reflect the impacts of climate change to ensure the safety of our food, reduce pesticide residue levels, maintain effective pest management, and support resilient and sustainable farming practices. The cooperation of researchers, policymakers, agricultural producers, and other stakeholders is essential in developing and implementing strategies that optimize PHI under changing climatic conditions.

6.4 Mitigation and Adaptation Strategies

The implications of climate change for pesticide usage and environmental fate are substantial, prompting a need for the development and application of robust mitigation and adaptation measures. Abbass et al. (2022) and Pathak (2023) have conducted comprehensive reviews of the global and local effects of climate change and sustainable mitigation measures, calling for a holistic approach to agricultural policy, scientific research, and practical farming methods. Mitigation and adaptation strategies for pesticide use to climate change are essential for safeguarding agricultural productivity and environmental health. By integrating these strategies, those involved in agriculture can more effectively navigate the challenges posed by climate change and pesticide application, thus promoting sustainable crop production and food security while reducing risks to the environment and human health.

Mitigation strategies for pesticide utilization encompass a combination of policy measures, technological advancements, and farm management practices. Möhring et al. (2020) presented approaches to improve pesticide policies by calling for stricter regulations, incentives for lessening pesticide application, and support for alternative pest control methods. Malhi et al. (2021) discussed a range of strategies for mitigation, including precision agriculture, organic farming, and the encouragement of biological control agents. Bibi and Rahman (2023) highlighted the importance of policy reinforcements and the involvement of relevant parties in fostering sustainable agricultural practices. Several strategies can be employed for mitigating the impacts of climate change on pesticide use and fate such as improving pesticide formulations and application techniques and enhancing surveillance and modeling efforts to assess and forecast pesticide fate under changing climatic conditions. Integrated pest management (IPM), breeding climate-resilient crop varieties, precision agriculture, and organic farming are essential mitigation measures. IPM approaches encompass a range of techniques, such as cultural practices, biological control, and the careful application of pesticides. Implementing IPM can decrease the reliance on pesticides and improve the efficiency of pest control, all while reducing harm to the environment. Breeding plant varieties that are tolerant to both pests and environmental stress factors can also diminish the need for pesticides and enhance crop resilience. This approach involves identifying and incorporating genes associated with resistance to pests and climate-related stresses into breeding programs. Precision agriculture, through utilizing advanced technologies, such as remote sensing, drones, and global positioning systems (GPSs), enables targeted and precise application of pesticides that is informed by immediate data and vigilant pest surveillance. This approach optimizes pesticide use, curbing excess and environmental pollution. Organic farming practices, which exclude synthetic pesticides, promote biodiversity, and enhance soil health, can serve as a sustainable alternative to conventional pesticide-dependent systems. The adoption of organic farming practices can diminish the ecological footprint linked to the use of pesticides.

Adapting agricultural practices in response to climate change is vital for ensuring food security and environmental sustainability. Grigorieva et al. (2023) demonstrated that IPM and the development of climate-resilient crops are key strategies for adapting agriculture to climate change. Anderson et al. (2020) emphasized the need for the incorporation of climate forecasting into pest management plans. Muchuru and Nhamo (2019) pointed out the significance of local knowledge and technology transfer in developing resilience to climate impacts. Additionally, Skendžić et al. (2021) provided insights into the effects of climate change on agricultural insect pests, underscoring the need for pest management strategies that are both responsive and proactive. Climate information services, educational programs for farmers, and policy interventions can aid farmers in adapting to changing pest dynamics while also reducing the ecological effects of pesticide application. Supplying farmers with accurate and localized climate data helps them make well-informed pest control choices. This encompasses predicting pest outbreaks, advising on the best times for pesticide application, and proposing other pest management options. It is essential to improve farmers' knowledge and understanding of the consequences of climate change, the biology of pests, and the principles of IPM for successful adaptation. Capacity-building programs can enable farmers to implement suitable pest management strategies in response to changing climatic conditions. Policy and regulatory measures are crucial in fostering the responsible use of pesticides and encouraging the adoption of climate-smart practices. Developing and implementing policies that incentivize sustainable pest management approaches can support the transition toward more sustainable agricultural practices.

6.5 Research Gap

Although this chapter provides an extensive review of the different impacts that climate change has on the usage and fate of pesticides in the environment, research gaps in this field, where sufficient research and knowledge are still lacking, remain. By filling these gaps, stakeholders can better understand the complexities of pesticide use and fate in a changing climate and formulate pest management approaches that mitigate risks to the environment and human health. There are several potential areas in this field that may require further research:

Effectiveness and Application of Pesticides: As climate change potentially alters pest behavior, physiology, and life cycle, the efficacy of current pesticides may be affected, leading to a need for research into how these changes influence pest resistance and the necessity for new pesticide application methods. It is essential to comprehend the varied effects that climate change has on pesticide utilization across different global regions, which requires targeted region-specific research that takes into account the specific local weather patterns, agricultural practices, and pest species. Investigations into how climate change will shape IPM practices are crucial, especially with respect to the role of biological control agents in minimizing pesticide dependence.

Alterations in Pest Populations: There is a necessity for studies to understand how changes in temperature and weather influence pest populations and distribution, including the possibility of the emergence of new pests in regions where they previously did not exist due to changing climates. There is also a need for the development of predictive instruments that support future-oriented planning and decision-making regarding pesticide management in the context of climatic changes.

Pesticide Degradation and Fate: Investigating the impact of rising temperatures and fluctuating humidity on the PHIs and degradation rates of pesticides in the environment is critical, in addition to understanding how extreme weather events affect the transport and fate of pesticides within the environment. The effects of changing precipitation on the movement of pesticides into water systems are an important research focus, along with the role of soil health and microbial activity in pesticide degradation under changing climate conditions. There is a necessity for enhanced modeling that integrates various climatic variables with pesticide fate.

Regulatory and Management Approaches: An urgent requirement exists to assess whether existing pesticide regulations are adequate in light of climate change and to develop adaptable management strategies for pesticide application that can handle the uncertainties associated with the climate.

Finally, additional investigation is required to comprehend how mitigation and adaptation measures are put into practice and their efficacy, which can be determined through long-term field studies and evaluations of their impacts.

6.6 Conclusion

This chapter examined the effects of climate change on the usage and environmental fate of pesticides, while also emphasizing an array of proposed measures for mitigation and adaptation to address these challenges. Alterations in climate conditions, such as temperature increases, elevated CO_2 concentrations, and altered precipitation patterns, have been observed to influence pesticide efficacy and the frequency of applications. Such climate-related elements can compromise the effectiveness of pesticides, resulting in intensified pest pressure and necessitating adjustments in pest management strategies. In addition, climate change has been found to impact pest dynamics, with consequences such as the expansion of pests' geographic ranges, altered pest life cycles, outbreaks of plant diseases, increased pest resistance, and changes in overwintering survival. These findings underline the necessity for adaptive and flexible approaches to pesticide application to address changing pest dynamics driven by climate change. Moreover, changes in precipitation patterns and temperature can alter runoff, leaching, volatilization, and sorption processes, which may result in pesticide contamination of water bodies, soil, and air. Additionally, climate change can interfere with pesticide degradation processes, such as photodegradation, microbial degradation, and chemical degradation, which may pose threats to ecosystems and potentially to human health. To tackle these issues, various mitigation and adaptation strategies should be adopted, such as implementing IPM, developing climate-resilient pesticide formulations, incorporating climate change considerations into pesticide policy and regulations, and promoting precision and organic agriculture techniques. By understanding these relationships and implementing appropriate mitigation and adaptation strategies, stakeholders can work toward sustainable and eco-friendly pest management practices in the face of a changing climate. Ongoing research and cooperative efforts between researchers, policymakers, and those involved in agriculture are crucial in tackling the difficulties and developing innovative solutions for the complex interactions between climate change and pesticide dynamics.

References

Abbass K, Qasim MZ, Song H, Murshed M, Mahmood H, Younis I (2022) A review of the global climate change impacts, adaptation, and sustainable mitigation measures. Environ Sci Pollut Res 29(28):42539–42559

Adil M, Abbas G, Khan RN, Abbas F (2023) Impact of climate change on environmental fate and ecological effects of pesticides. In: Strategizing agricultural management for climate change mitigation and adaptation. Springer International Publishing, Cham, pp 247–263

Ahmed SS, Liu D, Simon JC (2017) Impact of water-deficit stress on tritrophic interactions in a wheat-aphid-parasitoid system. PLoS One 12(10):e0186599

Altermatt F (2010) Climatic warming increases voltinism in European butterflies and moths. Proc R Soc B Biol Sci 277(1685):1281–1287

Anderson R, Bayer PE, Edwards D (2020) Climate change and the need for agricultural adaptation. Curr Opin Plant Biol 56:197–202

Audsley E, Stacey KF, Parsons DJ, Williams AG (2009) Estimation of the greenhouse gas emissions from agricultural pesticide manufacture and use. Cranfield University, UK

Bailey SW (2003) Climate change and decreasing herbicide persistence. Pest Manag Sci 60:158–162

Balbus JM, Boxall AB, Fenske RA, McKone TE, Zeise L (2013) Implications of global climate change for the assessment and management of human health risks of chemicals in the natural environment. Environ Toxicol Chem 32(1):62–78

Bedos C, Cellier P, Calvet R, Barriuso E, Gabrielle B (2002) Mass transfer of pesticides into the atmosphere by volatilization from soils and plants: overview. Agronomie 22:21–33. https://doi.org/10.1051/AGRO:2001003

Beresford DV, Sutcliffe J (2009) Sampling designs of insect time series data: are they all irregularly spaced? Oikos 118(1):115–121

Berg EE, Henry JD, Fastie CL, De Volder AD, Matsuoka SM (2006) Spruce beetle outbreaks on the Kenai Peninsula, Alaska and Kluane National Park and Reserve, Yukon Territory: relation-

ship to summer temperatures and regional differences in disturbance regimes. For Ecol Manag 227(3):219–232
Besil N, Pérez-Parada A, Bologna F, Cesio MV, Rivas F, Heinzen H (2019) Dissipation of selected insecticides and fungicides applied during pre-harvest on mandarin and orange trees in Uruguay. Sci Hortic 248:34–40
Bibi F, Rahman A (2023) An overview of climate change impacts on agriculture and their mitigation strategies. Agriculture 13(8):1508
Bloomfield JP, Williams RJ, Gooddy DC, Cape JN, Guha PM (2006) Impacts of climate change on the fate and behaviour of pesticides in surface and groundwater—a UK perspective. Sci Total Environ 369(1–3):163–177
Boggs CL, Inouye DW (2012) A single climate driver has direct and indirect effects on insect population dynamics. Ecol Lett 15(5):502–508
Cech R, Leisch F, Zaller JG (2022) Pesticide use and associated greenhouse gas emissions in sugar beet, apples, and viticulture in Austria from 2000 to 2019. Agriculture 12(6):879
Chen CC, McCarl BA (2001) An investigation of the relationship between pesticide usage and climate change. Clim Chang 50(4):475–487
Clover GRG, Smith HG, Azam-Ali SN, Jaggard KW (1993) The effects of drought on sugar beet growth in isolation and in combination with beet yellows virus infection. J Agric Sci 133:251–261
de Beeck LO, Verheyen J, Stoks R (2017) Integrating both interaction pathways between warming and pesticide exposure on upper thermal tolerance in high-and low-latitude populations of an aquatic insect. Environ Pollut 224:714–721
Delcour I, Spanoghe P, Uyttendaele M (2015) Literature review: impact of climate change on pesticide use. Food Res Int 68:7–15
Delorenzo ME, Wallace SC, Danese LE, Baird TD (2009) Temperature and salinity effects on the toxicity of common pesticides to the grass shrimp, Palaemonetes pugio. J Environ Sci Health B 44(5):455–460
Denlinger DL (2022) Insect diapause. Cambridge University Press
Duchenne-Moutien RA, Neetoo H (2021) Climate change and emerging food safety issues: a review. J Food Prot 84(11):1884–1897
Eid HM, El-Marsafawy SM, Ouda SA (2007) Assessing the economic impacts of climate change on agriculture in Egypt: a Ricardian approach. The World Bank policy research working paper, 4293
FAO (2022) Pesticides use, pesticides trade and pesticides indicators—Global, regional and country trends, 1990–2020. FAOSTAT Analytical Briefs, no. 46. Rome. https://doi.org/10.4060/cc0918en
Gomez-Zavaglia A, Mejuto JC, Simal-Gandara J (2020) Mitigation of emerging implications of climate change on food production systems. Food Res Int 134:109256
Gregory PJ, Johnson SN, Newton AC, Ingram JS (2009) Integrating pests and pathogens into the climate change/food security debate. J Exp Bot 60(10):2827–2838
Grigorieva E, Livenets A, Stelmakh E (2023) Adaptation of agriculture to climate change: a scoping review. Climate 11(10):202
Gustafson EJ, Miranda BR, Dreaden TJ, Pinchot CC, Jacobs DF (2022) Beyond blight: Phytophthora root rot under climate change limits populations of reintroduced American chestnut. Ecosphere 13:18
Gutierrez AP, Ponti L, d'Oultremont T, Ellis CK (2008) Climate change effects on poikilotherm tritrophic interactions. Clim Chang 87(1):167–192
Hatfield JL, Boote KJ, Kimball BA, Ziska LH, Izaurralde RC, Ort DR, Thomson AM, Wolfe D (2011) Climate impacts on agriculture: implications for crop production. Agron J 103(2):351
Heimpel GE, Yang Y, Hill JD, Ragsdale DW (2013) Environmental consequences of invasive species: greenhouse gas emissions of insecticide use and the role of biological control in reducing emissions. PLoS One 8(8):e72293. https://doi.org/10.1371/journal.pone.0072293
IPCC (Intergovernmental Panel on Climate Change) (2018) In: Masson-Delmotte V, Zhai P, Pörtner H-O, Roberts D, Skea J, Shukla PR, Pirani A et al (eds) Global warming of 1.5 °C: an

IPCC special report on the impacts of global warming of 1.5 °C above pre-industrial levels and related global greenhouse gas emission pathways, in the context of strengthening the global response to the threat of climate change, sustainable development, and efforts to eradicate poverty. IPCC, Geneva. 630 pp

IPPC Secretariat (2021) Scientific review of the impact of climate change on plant pests—a global challenge to prevent and mitigate plant pest risks in agriculture, forestry and ecosystems. FAO on behalf of the IPPC Secretariat, Rome. https://doi.org/10.4060/cb4769en

Jackson LE, Wheeler SM, Hollander AD, O'Geen AT, Orlove BS, Six J, Sumner DA, Santos-Martin F, Kramer JB, Horwath WR, Howitt RE (2011) Case study on potential agricultural responses to climate change in a California landscape. Clim Chang 109(1):407–427

Jager ME, Bourbon C, Levsen K (1998) Analysis of pesticides and their degradation products in rainwater: a probe into their atmospheric degradation. Int J Environ Anal Chem 70(1–4):149–162

Jönsson AM, Appelberg G, Harding S, Bärring L (2009) Spatio-temporal impact of climate change on the activity and voltinism of the spruce bark beetle, *Ips typographus*. Glob Change Biol 15(2):486–499

Kaka H, Opute P, Maboeta M (2021) Potential impacts of climate change on the toxicity of pesticides towards earthworms. J Toxicol 2021. https://doi.org/10.1155/2021/8527991

Kannan N (2023) An analysis of the climate change effects on pesticide vapor drift from ground-based pesticide applications to cotton. Sci Rep 13:9740. https://doi.org/10.1038/s41598-023-36941-4

Karuppaiah V, Sujayanad GK (2012) Impact of climate change on population dynamics of insect pests. World J Agric Sci 8(3):240–246

Kattwinkel M, Kühne JV, Foit K, Liess M (2011) Climate change, agricultural insecticide exposure, and risk for freshwater communities. Ecol Appl 21(6):2068–2081. https://doi.org/10.1890/10-1993.1

Keikotlhaile BM (2011) Influence of the processing factors on pesticide residues in fruits and vegetables and its application in consumer risk assessment (Doctoral dissertation, Ghent University)

Khan MR, Rizvi TF (2020) Effect of elevated levels of CO_2 on powdery mildew development in five cucurbit species. Sci Rep 10(1):4986

Koleva N, Schneider U (2009) The impact of climate change on the external cost of pesticide applications in US agriculture. Int J Agric Sustain 7:203–216. https://doi.org/10.3763/ijas.2009.0459

Komprda J, Komprdová K, Sánka M, Mozny M, Nizzetto L (2013) Influence of climate and land use change on spatially resolved volatilization of persistent organic pollutants (POPs) from background soils. Environ Sci Technol 47(13):7052–7059. https://doi.org/10.1021/es3048784

Lehmann P, Ammunet T, Barton M, Battisti A, Eigenbrode SD, Jepsen JU, Kalinkat G, Neuvonen S, Niemelä P, Økland B, Terblanche JS (2018) Complex responses of global insect pests to climate change. bioRxiv:425488

Leong WH, Teh SY, Hossain MM, Nadarajaw T, Zabidi-Hussin Z, Chin SY, Lai KS, Lim SHE (2020) Application, monitoring and adverse effects in pesticide use: the importance of reinforcement of Good Agricultural Practices (GAPs). J Environ Manag 260:109987

Li CX, Schuler MA, Berenbaum MR (2007) Molecular mechanisms of metabolic resistance to synthetic and natural xenobiotics. Annu Rev Entomol 52:231–253

Liang D, Lu X, Zhuang M, Shi G, Hu C, Wang S, Hao J (2021) China's greenhouse gas emissions for cropping systems from 1978–2016. Sci Data 8(1):171

Majed L, Hayar S, Zeitoun R, Maestroni BM, Dousset S (2021) The effects of formulation on imidacloprid dissipation in grapes and vine leaves and on required pre-harvest intervals under Lebanese climatic conditions. Molecules 27(1):252

Malhi GS, Kaur M, Kaushik P (2021) Impact of climate change on agriculture and its mitigation strategies: a review. Sustainability 13(3):1318

Marshall KE, Gotthard K, Williams CM (2020) Evolutionary impacts of winter climate change on insects. Curr Opin Insect Sci 41:54–62

Matzrafi M (2019) Climate change exacerbates pest damage through reduced pesticide efficacy. Pest Manag Sci 75(1):9–13

Miraglia M, Marvin HJP, Kleter GA, Battilani P, Brera C, Coni E, Cubadda F, Croci L, De Santis B, Dekkers S, Filippi L (2009) Climate change and food safety: an emerging issue with special focus on Europe. Food Chem Toxicol 47(5):1009–1021

Möhring N, Ingold K, Kudsk P, Martin-Laurent F, Niggli U, Siegrist M, Studer B, Walter A, Finger R (2020) Pathways for advancing pesticide policies. Nat Food 1(9):535–540

Muchuru S, Nhamo G (2019) A review of climate change adaptation measures in the African crop sector. Clim Dev 11(10):873–885

Netherer S, Matthews B, Katzensteiner K, Blackwell E, Henschke P, Hietz P et al (2015) Do water-limiting conditions predispose Norway spruce to bark beetle attack? New Phytol 205(3):1128–1141

Netherer S, Panassiti B, Pennerstorfer J, Matthews B (2019) Acute drought is an important driver of bark beetle infestation in Austrian Norway spruce stands. Front For Glob Change 2(39). https://doi.org/10.3389/ffgc.2019.00039

Nicholls P (1988) Factors influencing entry of pesticides into soil water. Pesticide Science 22:123–137. https://doi.org/10.1002/PS.2780220204

Noyes PD, Lema SC (2015) Forecasting the impacts of chemical pollution and climate change interactions on the health of wildlife. Curr Zool 61(4):669–689

Noyes PD, McElwee MK, Miller HD, Clark BW, Van Tiem LA, Walcott KC, Erwin KN, Levin ED (2009) The toxicology of climate change: environmental contaminants in a warming world. Environ Int 35(6):971–986

Oerke EC (2006) Crop losses to pests. J Agric Sci 144:31–43

Otieno PO, Owuor PO, Lalah JO, Pfister G, Schramm KW (2013) Impacts of climate-induced changes on the distribution of pesticides residues in water and sediment of Lake Naivasha, Kenya. Environ Monit Assess 185(3):2723–2733

Parikka P, Hakala K, Tiilikkala K (2012) Expected shifts in *Fusarium* species' composition on cereal grain in Northern Europe due to climatic change. Food Addit Contam Part A Chem Anal Control Expo Risk Assess 29:1543–1555

Pathak H (2023) Impact, adaptation, and mitigation of climate change in Indian agriculture. Environ Monit Assess 195(1):52

Pathak H, Aggarwal PK, Singh S (2012) Climate change impact, adaptation and mitigation in agriculture: methodology for assessment and applications. Indian Agricultural Research Institute, New Delhi, p 302

Patterson DT, Westbrook JK, Joyce RJV, Lingren PD, Rogasik J (1999) Weeds, insects, and diseases. Clim Chang 43(4):711–727

Porretta D, Mastrantonio V, Bellini R, Somboon P, Urbanelli S (2012) Glacial history of a modern invader: phylogeography and species distribution modelling of the Asian Tiger Mosquito *Aedes albopictus*. PLoS One 7(9):e44515

Roos J, Hopkins R, Kvarnheden A, Dixelius C (2011) The impact of global warming on plant diseases and insect vectors in Sweden. Eur J Plant Pathol 129(1):9–19

Schiedek D, Sundelin B, Readman JW, Macdonald RW (2007) Interactions between climate change and contaminants. Mar Pollut Bull 54(12):1845–1856

Schneider M, Endo S, Goss K (2013) Volatilization of pesticides from the bare soil surface: evaluation of the humidity effect. J Environ Qual 42(3):844–851. https://doi.org/10.2134/jeq2012.0320

Sharma H, Srivastava C, Durairaj C, Gowda C (2010) Pest management in grain legumes and climate change. In: Climate change and management of cool season grain legume crops. Springer, Dordrecht, pp 115–139

Singh BK, Delgado-Baquerizo M, Egidi E, Guirado E, Leach JE, Liu H, Trivedi P (2023) Climate change impacts on plant pathogens, food security and paths forward. Nat Rev Microbiol 21:640–656

Skendžić S, Zovko M, Živković IP, Lešić V, Lemić D (2021) The impact of climate change on agricultural insect pests. Insects 12(5):440

Steffens K, Larsbo M, Moeys J, Kjellström E, Jarvis N, Lewan E (2013a) Modelling pesticide leaching under climate change: parameter vs. climate input uncertainty. Hydrol Earth Syst Sci 18:479–491. https://doi.org/10.5194/HESS-18-479-2014

Steffens K, Larsbo M, Moeys J, Jarvis N, Lewan E (2013b) Predicting pesticide leaching under climate change: importance of model structure and parameter uncertainty. Agric Ecosyst Environ 172:24–34. https://doi.org/10.1016/J.AGEE.2013.03.018

Steffens K, Jarvis N, Lewan E, Lindström B, Kreuger J, Kjellström E, Moeys J (2015) Direct and indirect effects of climate change on herbicide leaching—a regional scale assessment in Sweden. Sci Total Environ 514:239–249. https://doi.org/10.1016/j.scitotenv.2014.12.049

Tirado MC, Clarke R, Jaykus LA, McQuatters-Gollop A, Frank JM (2010) Climate change and food safety: a review. Food Res Int 43(7):1745–1765

Tudi M, Daniel Ruan H, Wang L, Lyu J, Sadler R, Connell D, Chu C, Phung DT (2021) Agriculture development, pesticide application and its impact on the environment. Int J Environ Res Public Health 18(3):1112. https://doi.org/10.3390/ijerph18031112

Var Z, Mullins E, McElwain JC, Doohan FM (2015) The severity of wheat diseases increases when plants and pathogens are acclimatized to elevated carbon dioxide. Glob Chang Biol 21:2661–2669

Velasquez AC, Castroverde CDM, He SY (2018) Plant–pathogen warfare under changing climate conditions. Curr Biol 28:R619–R634

Williams CM, Hellmann J, Sinclair BJ (2012) Lepidopteran species differ in susceptibility to winter warming. Clim Res 53(2):119–130

Yamamura K, Kiritani K (1998) A simple method to estimate the potential increase in the number of generations under global warming in temperate zones. Appl Entomol Zool 33(2):289–298

Zayan SA (2019) Impact of climate change on plant diseases and IPM strategies. In: Plant diseases-current threats and management trends. https://doi.org/10.5772/intechopen.87055

Zhang ZY, Zhang CZ, Liu XJ, Hong XY (2006) Dynamics of pesticide residues in the autumn Chinese cabbage (Brassica chinensis L.) grown in open fields. Pest Manag Sci 62(4):350–355

Zhang YW, McCarl BA, Luan Y, Kleinwechter U (2018) Climate change effects on pesticide usage reduction efforts: a case study in China. Mitig Adapt Strateg Glob Chang 23:685–701

7 Climate Change Impacts of Pesticide Toxicity

El-Sayed A. El-Sheikh, Abdel-Tawab H. Mossa, and Moustafa A. Abbassy

Abstract

Climate change has a significant impact on the ecosystem and will therefore impact the chemicals used in the agricultural sector. This relationship represents an area of growing concern as the delicate balance between environmental conditions, agricultural practices, and chemical compound behavior is disturbed. This imbalance leads to harmful effects on the environment and humans. One of the main ways climate change affects pesticide toxicity is through changes in temperature and precipitation patterns, which affect groundwater contamination and pesticide stability. Many pesticides become more easily evaporated in warmer conditions, increasing the likelihood of inhalation by humans and wildlife. Climate change also impacts agricultural pests' life cycles and population dynamics. As temperatures rise and precipitation patterns shift, many pest species may expand their geographic ranges, adapt to new climatic conditions, and develop resistance to commonly used pesticides. This can lead to the increased application of pesticides, further growing their toxicity effects. Climate change will likely increase the movement of pesticides away from their intended targets, polluting the environment and endangering public health. Increased temperatures are anticipated to result in more pesticide volatilization, meaning more pesticides will end up in air rather than on their application targets.

E.-S. A. El-Sheikh (✉)
Department of Plant Protection, Faculty of Agriculture, Zagazig University, Zagazig, Egypt
e-mail: eaelsheikh@agri.zu.edu.eg

A.-T. H. Mossa
Pesticide Chemistry Department, Chemical Industries Research Institute, National Research Centre, Giza, Egypt

M. A. Abbassy
Department of Plant Protection, Damanhour University, Damanhur, Egypt

A. Saad-Hussein et al. (eds.), *Climate Change Impacts on Toxins and Health Effects*,
https://doi.org/10.1007/978-981-96-3416-3_7

Keywords

Climate change · Pesticide toxicity · Risks · Mitigation · Adaptation

Abbreviations

AChE	Acetylcholine esterase
CaE	Carboxylesterase
GST	Glutathione S-transferase
IGR	Insect growth regulators
IPM	Integrated pest management
OCs	Organochlorines
PCBs	Polychlorinated biphenyls
PCDDs	Polychlorinated dibenzodioxins
PM	Particulate matter
PSU	Practical salinity unit
PUFA	Polyunsaturated fatty acids

7.1 Introduction

From 2006 to 2015, the Intergovernmental Panel on Climate Change estimated that global warming reached 87 °C, above the 1850–1900 average (Kogan 2023). According to current estimates, the earth is warming by 0.2 °C every ten years due to human activities (Fang et al. 2011; Hoegh-Guldberg et al. 2019; Hoegh Guldberg et al. 2018). However, climate change describes long-term variations in wind, precipitation, and temperature (Mote et al. 2013). Increased extreme weather occurrences, like droughts and floods, correlate with these changes. A key component of ecosystem services and functions is soil variety, which comprises species ranging in size from microscopic to giant animals and plants (Bardgett 2005; Coleman et al. 2024). However, the increase in soil pollution and decreased soil biodiversity are caused by human activities like mining and agriculture (Cachada et al. 2018). Global pesticide use in agriculture increased from less than 3 million tons in 2010 to 3.5 million tons in 2021 (Zhang 2018). This rise in amounts can be attributed to the protection of plants from pests and increased yields, which may be partly due to climate change that causes an increase in pesticide usage (Delcour et al. 2015). The degradation of pesticides results from climate change, which significantly impacts distribution, availability, and toxicity. Temperature fluctuations and rainfall changes affect the chemical structure of pesticides, leading to increased toxicity or decreased efficacy (Matzrafi 2019; Tudi et al. 2021). For example, rising temperatures cause pesticides to break down more quickly, reducing their accumulation in the environment. Nevertheless, it can generate other toxic substances that negatively affect non-target species. Therefore, it is imperative to gain more knowledge about the impact of climate change on the toxification/detoxification processes and pesticides on ecosystems, as these are complex interactions with climate variables. Thus, the effects of climate change on pesticide toxicity are highlighted in this chapter.

7.2 Pesticides and Climate Change

While pesticides are crucial to agricultural production, they also significantly threaten the environment and human health. Pesticide residues in ecosystems such as soil, water, and food have been extensively studied to assess their potential health and environmental risks (Abbassy et al. 2021; El-Sheikh et al. 2022). A study by Huang and Li (2024) evaluated the extent of atmospheric pollution with pesticides in numerous countries and analyzed the associated health risks. The findings of this study revealed stark disparities in the levels of air pollution with pesticides, with Asian countries registering the highest levels compared to other countries like America and Europe. However, it was also noted that European countries were among the highest users of pesticides. This underscores the fact that excessive exposure to pesticides in the atmosphere can lead to chronic health risks, including neurological and psychological disorders and even cancer (Kaur et al. 2024). Choudhury and Saha (2020) found that the declining efficacy of pesticides and increased pest pressures associated with a changing climate will likely increase synthetic pesticide use in conventional agriculture.

The quantity and quality of agricultural production depend on measures to protect crops from pests and diseases. Applying various types of pesticides (insecticides, fungicides, and herbicides) is essential in agricultural production to increase crop productivity, maintain food security, and raise the standard of living (Tudi et al. 2021). It is expected that the use of pesticides will be affected by climate changes due to the expectation that pests will increase with rising temperatures and changes in rainfall patterns, which will require an increase in the application of pesticides. Climate change is also expected to increase the volatility of pesticides and their decomposition, increasing the doses and frequency of pesticide application. The increase in pesticide application will negatively affect the consumer's exposure to large quantities of their residues on crops and, thus, the health of the consumer at the end of the food chain (Delcour et al. 2015; El-Sheikh et al. 2023).

Climate change has emerged as a formidable threat to global food security, primarily by altering the distribution of pests and their control. The expansion of the geographic range of pests due to climate variability has raised concerns about the efficacy of current pest control strategies. Research by Ma et al. (2021) indicates that over the past 50 years, climate change has increased the overwintering range of the diamondback moth by approximately 2.4 million km^2 worldwide. This expansion underscores the direct impact of climate change on pest behavior and distribution. Furthermore, studies have shown that warmer temperatures resulting from climate change can hasten the dissipation of pesticides, thereby affecting their efficacy in pest management (Zinyemba et al. 2021). This interplay between climate change and pesticide resistance underscores the importance of holistic approaches that consider both factors when devising sustainable pest management strategies

(Muhie 2022). While pesticide usage has surged in regions experiencing expanded pest ranges, there is a lack of understanding regarding how these shifts will influence pesticide resistance development under future climatic conditions (Ma et al. 2021). Therefore, it is crucial to explore how changes in pest distribution induced by climate change will affect the efficacy of pesticides in controlling these evolving pest populations.

7.3 Impact of Climate Change on Pesticide Use and Ecological Risk

7.3.1 Effect of Climate Change on Pesticide Use

The United Nations Food and Agriculture Organization (FAO) has compiled extensive data on the use of pesticides in agriculture around the globe. Based on their records, there has been a noticeable increase in the use of pesticides over time (FAOSTAT 2024). In total, 2,989,646.99 tons of active ingredients, or less than 3 million tons, were used in 2010. This number, however, rose to over 3.5 million tons by 2021, with 3,531,959.08 tons of active ingredients consumed (Table 7.1). When agricultural practices in 2010 and 2021 are compared, notable trends in pesticide use are shown. In addition to a 14.1% increase in the consumption rate per hectare, there was an overall increase of 18.1% in pesticide use in agriculture. By the end of 2021, the average amount of pesticides used per person had increased by 4.65%. These shifts might result from growing problems of pests and plant diseases, which could have been exacerbated by how climate change affects insect and natural predator populations. However, compared to 2010 data, the economic value of pesticides used in agricultural production dropped dramatically by 5.49%. The information up to this point covers a variety of pesticide classes that are frequently used on crops and seeds. These categories include herbicides, fungicides, insect growth regulators (IGRs), rodenticides, and other substances.

Climate change is one of the main factors contributing to the growing use of synthetic pesticides in agriculture today. It is greatly influencing farming systems, especially when it comes to pest management (Skendžić et al. 2021; Subedi et al. 2023). Pests adapt to changes in global temperature by moving to new areas where they were previously unknown. Climate changes significantly impact the dynamics of insects. These changes have the potential to directly impact the life-table characteristics of insects, including growth rates, fecundity, and mortality (Kocmánková et al. 2010; Nyamukondiwa et al. 2022; Prakash et al. 2014; Robinet and Roques 2010; Skendžić et al. 2021; Subedi et al. 2023). Simultaneously, they may exert indirect influence by altering host interactions, competitive scenarios, and predation levels. Understanding these impacts is crucial for predicting insect population trends and implementing effective management strategies. This shift is not only changing pest behavior but also exacerbating the spread of disease and forcing farmers to increasingly rely on synthetic pesticides (Delcour et al. 2015).

Table 7.1 Estimated value of total pesticides used as active ingredients and their global distribution during 2010–2021

	Pesticides active ingredients							
	Agricultural use		Use per area of cropland		Use per capita		Use per value of agricultural production	
Year	Value (million tons)	% of change	Value (kg/hectare)	% of change	Value (kg/cap)	% of change	value (g/Int$)	% of change
2010	2.990	0.00	1.98	0.00	0.43	0.00	0.91	0.00
2011	3.016	0.88	1.98	0.00	0.43	0.00	0.88	−3.30
2012	3.170	6.02	2.08	5.05	0.45	4.65	0.92	1.10
2013	3.210	7.38	2.10	6.06	0.45	4.65	0.89	−2.20
2014	3.220	7.71	2.1	6.06	0.45	4.65	0.88	−3.30
2015	3.255	8.88	2.11	6.57	0.45	4.65	0.87	−4.40
2016	3.379	13.03	2.19	10.61	0.46	6.98	0.90	−1.10
2017	3.317	10.95	2.14	8.08	0.44	2.33	0.86	−5.49
2018	3.292	10.11	2.12	7.07	0.44	2.33	0.84	−7.69
2019	3.382	13.14	2.18	10.10	0.44	2.33	0.86	−5.49
2020	3.399	13.69	2.18	10.10	0.44	2.33	0.85	−6.59
2021	3.532	18.14	2.26	14.14	0.45	4.65	0.86	−5.49

Source: Food and Agriculture Organization of the United Nations (http://www.fao.org/faostat/en/#data, accessed 12 March 2024). % of change = [(value at selected year – value at year 2010)/value at year 2010] × 100

To formulate their commercial pesticide products, the pesticides consist of active ingredients and additives (Cox and Surgan 2006; Hassaan and El Nemr 2020; US EPA 2021). The active ingredients in pesticides are categorized and described based on the type of pest they control or how they work (Yadav et al. 2015). Although insecticides are the only products commonly referred to as "pesticides", the term encompasses all active ingredients used to control pests. Consequently, insecticides, molluscicides, herbicides, acaricides, nematicides, rodenticides, and fungicides are the different categories of pesticides, depending on which pests they control (US EPA 2021). In addition to algaecides, disinfectants, plant defoliants, attractants, and swimming pool treatment products (SPT), other pesticides are less well known. However, they are still critically important for protecting against disease or controlling major pests. Based on the data shown in Table 7.2, there was a notable increase in the usage percentages of various types of pesticides in 2021 compared to the data recorded in 2010. Specifically, the usage percentages of insecticides, herbicides, fungicides and bactericides, and rodenticides increased by 7.21%, 21.26%, 26.34%, and 14.84%, respectively. The most significant increases were observed in herbicides, fungicides and bactericides, and insecticides. In overall usage, herbicides were the most prevalent pesticides in 2021, comprising 48.99% of the total volume. Fungicides and bactericides were the second most used at 23.1%, followed by insecticides at 21.43% and rodenticides at only 0.5%. Such dependence poses challenges for sustainable agriculture and impacts ecosystems and human health. FAO

Table 7.2 The distribution of the most common types of pesticides used in agriculture between 2010 and 2021

	Type of pesticide use (million tons)							
	Insecticides		Herbicides		Fungicides and bactericides		Rodenticides	
Year	Value	% of change	Value	% of change	Value	% of change	Value	% of change
2010	0.706	0.00	1.427	0.00	0.646	0.00	0.0155	0.00
2011	0.657	−6.91	1.498	5.02	0.658	1.96	0.0120	−22.58
2012	0.676	−4.30	1.605	12.49	0.667	3.29	0.0132	−14.84
2013	0.738	4.59	1.576	10.46	0.668	3.39	0.0152	−1.94
2014	0.727	3.02	1.557	9.11	0.716	10.96	0.0136	−12.26
2015	0.736	4.19	1.575	10.40	0.706	9.41	0.0158	1.94
2016	0.760	7.67	1.641	15.03	0.737	14.18	0.0170	9.68
2017	0.721	2.04	1.629	14.19	0.734	13.64	0.0153	−1.29
2018	0.701	−0.74	1.604	12.40	0.760	17.67	0.0150	−3.23
2019	0.737	4.32	1.655	16.01	0.762	18.08	0.0168	8.39
2020	0.694	−1.68	1.727	21.03	0.766	18.59	0.0153	−1.29
2021	0.757	7.21	1.730	21.26	0.816	26.34	0.0178	14.84
% of total pesticides	21.43		48.99		23.10		0.50	

Source: Food and Agriculture Organization of the United Nations (http://www.fao.org/faostat/en/#data, accessed 12 March 2024). % of change = [(value at selected year – value at year 2010)/value at year 2010] × 100. % of total pesticides = (type pesticide value at 2021/total value of pesticides at 2021) × 100

data is critical to mapping these trends and stimulating discussions that lead to environmentally friendly pest control methods (FAO 2023; US EPA 2021).

Evidence shows that changes in phenology and regional distribution occur in various ecosystems due to climate change (Greco et al. 2011; Scherm 2004; Visser and Both 2005). It is anticipated that altered temporal synchrony of pest, host, and bio-control agents would occur, resulting in associated yield losses, as a result of the likely reaction of individual species and ecosystems to climate change (Fontaine et al. 2009; Müller et al. 2010; Perarnaud et al. 2005). Studies reveal that pest infestations frequently follow climate changes, indicating that these changes are species- and region-specific (Gullino et al. 2022; Noyes et al. 2009; Perarnaud et al. 2005). These findings are consistent with the recognition that temperature plays a crucial role in influencing various ecological factors that affect insect life (Rosenzweig et al. 2001). It is interesting to note that there is evidence that warmer temperatures can reduce the impact of fungal pathogens on insects, which may change the method by which the insects interact with their environment and alter their survival strategies (de Lorenzo et al. 2009; Rosenzweig et al. 2001).

7.3.2 Effect of Climate Change on Ecological Risk

The interplay of climate change with various environmental stressors is poised to alter the dynamics of ecosystem operations, potentially leading to significant shifts in the provision of vital ecosystem services (Malhi et al. 2020). These changes are not solely a consequence of shifting temperatures but are also intricately linked to alterations in rain patterns, atmospheric carbon dioxide levels, hydrological dynamics, oceanic composition, and an escalation in the frequency and intensity of extreme weather events (Malhi et al. 2020; Pandit and Sharma 2023; Shivanna 2022; USEPA 2017). The resilience and adaptability of ecosystems to these changes are highly variable, influenced by a web of interactions among living organisms, environmental disturbances, and additional stress factors. This complexity necessitates a nuanced understanding of ecological processes to effectively predict and manage climate change's implications (Mantyka-Pringle et al. 2015; Weiskopf et al. 2020; Zarrineh et al. 2020). It is imperative that future evaluations comprehensively integrate the multifaceted risks posed by altered ecosystem functions to ensure that policy frameworks can effectively mitigate adverse outcomes for both natural habitats and human communities (Hurlbert et al. 2019; Kabisch et al. 2016).

Ecological risk is anchored in two fundamental components: hazard and exposure. A hazard represents the intrinsic potential for the harm a stressor poses, indicating the probability of adverse effects on organisms or ecosystems (Bartell 2008; USEPA 2011). Conversely, exposure denotes the degree to which an organism or ecosystem is subject to this stressor. In public health, for example, the intricate relationship between climate change and the prevalence of certain diseases is a critical focus within ecological studies (Hileman 2003; Neira et al. 2023). In addition, the interaction of ecosystems and climate significantly impacts the development and transmission of severe illnesses like malaria, West Nile virus, and Lyme disease (Caminade et al. 2019; Mojahed et al. 2022).

Global temperature changes can alter habitats, affecting biodiversity and species distribution (Weiskopf et al. 2020). It may enable invasive species to thrive, outcompeting native flora and fauna, thus increasing ecological risk (Koh et al. 2013; Sorte 2016). Climate change intensifies natural disasters like wildfires and hurricanes, which can devastate ecosystems and reduce their ability to recover (Shivanna 2022). Likewise, changes in precipitation patterns and water availability pose severe risks to aquatic ecosystems, potentially leading to reduced water quality and scarcity of resources for species dependent on these habitats (Pletterbauer et al. 2018; USEPA 2023).

7.4 Climate Change and Alter in Pesticide Toxicity to Mammals

7.4.1 Effect of Temperature on the Toxicity of Pesticides

For environmental scientists and policymakers, the interaction of pesticide toxicity and climate change poses a challenging problem. The behavior of chemicals, especially pesticides, can change significantly with rising global temperatures. Warmer climates may cause pesticides to be more volatile, which could result in higher environmental concentrations and possibly increase exposure for mammals, including humans (Levy and Roelofs 2019). Furthermore, variations in precipitation patterns may impact these chemicals' persistence and distribution within ecosystems (Nadal et al. 2015).

7.4.2 Effect of Climate Change on the Metabolic Systems of Mammals

The relationship between pesticide exposure, climate change, and mammalian metabolism is intricate. Climate change affecting the metabolic systems of mammals can have diverse effects on all living beings worldwide, including animals and plants (Kaka et al. 2021; Koleva and Schneider 2010). For example, temperature fluctuations in mammals can alter the physiology of thermoregulatory systems, leading to fluctuations in metabolic rates (Mota-Rojas et al. 2021; Norin and Metcalfe 2019; Seebacher 2009). These physiological adaptations are crucial to maintaining homeostasis and ensuring survival in a dynamic environment. However, the widespread use of pesticides in agriculture and industry presents potential metabolic problems for mammals (Abbas et al. 2014; Abbott 1925; Becker et al. 2003). These chemical compounds can inhibit immunological reactions, interfere with lipid and carbohydrate metabolism, and even affect endocrine processes. As we grapple with the consequences of climate change and pesticide exposure, it becomes increasingly important to understand their complex effects on mammalian metabolism (De Coster and Van Larebeke 2012; Kaka et al. 2021; Norin and Metcalfe 2019). The metabolic responses of mammals to exposure to pesticides and climate change are now being investigated. Different studies aim to develop prediction models to forecast how mammals respond to various environmental stresses. These models should be considered when making policy decisions and managing systems that aim to reduce negative impacts on biodiversity conservation and wildlife.

7.4.3 Effect of Climate Changes on Detoxification Processes

Climate change significantly impacts the detoxification processes of pesticides, affecting their distribution, bioavailability, and toxicity (Noyes et al. 2009; Rajmohan et al. 2020). Increasing temperatures and changes in rain patterns can alter the chemical structure of pesticides, potentially increasing their toxicity or reducing their effectiveness (Kaka et al. 2021). For instance, increased temperatures can accelerate the degradation of pesticides, leading to a decrease in their accumulation in the environment (Arbeli and Fuentes 2007; Noyes et al. 2009). For example, heat stress can impede the liver's ability to function, where most detoxification processes occur, causing the body to accumulate harmful substances. Furthermore, the dynamics of pesticide exposure and effects may change due to changes in biodiversity and mammal populations brought on by climate change-related ecosystem shifts (Gatto et al. 2016). In light of these variables, reviewing the legal frameworks controlling pesticide use and considering climate change effects on their toxicity is imperative. This could entail creating new models to forecast how pesticides behave in climate change scenarios and adjusting safety thresholds to consider mammals' heightened susceptibility to pesticide exposure.

7.5 Impact of Climate Change on Pesticide Toxicity to Ecosystem and Wildlife

The interplay between climate change and pesticide toxicity forms a complex web of environmental challenges (de Lorenzo et al. 2009; Tudi et al. 2021). Climate change can alter the distribution and toxicity of chemical contaminants, including pesticides, which can profoundly affect ecosystems and wildlife (Noyes et al. 2009; Ritter et al. 2002). For instance, increased temperatures and changes in precipitation patterns can affect the bioavailability of pesticides in the soil, potentially increasing their toxicity. This can lead to various detrimental effects on multiple species, from earthworms, which are crucial for soil health, to bees, essential for pollination (Kaka et al. 2021). Pesticides can impact wildlife through direct application, drift, secondary poisoning, runoff into local water bodies, or groundwater contamination (Cope 1971; Kumar et al. 2021; Mishra et al. 2023; Moriarty 1972). These pesticides have been linked to severe health issues in wildlife, such as cancer, endocrine disruption, reproductive effects, neurotoxicity, and developmental changes (Encarnação et al. 2019; Köhler and Triebskorn 2013; Sabarwal et al. 2018). The behavioral alterations in wildlife due to pesticide exposure can also significantly affect their survival and reproduction. For example, birds exposed to certain pesticides may experience difficulties singing, vital for attracting mates or caring for their offspring, leading to population declines (Newton 1998; Ottinger and Dean 2022).

7.6 Climate Change and Pesticide Toxicity to Aquatic Organisms

Environmental pollutants lead to widespread disease effects, as climate changes affect the distribution and toxicity of these chemical pollutants—including pesticides—in the aquatic environment. These environmental impacts represent a potential problem for aquatic species in the short and long term, ultimately negatively affecting human health (Manciocco et al. 2014). The distribution of environmental pollutants in the atmosphere, water, and soil depends mainly on temperature changes (Noyes et al. 2009), which leads to the accumulation of many of these pollutants, including chemical pesticides, in the food chain and their persistence in the environment for many years (European Food Safety Authority 2005; UNEP/AMAP 2011).

Understanding the interactions between climate-induced habitat processes and the effects of pollution (particularly pesticides) is becoming increasingly crucial in anticipating and minimizing threats to aquatic communities (Hooper et al. 2013). Environmental toxicants such as pesticides, polychlorinated biphenyls (PCBs), polychlorinated dibenzodioxins (PCDDs), and heavy metals are widely found in the ambient water and transferred to aquatic vertebrates (Borghesi et al. 2009; Webster et al. 2011). These pollutants are distributed across the aquatic environment and abundantly accumulate in edible wild fish in the Mediterranean and Arctic and temperate marine food webs (Serrano et al. 2008; Sobek et al. 2010). The concentrations of environmental toxicants detected in aquatic organisms depend on several factors, including feeding nature, lifespan, and body size (Alleva et al. 2006; Michelutti et al. 2010), confirming that larger and longer-living fish have higher pesticide and other toxicant concentrations than smaller and shorter-living species. Research conducted by Laetz et al. (2014) found that exposure to malathion-ethoprop mixture leads to loss of locomotor function in salmonids when the temperature changes from 12 to 18 °C. Their study also showed a slight decrease in AChE activity due to salmon being exposed to the pesticide mixture at a temperature of 12 °C, and this reduction in enzyme activity increases to 65% at a temperature of 18 °C. It has been found that significant inhibition in the enzyme level results in salmon being unable to control their position in moving water or avoid predators in the aquatic environment compared to fish with an average level of the enzyme (Labenia et al. 2007). Different agricultural practices contaminate the aquatic environment with various toxic substances. Compared to normal environmental dynamics, climate change has been reported to play a vital role in potentially increasing toxic interactions between living organisms and chemical compounds. Precipitation is expected to increase the rate of discharge of nutrients and pollutants into groundwater in regions that face climate alterations (IPCC 2007). The Mondego estuary basin is found to be contaminated with low- to moderate-levels of heavy metals (Coelho et al. 2006), PCDD/Fs (Nunes et al. 2011), and PCBs (Baptista et al. 2013) due to the Mondego basin being located in areas surrounded by vast agriculture, industrial, and aquaculture (Ferreira dos Santos and Freitas 2012). Also, significant amounts of different groups of pesticides (insecticides, fungicides, and herbicides) are detected in the Mondego estuary basin. Most detected pesticides exceeded the maximum levels established by the European Union (Cruzeiro et al. 2016). The increase of pesticides and their mixtures in fish communities in the Mondego estuary during periods of higher runoff was reported to highlight/elevate the risk of toxicity (Silva and Cerejeira 2015). According to the study of Vieira et al. (2018), exposure to pesticides with high synergetic potential (e.g., dichlorvos, endosulfan, dithiocarbamates, fenthion, vilogen, hexachloro-benzene) induced depletion of enzymatic activity in the brain of several fish species, such as catfish (Hai et al. 1997), Atlantic salmon (Berntssen et al. 2003), rainbow trout and European eels (Peña-Llopis et al. 2003), and grey mullet (Mieiro et al. 2010). After pesticide exposure, the brain showed the highest sensitivity and marked reductions in enzyme activities (glutathione reductase, glutathione S-transferase (GST), glutathione peroxidase) (Song et al. 2006). The high effects on the brain may be associated with the hypothesis that the brain accumulates higher concentrations of pesticides than other organs (i.e., liver and muscle) for some species. Additionally, the brain's high content of polyunsaturated fatty acids (PUFAs), which facilitates the binding of pesticides, may significantly increase the risk of neurotoxicity and make it more susceptible to enzyme inhibition and oxidative damage (Ballesteros et al. 2009), particularly when exposure occurs during high temperatures (Ruszkiewicz et al. 2019).

The application of pesticides and agricultural chemicals has a role in causing toxic effects on the environment and ecosystems, thus harming the environment, mammals, and humans (El-Sheikh and Galal 2015; Hassan et al. 2022; El-Sheikh and Ashour 2022; Abbassy et al. 2019, 2021). Osterauer and Köhler (2008) determined that pesticides affect fish due to climate change and temperature increases. It was found that imidacloprid increases the heart rate in zebrafish when tested at concentrations ranging from 1 to 20 mg/L (Table 7.3). There was a significant increase in the mortality rate of zebrafish when exposed to diazinon, especially with the increase in temperature, because the increase in the solubility of the chemicals is linked to the increase in temperature, thus increasing their toxicity and increasing the rate of their spread and active absorption through the gills and the surface of the body (Heugens et al. 2001). This study also showed that zebrafish hatch earlier than usual when the temperature rises, which strongly impacts embryonic development in the fish (Muller 2005; Radonic et al. 2005). It was also found that one of the harmful effects of climate change on the aquatic environment is that the rise in temperature leads to a decrease in the degree of solubility of oxygen in the water, thus reducing its quantity. In contrast to this, metabolic rates increase, which requires the consumption of high amounts of oxygen. Thus, the heart rate increases due to the imbalance in these biological processes (Mortimer 2003).

Climate change has been identified as a significant factor influencing the susceptibility of fish to pesticide toxicity. Alterations in salinity levels, driven by climate change-induced sea level rise and changing rainfall patterns, have been found to affect the interactions of aquatic pollutants and alter their toxicity profiles. Research by Hutton et al. (2021) indicates that changes in salinity levels can significantly affect fish's physiological responses and enzyme metabolism, potentially altering how they process and respond to pesticides. Furthermore, variations in salinity levels have been linked to changes in the toxic effects of commonly used pesticides on fish species like Menidia beryllia, leading to increased toxicity levels for certain compounds (Hutton et al. 2021).

This interaction between climate change-induced alterations in environmental conditions like salinity and pesticide toxicity highlights the complex relationship

Table 7.3 The effect of different climatic factors on pesticides toxicity for aquatic organisms

Organism	Pesticide	Concentration	Effect	Factor	Reference
Zebrafish	Thiacloprid, diazinon	5, 10, 15 mg L^{-1} (thiacloprid); 0.1 mg L^{-1} (diazinon)	Increase heart rate	Temperature (26 – 33°C)	Osterauer and Köhler (2008)
Zebrafish	diazinon	0.5–3 mg L^{-1}	Decrease hatching rate—high mortality	Temperature (26 – 33 °C)	Osterauer and Köhler (2008)
Salmon	Mixture of ethoprop and malathion	1.1+1.4 μg L^{-1}	Decrease in AChE and CaE	Temperature (18 °C)	Laetz et al. (2014)
Inland silverside (Menidia beryllina)	Bifenthrin, chlorpyrifos, dicloran, myclobutanil, paraquat dichloride hydrate, penconazole, triadimefon	NA	High mortality rate	Salinity (5, 15 PSU)	Hutton et al. (2021)
Zebrafish	Different pesticide groups	NA	Developmental-, genetic-, endocrine-, and reproductive-toxicity	NA	Cui et al. (2023)
Fasting high arctic common eiders	Organochlorines (OCs) (HCB, p,ṕ-DDE, Chlordanes)	NA	Detection of high concentrations of OCs with negative effects on body mass and clutch size	When temperature increases by >1 °C	Bustnes et al. (2024)
Tambaqui (Colossoma macropomum)	Mixture of chlorpyrifos, malathion, carbandazim, and atrazine	0.7, 0.535, 0.214, and 0.025 μg L^{-1}	Liver damage—alter in biochemical responses—Genetic damage	Different climatic factors (Temperature, 29.9 °C; O_2, 6.89 mg L^{-1}; pH, 6.89; Co_2, 19.50 ppm)	de Souza et al. (2024)

between human activities and ecosystem health. Understanding these dynamics is crucial for assessing the risks posed by pesticide exposure to fish populations, particularly those inhabiting estuarine and marine environments undergoing rapid changes due to climate change (Werner et al. 2021).

7.7 Effect of Climate Change on Pesticide Toxicity and Risks to Human Health

While some may argue that climate change-induced shifts in temperature negatively impact pesticide breakdown rates and environmental stability (Op de Beeck et al. 2017), it is essential to recognize that these changes also bring about increased biodegradation activities that can expedite the dissipation of harmful chemicals from agricultural ecosystems (Kattwinkel et al. 2011; Pathak et al. 2022). As highlighted by research on rice paddy soils (Tudi et al. 2023), an elevation in temperature drives enhanced pesticide metabolism rates—an aspect that needs acknowledgment when considering the overall impact of climate change on pesticide persistence. Therefore, rather than solely focusing on the potential drawbacks of rising temperatures, it is crucial to understand how these changes could also positively influence pesticide degradation processes for a more balanced perspective (Bolan et al. 2024). Temperature variation plays a crucial role in influencing the breakdown of pesticides in the environment and their subsequent toxicity to humans. Research has shown that higher temperatures can accelerate the degradation process of pesticides, leading to faster breakdown rates and the formation of more toxic metabolites. For example, studies have demonstrated that bacterial genera like Bacillus, Pseudomonas, and Flavobacterium are more proficient in breaking down pesticide compounds at elevated temperatures (Kafilzadeh et al. 2015).

Moreover, as pesticides degrade more rapidly in warmer conditions, there is a concern regarding releasing by-products from these breakdown processes into the environment. These by-products may exhibit varying levels of toxicity and persistence, posing risks to human health upon exposure. Higher temperatures can also increase human susceptibility to pesticide-related health complications due to enhanced absorption rates through inhalation or skin contact (Thredgold et al. 2019). Understanding how temperature variations impact pesticide breakdown is essential for assessing environmental risks and implementing effective mitigation strategies. By elucidating the relationship between temperature fluctuations and pesticide degradation pathways, we can develop targeted interventions to minimize environmental contamination and mitigate potential health hazards for humans and ecosystems (Ruomeng et al. 2023).

Increased precipitation can have significant implications for the leaching of pesticides into groundwater sources (Lewan et al. 2009). As rainfall intensifies, it can lead to higher levels of soil erosion, facilitating the movement of water carrying various chemicals, including pesticides, into underground water reservoirs (Tudi et al. 2023). This heightened leaching process poses a substantial risk to groundwater quality and human health as these chemicals contaminate drinking water sources.

Therefore, the direct dangers of floods resulting from climate change constitute a source of great concern. In addition, floodwaters transport various pollutants, such as sewage, chemicals, and waste, into the groundwater, polluting waterways and concentrating pollutants in the food chain, leading to harmful effects on biodiversity and human health (Song et al. 2019; Sulaiman et al. 2019; Wang et al. 2023).

Increased precipitation plays a significant role in transporting pesticides to water sources, subsequently elevating human exposure and toxicity levels. Heavy rainfall leads to the runoff of pesticides from agricultural fields into nearby water bodies, contaminating drinking water sources, surface water, and groundwater, potentially exacerbating their toxicity to aquatic ecosystems and human health (Ahirwar 2023). This process poses serious health risks, as these pesticides can persist in the environment, accumulate in food chains, and ultimately affect non-target populations, including humans. The combination of pesticide contamination with increased precipitation heightens concerns about human health implications. The pattern of increases in temperature, precipitation, and flood on pesticide stability and efficacy is highlighted in Fig. 7.1. Exposure to these toxic pesticides through contaminated water and other sources has been linked to various adverse health effects ranging from mild symptoms like nausea to severe conditions such as respiratory distress and even cancer (de-Assis et al. 2021; Ahmad et al. 2024; Kaur et al. 2024). Furthermore, hormonal imbalances, reproductive difficulties, and heightened sensitivity to allergens are among the reported health issues associated with pesticide exposure (Ahirwar 2023). Understanding the impact of increased precipitation on pesticide transport is crucial for implementing effective strategies to minimize environmental contamination and reduce human health risks related to pesticide exposure. Furthermore, climate change-induced stress on crop plants can influence their pesticide interactions. Drought, extreme temperatures, and other environmental stressors can alter the plant's physiology, affecting pesticide uptake, translocation, and metabolism. This, in turn, can lead to changes in the concentration and distribution of these chemicals within the plant, potentially increasing their toxicity to consumers (Alengebawy et al. 2021).

Consumers who depend on seafood in their diet are vulnerable to poisoning due to the impact of climate change on the accumulation of environmental pollutants in marine organisms. Examples include the intensification of mercury methylation and its increased absorption in fish and marine organisms due to rising ocean temperatures (Booth and Zeller 2005; Noyes et al. 2009; Schiedek et al. 2007). The use of pesticides, herbicides, artificial fertilizers, and other chemicals is also increasing due to the impact of climate change on the increased growth of weeds harmful to crops and the increase in the number and duration of survival of plant and animal pests and diseases. High temperatures lead to agricultural soil erosion and increase the chances of these environmental pollutants, especially pesticides, reaching groundwater (Boxall et al. 2009; Myers et al. 2017) and magnifying their toxic effects on humans. Exposure to pesticides with higher temperatures increases the chance of magnifying their toxicity due to many factors (Fig. 7.2).

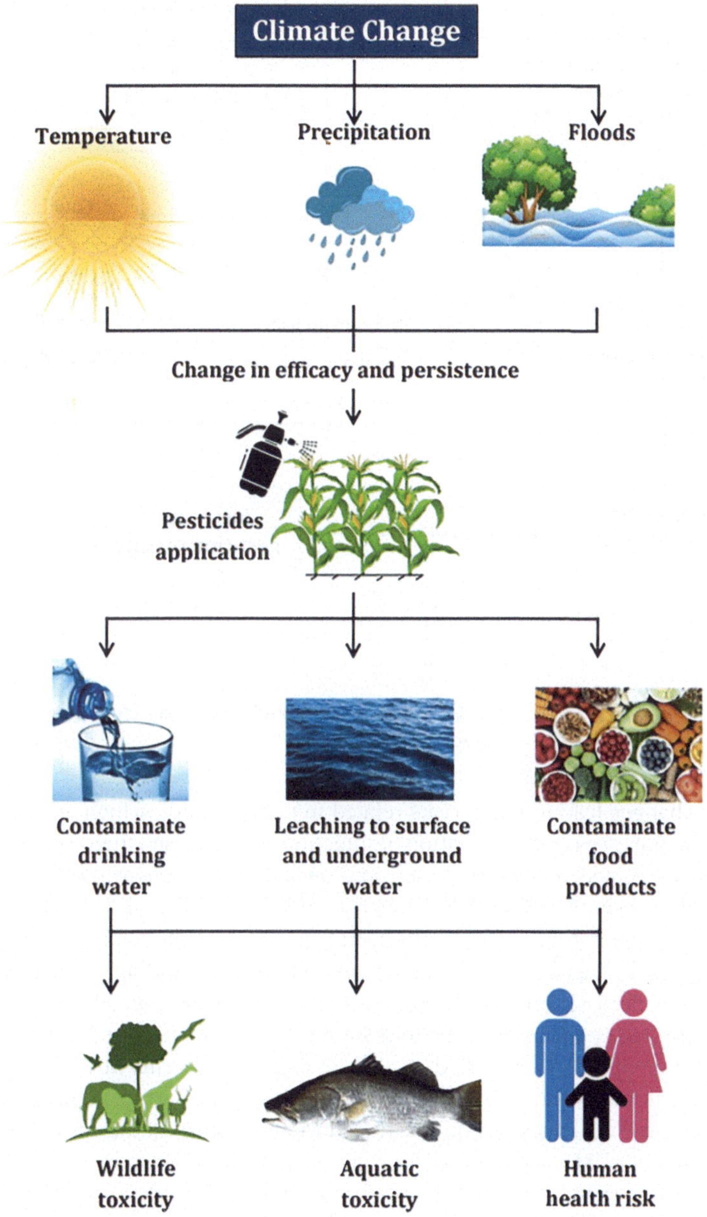

Fig. 7.1 The effect of climate change on pesticide toxicity to humans and the environment

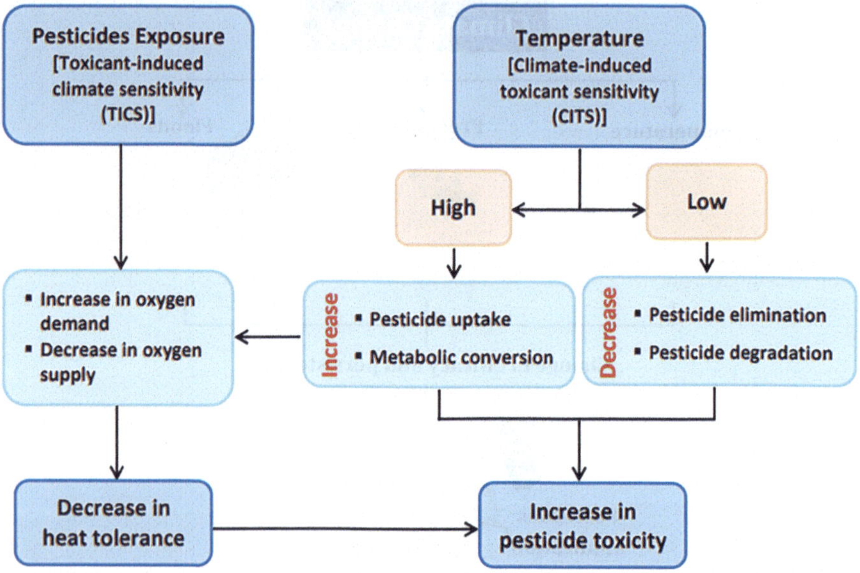

Fig. 7.2 The effect of temperature and pesticide exposure on the increase in pesticide toxicity

7.8 Climate Change Adaptation Strategies to Reduce the Toxic Effects of Pesticides

Climate changes lead to many risks related to public health, which are represented by shifts in the types and distribution of infectious diseases and their causes, changes in the quantity and quality of water, and an increase in environmental pollution rates, which severely affect the production and availability of food (Ruszkiewicz et al. 2019). Adaptation and mitigation strategies for climate impacts may also lead to brain-related risks due to environmental pollutants (including pesticides), which have brain-neurotoxic effects (Burke et al. 2017; Cassereau et al. 2017). The increase in environmental pollutants is due to natural or human factors, and their accumulation in the environment is linked to climate changes in terms of their distribution and concentration in the air. Increased temperature, humidity, changing weather conditions, and decreased rainfall will lead to an increase in the formation of particulate matter (PM) particles and an increase in the release of fine particles of 2.5 μM or less (PM2.5), which is dangerous as they can adversely affect a broad range of the population, causing human impacts due to traveling hundreds of kilometers (Xu et al. 2023). PM contains proinflammatory and highly oxidative components (e.g., aldehydes and nitrogen oxides as proinflammatory, quinones, and oxygenated polycyclic aromatic hydrocarbons as oxidative substances) with substantial toxicity (Xu et al. 2020).

Heavy metals (such as mercury, zinc, and lead) are also increased in the environment due to the increase in forest fire rates due to a temperature rise, which is reflected in increased changes in ozone (O_3) levels (Doherty et al. 2017; Kinney 2018; Cascio 2018; Liu et al. 2016), which in turn increases the temperature of the atmosphere. Reducing global warming to reduce health risks resulting from climate change is one of the goals of the Paris Agreement 2015 so that temperature increases are controlled to no more than two degrees Celsius by the year 2100. Mitigation strategies must be adopted to avoid health risks. Adaptation to climate change, as the primary strategy for mitigating it, is to control emissions of gases harmful to the environment, especially those resulting from greenhouses worldwide, by reducing energy and fossil fuel consumption in exchange for increasing the use of alternative energy (Tong and Ebi 2019; Woodward 2019). In addition, capturing and storing carbon, increasing green areas, and adopting measures to preserve ecosystems and biological diversity will be required. These measures may contribute to reducing climate impacts on the environment (Fig. 7.3). For example, reducing the use of fossil fuels may contribute to reducing emissions of neurotoxic substances such as arsenic, mercury, and other chemicals into the environment. In contrast, adaptation strategies may lead to increased environmental pollution and human harm, such as increased use of pesticides to control vector disease (VBZD) and plant pests and diseases (Gustin and Ladwig 2004; Hu and Cheng 2016; Ito et al. 2006; Noyes et al. 2009). One of the things necessary to mitigate the impact of climate change on pesticide poisoning is the need to establish a system to monitor pesticides in the environment and regulate their use by standards to preserve the environment by strengthening international cooperation (Reichenberger et al. 2007; Goh et al. 2011; Leskovac and Petrović 2023).

When considering the impact of climate change adaptation strategies on the toxicity of pesticides, it is crucial to understand how environmental changes can influence the dynamics between chemical contaminants and soil organisms. Research by Kaka et al. (2021) highlights that climate change variables, such as temperature and soil moisture, significantly affect the bioavailability of pesticides in the soil. Thereby, impacting the growth and reproduction of soil organisms such as earthworm species. Moreover, as highlighted in the study, indirect effects of climate change on pesticides include shifts in cultivation practices toward higher latitudes and extended cultivation periods, potentially enhancing pesticide volatility and degradation processes. These alterations may increase pesticide application levels to maintain effectiveness against pests. Several strategies can be implemented to mitigate these toxic effects in the face of climate change. Firstly, promoting integrated pest management (IPM) practices can help reduce pesticide reliance. Ali et al. (2022) reported that the application of a botanical extract of neem on cotton plants as an alternative to chemical pesticides was effective in reducing the number of sucking pests, reducing the number of applications of chemical pesticides, and reducing the total number of pesticides used in experimental plots compared to control ones. These experiments were conducted to find a viable approach that would ultimately play a role in mitigating climate change by reducing the number of applied chemical pesticides and their residues in water, soil, and the environment and consequently maintaining the health of farmers (Ali et al. 2022). Accordingly, IPM involves utilizing biological control methods, crop rotation, and resistant plant

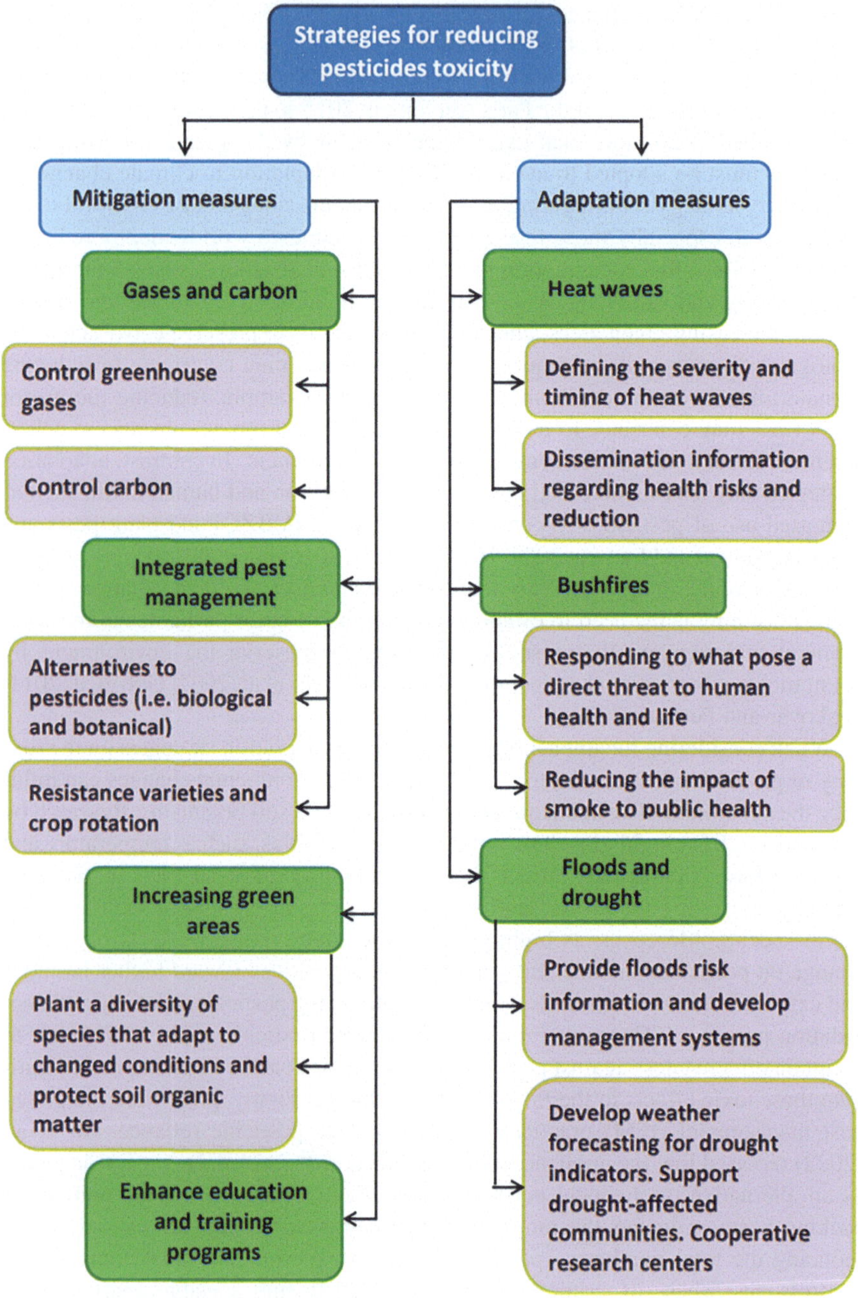

Fig. 7.3 Adaptation and mitigation strategies for reducing the effect of climate change

varieties to manage pests sustainably without solely relying on chemical interventions.

This strategy aligns with the findings that shifts in cultivation practices due to climate change may lead to increased pesticide use. Furthermore, enhancing education and training programs for farmers on judicious pesticide application can minimize overuse and reduce environmental contamination. By raising awareness about alternative pest control methods and emphasizing proper pesticide handling techniques, farmers can minimize adverse impacts on soil organisms while maintaining effective pest management.

Resource conservation strategies, cropping systems technologies, and social, economic, and political systems are essential in adapting to climate change (Ventakeswarlu and Shanker 2009). For example, small farmers may be unable to confront climate change due to a lack of awareness of the side effects on the agricultural sector resulting from rising temperatures and changes in rainfall patterns, making them more vulnerable to losses in crop production (Baul and McDonald 2015). This also applies to farmers in various continents, especially African countries, due to the shortage of financial resources and management strategies (Biber-Freudenberger et al. 2016). In contrast, the agroforestry sector can help small farmers in Africa adapt to climate change by mitigating the accumulation of greenhouse gases (GHG) in the atmosphere (Malhi et al. 2021). The crop production losses become minimal when farmers adjust planting dates according to the climate, adapt sequential cropping systems, and use appropriate varieties, effectively reducing climate change's impact (Verchot et al. 2007). In addition, simple approaches such as alternate drying and mid-rice drainage, improving livestock diet, and more efficient use of nitrogen and soil carbon all reduce the impact of climate change (Aggarwal 2008).

7.9 Conclusion

The environment and human health are matters that must be given priority in preservation. Many human practices (i.e., industrial and agricultural) lead to climate changes that harm the environment and health. Continuing these irrational practices will exacerbate the problem of climate change, which in turn will lead to health impacts that may be uncontrollable. However, potential solutions can be identified and negative impacts mitigated by understanding the interaction between climate change and its causative factors, including pesticide behaviour. Pesticides are one of the essential elements in agricultural production due to their vital role in combating many plant pests and diseases to ensure an increase in the quantity and quality of food products to meet the global population increase. Pesticides are affected by climate change in terms of changes in their efficiency and their toxic effects toward environment and human health. As the world is concerned with the long-term consequences of climate change, the interplay between ecological conditions and pesticide behavior must be at the forefront of our collective efforts to ensure food security, protect public health, and maintain the delicate balance between natural resources and ecosystems.

Temperature changes are not just affecting the behavior of pesticides but also posing a significant and immediate threat to human health. As temperatures rise, some pesticides become more volatile, increasing their concentrations in the air. This is a particular concern for agricultural workers and neighboring communities who may inhale these chemicals. Organophosphate and carbamate pesticides, in particular, are becoming more volatile at higher temperatures. If global temperatures continue to rise, these pesticides could be released more easily into the atmosphere, increasing the risk of accidental poisoning and chronic health effects for those who come into contact with them.

Changes in rainfall patterns are another critical factor that can lead to water runoff and the leakage of pesticides into water bodies. This can lead to high concentrations of pesticides in drinking water, surface water, and groundwater, potentially exposing humans and aquatic life to harmful levels of these pesticide chemicals. Climate change also affects agricultural pests' life cycles, behaviors, and spread. As temperatures rise and rainfall regimes change, many pest species are expanding their geographic ranges, adapting to new conditions, and developing resistance to commonly used insecticides. This could lead to increased use of pesticides, increasing their spread and toxic environmental and health effects. In addition, climate-induced stress on crop plants can alter their interactions with pesticides. Drought, heat waves, and other extreme events can change how plants absorb, distribute, and metabolize these chemicals, potentially leading to higher concentrations of toxic residues in harvested produce. To address these challenges, researchers are exploring different strategies, such as developing more climate-resilient crop varieties, implementing an IPM approach, and advocating for policy changes that promote sustainable agricultural practices and the responsible use of pesticides in light of changing environmental conditions. Ultimately, the interaction between climate change and pesticide toxicity is a complex and multifaceted issue that requires a comprehensive and multidisciplinary approach. By understanding these dynamics and taking decisive action, human health and the environment can be protected, as well as ensuring the sustainability of agricultural systems in the face of rapid climate change.

References

Abbas N, Khan HAA, Shad SA (2014) Resistance of the house fly Musca domestica (Diptera: Muscidae) to lambda-cyhalothrin: mode of inheritance, realized heritability, and cross-resistance to other insecticides. Ecotoxicology 23(5):791–801. https://doi.org/10.1007/s10646-014-1217-7

Abbassy MA, Abdel-Halim KY, El-Meseiry MA (2019) Bioaccumulation of pesticides in the aquatic system of Edko Lake, Egypt: an approach for risk associated with fish. Acad J Life Sci 5(9):61–69

Abbassy MA, Khalifa M, Nassar AK et al (2021) Analysis of organochlorine pesticides (OCPS) residues in fish from Edko Lake (North of Egypt) using eco-friendly methods and their health implications for humans. Toxicol Res 37(4):495–503

Abbott WS (1925) A method of computing the effectiveness of an insecticide. J Econ Entomol 18(2):265–267. https://doi.org/10.1093/jee/18.2.265a

Aggarwal PK (2008) Global climate change and Indian agriculture: impacts, adaptation and mitigation. Indian J Agric Sci 78:911–919

Ahirwar. (2023) Advanced pesticide nanoformulations and understanding their breakdown by Bacteria. Indian J Microbiol Res 10:193–201

Ahmad MF, Ahmad FA, Alsayegh AA et al (2024) Pesticides impact human health and the environment with their mechanisms of action and possible countermeasures. Heliyon 10(7):e29128

Alengebawy A, Abdelkhalek ST, Qureshi SR et al (2021) Heavy metals and pesticides toxicity in agricultural soil and plants: ecological risks and human health implications. Toxics 9:42

Ali MY, Saleem S, Nasir M et al (2022) Use of botanical spray to delay the application of the first pesticide against sucking pests of cotton which ultimately mitigate climate change. J King Saud Univ Sci 34:101903

Alleva E, Francia N, Pandolfi M et al (2006) Organochlorine and heavy metal contaminants in wild mammals and birds of Urbino-Pesaro province, Italy: an analytic overview for potential bioindicators. Arch Environ Contam Toxicol 51:123–134

Arbeli Z, Fuentes CL (2007) Accelerated biodegradation of pesticides: an overview of the phenomenon, its basis and possible solutions; and a discussion on the tropical dimension. Crop Prot 26(12):1733–1746

Ballesteros ML, Wunderlin DA, Bistoni MA (2009) Oxidative stress responses in different organs of Jenynsia multidentata exposed to endosulfan. Ecotoxicol Environ Saf 72:199–205

Baptista J, Pato P, Tavares S et al (2013) PCB bioaccumulation in three mullet species—a comparison study. Ecotoxicol Environ Saf 94:147–152

Bardgett R (2005) The biology of soil: a community and ecosystem approach. Oxford University Press

Bartell SM (2008) Ecological risk assessment. In: Encyclopedia of ecology, five-volume set, vol 1–5. Academic Press, pp 1097–1101. https://doi.org/10.1016/B978-008045405-4.00387-6

Baul TK, McDonald M (2015) Integration of indigenous knowledge in addressing climate change. Indian J Tradit Knowl 1:20–27

Becker N, Petrić D, Boase C et al (2003) Mosquitoes and their control. Springer US. https://doi.org/10.1007/978-1-4757-5897-9

Berntssen MHG, Aatland A, Handy RD (2003) Chronic dietary mercury exposure causes oxidative stress, brain lesions, and altered behavior in Atlantic salmon (Salmo salar). Aquat Toxicol 65:55–72

Biber-Freudenberger L, Ziemacki J, Tonnang HEZ et al (2016) Future risks of pest species under changing climatic conditions. PLoS One 11:e0153237

Bolan S, Padhye LP, Jasemizad T et al (2024) Impacts of climate change on the fate of contaminants through extreme weather events. Sci Total Environ 909:168388

Booth S, Zeller D (2005) Mercury, food webs, and marine mammals: implications of diet and climate change for human health. Environ Health Perspect 113:521–526

Borghesi N, Corsolini S, Leonards P et al (2009) Polybrominated diphenyl ether contamination levels in fish from the antarctic and the mediterranean sea. Chemosphere 77:693–698

Boxall AB, Hardy A, Beulke S et al (2009) Impacts of climate change on indirect human exposure to pathogens and chemicals from agriculture. Environ Health Perspect 117:508–514

Burke RD, Todd SW, Lumsden E et al (2017) Developmental neurotoxicity of the organophosphorus insecticide chlorpyrifos: from clinical findings to preclinical models and potential mechanisms. J Neurochem 142:162–177

Bustnes JO, Bårdsen BJ, Moe B et al (2024) Impacts of a warming climate on concentrations of organochlorines in a fasting high arctic marine bird: direct vs. indirect effects? Sci Total Environ 908:168096

Cachada A, Rocha-Santos T, Duarte AC (2018) Soil and pollution: an introduction to the main issues. In: Soil pollution. Elsevier, pp 1–28
Caminade C, McIntyre KM, Jones AE (2019) Impact of recent and future climate change on vector-borne diseases. Ann N Y Acad Sci 1436(1):157–173. https://doi.org/10.1111/nyas.13950
Cascio WE (2018) Wildland fire smoke and human health. Sci Total Environ 624:586–595
Cassereau J, Ferré M, Chevrollier A et al (2017) Neurotoxicity of insecticides. Curr Med Chem 24:2988–3001
Choudhury PP, Saha S (2020) Dynamics of pesticides under changing climatic scenario. Environ Monit Assess 192(1):1–3
Coelho JP, Pimenta J, Gomes R et al (2006) Can Nassarius reticulatus be used as a bioindicator for Hg contamination? Results from a longitudinal study of the Portuguese coastline. Mar Pollut Bull 52:674–680
Coleman DC, Geisen S, Wall DH (2024) Soil fauna: Occurrence, biodiversity, and roles in ecosystem function. In: Soil microbiology, ecology and biochemistry. Elsevier, pp 131–159
Cope OB (1971) Interactions between pesticides and wildlife. Annu Rev Entomol 16:325–364. https://doi.org/10.1146/annurev.en.16.010171.001545
Cox C, Surgan M (2006) Unidentified inert ingredients in pesticides: implications for human and environmental health. Environ Health Perspect 114(12):1803–1806. https://doi.org/10.1289/ehp.9374
Cruzeiro C, Rocha E, Pardal MA et al (2016) Environmental assessment of pesticides in the Mondego River Estuary (Portugal). Mar Pollut Bull 103:240–246
Cui J, Tian S, Gu Y et al (2023) Toxicity effects of pesticides based on zebrafish (Danio rerio) models: advances and perspectives. Chemosphere 340:139825
De Coster S, Van Larebeke N (2012) Endocrine-disrupting chemicals: associated disorders and mechanisms of action. J Environ Public Health 2012:713696
De Lorenzo ME, Wallace SC, Danese LE, Baird TD (2009) Temperature and salinity effects on the toxicity of common pesticides to the grass shrimp, Palaemonetes pugio. J Environ Sci Heal Part B Pestic Food Contam Agric Wastes 44(5):455–460. https://doi.org/10.1080/03601230902935121
de Souza SS, Bruce KHR, da Costa JC et al (2024) Effects of climate change and mixtures of pesticides on the Amazonian fish *Colossoma macropomum*. Sci Total Environ 922:171379
De-Assis MP, Barcella RC, Padilha JC et al (2021) Health problems in agricultural workers occupationally exposed to pesticides. Revista Brasileira de Medicina do Trabalho 18(3):352–363
Delcour I, Spanoghe P, Uyttendaele M (2015) Literature review: impact of climate change on pesticide use. Food Res Int 68:7–15
Doherty RM, Heal MR, O'Connor FM (2017) Climate change impacts on human health over Europe through its effect on air quality. Environ Health 16:118
EFSA (European Food Safety Agency) (2005) Opinion of the scientific panel on contaminants in the food chain on the request from the European parliament related to the safety assessment of wild and farmed fish. EFSA 2005. http://cfpub.epa.gov/ncea/risk/recordisplay.cfm?deid=12486
El-Sheikh EA, Ashour M-BA (2022) Diamide insecticides: efficacy, toxicity and analytical methods for residue monitoring in food samples. Egypt J Chem 65(5):165–177
El-Sheikh EA, Galal AA (2015) Toxic effects of sub-chronic exposure of male albino rats to emamectin benzoate and possible ameliorative role of *Foeniculum vulgare* essential oil. Environ Toxicol Pharmacol 39(3):1177–1188
El-Sheikh E-SA, Ramadan MM, El-Sobki AE et al (2022) Pesticide residues in vegetables and fruits from farmer markets and associated dietary risks. Molecules 27:8072
El-Sheikh E-SA, Li D, Hamed I et al (2023) Residue analysis and risk exposure assessment of multiple pesticides in tomato and strawberry and their products from markets. Foods 12:1936
Encarnação T, Pais AACC, Campos MG, Burrows HD (2019) Endocrine disrupting chemicals: impact on human health, wildlife and the environment. Sci Prog 102(1):3–42
Fang J, Zhu J, Wang S, Yue C, Shen H (2011) Global warming, human-induced carbon emissions, and their uncertainties. Sci China Earth Sci 54:1458–1468
FAO (2023) Pesticides use. In: FAOSTAT. Rome. Published 2023. http://www.fao.org/faostat/en/#data/RP. Accessed 13 Mar 2024

FAOSTAT (2024) Pesticides use. Published 2024. https://www.fao.org/faostat/en/#data/RP. Accessed 13 Mar 2024

Ferreira dos Santos J, Freitas V (2012) Aproveitamento hidroagrícola do Baixo Mondego-actualidade e desafios futuros. In: aph (ed) IV Congresso nacional de rega e drenagem. Direcção-Geral de Agricultura e Desenvolvimento Rural (DGADR), Coimbra, Portugal, p 14

Fontaine JJ, Decker KL, Skagen SK, van Riper C (2009) Spatial and temporal variation in climate change: a bird's eye view. Clim Chang 97(1):305–311. https://doi.org/10.1007/s10584-009-9644-9

Gatto MP, Cabella R, Gherardi M (2016) Climate change: the potential impact on occupational exposure to pesticides. Ann Ist Super Sanita 52(3):374–385

Goh K, Bret B, Potter T et al (2011) Pesticide mitigation strategies for surface water quality, vol 1075. American Chemical Society

Greco L, Pellerin J, Capri E et al (2011) Physiological effects of temperature and a herbicide mixture on the soft-shell clam Mya arenaria (Mollusca, Bivalvia). Environ Toxicol Chem 30(1):132–141. https://doi.org/10.1002/etc.359

Gullino ML, Albajes R, Al-Jboory I et al (2022) Climate change and pathways used by pests as challenges to plant health in agriculture and forestry. Sustainability 14(19):12421. https://doi.org/10.3390/su141912421

Gustin MS, Ladwig K (2004) An assessment of the significance of mercury release from coal fly ash. J Air Waste Manag Assoc 54:320–330

Hai DQ, Varga SI, Matkovics B (1997) Organophosphate effects on antioxidant system of carp (Cyprinus carpio) and catfish (Ictalurus nebulosus). Comp Biochem Physiol C: Pharmacol Toxicol Endocrinol 117:83–88

Hassaan MA, El Nemr A (2020) Pesticides pollution: classifications, human health impact, extraction and treatment techniques. Egypt J Aquat Res 46(3):207–220. https://doi.org/10.1016/J.EJAR.2020.08.007

Hassan AA, Bel Hadj Salah K, Fahmy EM et al (2022) Olive leaf extract Attenuates Chlorpyrifos-induced neuro- and reproductive toxicity in male Albino Rats. Life 12(10):1500

Heugens EHW, Hendriks AJ, Dekker T et al (2001) A review of the effects of multiple stressors on aquatic organisms and analysis of uncertainty factors for use in risk assessment. Crit Rev Toxicol 31:247–284

Hileman B (2003) Climate change. Chem Eng News 81(50):27–37. https://doi.org/10.1021/cen-v081n050.p027

Hoegh Guldberg O, Jacob D, Taylor M et al (2018) Impacts of 1.5 °C global warming on natural and human systems. Published online 2018.

Hoegh-Guldberg O, Jacob D, Taylor M et al (2019) The human imperative of stabilizing global climate change at 1.5 °C. Science (80-) 365(6459):eaaw6974

Hooper MJ, Ankley GT, Cristol DA et al (2013) Interactions between chemical and climate stressors: a role for mechanistic toxicology in assessing climate change risks. Environ Toxicol Chem 32:32–48

Hu Y, Cheng H (2016) Control of mercury emissions from stationary coal combustion sources in China: current status and recommendations. Environ Pollut 218:1209–1221

Huang Y, Li Z (2024) Assessing pesticides in the atmosphere: a global study on pollution, human health effects, monitoring network and regulatory performance. Environ Int 187:108653

Hurlbert M, Krishnaswamy J, Johnson FX, Rodríguez-Morales JE, Zommers Z (2019) Risk management and decision making in relation to sustainable development. In: Shukla PR, Skea J, Buendia EC, Masson-Delmotte V, Pörtner H-O, Roberts DC, Zhai P, Slade R, Connors S, van Diemen R, Ferrat M, Haughey E, Luz S, Neogi S, Pathak M, Petzold J, Pereira JP, Vyas P, Huntley E, Kissick K, JM M (eds) Climate Change and land: an IPCC special report on climate change, desertification, land degradation, sustainable land management, food security, and greenhouse gas fluxes in terrestrial ecosystems. Intergovernmental Panel on Climate Change

Hutton SJ, St Romain SJ, Pedersen EI et al (2021) Salinity alters toxicity of commonly used pesticides in a model Euryhaline fish species (Menidia beryllina). Toxics 9(5):114

IPCC (2007) Climate change: synthesis report. Contribution of working groups I, II and III to the fourth assessment report of the intergovernmental panel on climate change. Geneva

Ito S, Yokoyama T, Asakura K (2006) Emissions of mercury and other trace elements from coal-fired power plants in Japan. Sci Total Environ 368:397–402

Kabisch N, Frantzeskaki N, Pauleit S et al (2016) Nature-based solutions to climate change mitigation and adaptation in urban areas: perspectives on indicators, knowledge gaps, barriers, and opportunities for action. Ecol Soc 21(2)

Kafilzadeh F, Ebrahimnezhad M, Tahery Y (2015) Isolation and identification of endosulfan-degrading bacteria and evaluation of their bioremediation in kor river, Iran. Osong Public Health Res Perspect 6(1):39–46

Kaka H, Opute PA, Maboeta MS (2021) Potential impacts of climate change on the toxicity of pesticides towards earthworms. J Toxicol 20:8527991. Published online 2021

Kattwinkel M, Kuhne J-V, Foit K et al (2011) Climate change, agricultural insecticide exposure, and risk for freshwater communities. Ecol Appl 21(6):2068–2081

Kaur R, Choudhary D, Bali S et al (2024) Pesticides: an alarming detrimental to health and environment. Sci Total Environ 915:170113

Kinney PL (2018) Interactions of climate change, air pollution, and human health. Curr Environ Health Rep 5:179–186

Kocmánková E, Trnka M, Eitzinger J et al (2010) Estimating the impact of climate change on the occurrence of selected pests in the central European region. Clim Res 44(1):95–105. https://doi.org/10.3354/cr00905

Kogan F (2023) Global warming impacts on earth systems. In: Remote sensing land surface changes: the 1981–2020 intensive global warming. Springer, pp 21–66

Koh LP, Kettle CJ, Sheil D et al (2013) Biodiversity state and trends in Southeast Asia. In: Encyclopedia of biodiversity, 2nd edn. Elsevier Inc, pp 509–527. https://doi.org/10.1016/B978-0-12-384719-5.00357-9

Köhler HR, Triebskorn R (2013) Wildlife ecotoxicology of pesticides: can we track effects to the population level and beyond? Science (80-) 341(6147):759–765

Koleva NG, Schneider UA (2010) The impact of climate change on aquatic risk from agricultural pesticides in the US. Int J Environ Stud 67(5):677–704

Kumar R, Sankhla MS, Kumar R, Sonone SS (2021) Impact of pesticide toxicity in aquatic environment. Biointerface Res Appl Chem 11(3):10131–10140

Labenia JS, Baldwin DH, French BL et al (2007) Behavioral impairment and increased predation mortality in cutthroat trout exposed to carbaryl. Mar Ecol Prog Ser 329:1–11

Laetz CA, Baldwin DH, Hebert VR, Stark JD, Scholz NL (2014) Elevated temperatures increase the toxicity of pesticide mixtures to juvenile coho salmon. Aquat Toxicol 146:38–44

Leskovac A, Petrović S (2023) Pesticide use and degradation strategies: food safety, challenges and perspectives. Foods 12(14):2709

Levy BS, Roelofs C (2019) Impacts of climate change on workers' health and safety. Oxford Res Encycl Glob Public Heal (March):1–15. https://doi.org/10.1093/acrefore/9780190632366.013.39

Lewan E, Kreuger J, Jarvis N (2009) Implications of precipitation patterns and antecedent soil water content for leaching of pesticides from arable land. Agric Water Manag 96(11):1633–1640

Liu JC, Mickley LJ, Sulprizio MP et al (2016) Particulate air pollution from wildfires in the western US under climate change. Climate Change 2016(138):655–666

Ma CS, Zhang W, Peng Y et al (2021) Climate warming promotes pesticide resistance through expanding the overwintering range of a global pest. Nat Commun 12:5351

Malhi Y, Franklin J, Seddon N et al (2020) Climate change and ecosystems: threats, opportunities and solutions. Philos Trans R Soc B Biol Sci 375(1794). https://doi.org/10.1098/rstb.2019.0104

Malhi GS, Kaur M, Kaushik P (2021) Impact of climate change on agriculture and its mitigation strategies: a review. Sustainability 13:1318

Manciocco A, Calamandrei G, Alleva E (2014) Global warming and environmental contaminants in aquatic organisms: the need of the etho-toxicology approach. Chemosphere 100:1–7

Mantyka-Pringle CS, Visconti P, Di Marco M, Martin TG, Rondinini C, Rhodes JR (2015) Climate change modifies risk of global biodiversity loss due to land-cover change. Biol Conserv 187:103–111

Matzrafi M (2019) Climate change exacerbates pest damage through reduced pesticide efficacy. Pest Manag Sci 75(1):9–13

Michelutti N, Blais JM, Mallory ML et al (2010) Trophic position influences the efficacy of seabirds as metal vectors. Proc Natl Acad Sci USA 107:10543–10548

Mieiro CL, Ahmad I, Pereira ME et al (2010) Antioxidant system breakdown in the brain of feral golden grey mullet (Liza aurata) as an effect of mercury exposure. Ecotoxicology 19:1034–1045

Mishra S, Sahu A, Dungdung M, Ahmed S, Baitharu I (2023) Pesticide pollution in freshwater and its impact on community health. In: Current developments in biotechnology and bioengineering. Elsevier, pp 33–52

Mojahed N, Mohammadkhani MA, Mohamadkhani A (2022) Climate crises and developing vector-borne diseases: a narrative review. Iran J Public Health 51(12):2664–2673. https://doi.org/10.18502/ijph.v51i12.11457

Moriarty F (1972) The effects of pesticides on wildlife: exposure and residues. Sci Total Environ 1(3):267–288. https://doi.org/10.1016/0048-9697(72)90023-X

Mortimer M (2003) Chemie: Das Basiswissen der Chemie uberarb. u. erw. ¨ Aufl. 2003. vol. 8. Georg Thieme Verlag

Mota-Rojas D, Titto CG, Orihuela A et al (2021) Physiological and behavioral mechanisms of thermoregulation in mammals. Animals 11(6):1733

Mote PW, Abatzoglou JT, Kunkel KE (2013) Climate: variability and change in the past and the future. Clim Chang Northwest Implic Landsc Waters Commun. Published online 2013:25–40

Muhie SH (2022) Novel approaches and practices to sustainable agriculture. J Agri Food Res 22:100446

Muller S (2005) Der Einfluss von Paracetamol, pH-Wert und Temperatur auf ¨ die Embryonalentwicklung von Danio rerio (Teleostei): Entwicklungsbiologische, biochemische und rasterelektronenmiktoskopische Untersuchungen. Diploma Thesis 2005. Eberhard-Karls-University Tubingen, FRG

Müller R, Bandow C, Seeland A et al (2010) BiKF AdaMus: a novel research project studying the response and adaptive potential of single species and communities to climate change in combination with other stressors. J Soils Sediments 10(4):718–721. https://doi.org/10.1007/s11368-010-0233-3

Myers SS, Smith MR, Guth S et al (2017) Climate change and global food systems: potential impacts on food security and undernutrition. Annu Rev Public Health 38:259–277

Nadal M, Marquès M, Mari M, Domingo JL (2015) Climate change and environmental concentrations of POPs: a review. Environ Res 143:177–185. https://doi.org/10.1016/j.envres.2015.10.012

Neira M, Erguler K, Ahmady-Birgani H et al (2023) Climate change and human health in the Eastern Mediterranean and middle east: literature review, research priorities and policy suggestions. Environ Res:216. https://doi.org/10.1016/j.envres.2022.114537

Newton I (1998) Population limitation in birds. Academic

Norin T, Metcalfe NB (2019) Ecological and evolutionary consequences of metabolic rate plasticity in response to environmental change. Philos Trans R Soc B 374(1768):20180180

Noyes PD, McElwee MK, Miller HD et al (2009) The toxicology of climate change: environmental contaminants in a warming world. Environ Int 35(6):971–986. https://doi.org/10.1016/J.ENVINT.2009.02.006

Nunes M, Marchand P, Vernisseau A et al (2011) PCDD/Fs and dioxin-like PCBs in sediment and biota from the Mondego estuary (Portugal). Chemosphere 83:1345–1352

Nyamukondiwa C, Machekano H, Chidawanyika F, Mutamiswa R, Ma G, Sen MC (2022) Geographic dispersion of invasive crop pests: the role of basal, plastic climate stress tolerance and other complementary traits in the tropics. Curr Opin Insect Sci 50:100878. https://doi.org/10.1016/j.cois.2022.100878

Op de Beeck L, Verheyen J, Olsen K et al (2017) Negative effects of pesticides under global warming can be counteracted by a higher degradation rate and thermal adaptation. J Appl Ecol 54:1847–1855

Osterauer R, Köhler HR (2008) Temperature-dependent effects of the pesticides thiacloprid and diazinon on the embryonic development of zebrafish (Danio rerio). Aquat Toxicol 86(4):485–494

Ottinger MA, Dean KDM (2022) Actions of toxicants and endocrine disrupting chemicals in birds. In: Sturkie's Avian physiology. Elsevier, pp 1373–1404

Pandit J, Sharma AK (2023) A comprehensive review of climate change's imprint on ecosystems. J Water Clim Chang 14(11):4273–4284

Pathak VM, Verma VK, Rawat BS et al (2022) Current status of pesticide effects on the environment, human health and its eco-friendly management as bioremediation: a comprehensive review. Front Microbiol 13:962619

Peña-Llopis S, Fernando MD, Peña JB (2003) Fish tolerance to organophosphateinduced oxidative stress is dependent on the glutathione metabolism and enhanced by N-acetylcysteine. Aquat Toxicol 65:337–360

Perarnaud V, Seguin B, Malezieux E, Deque M, Loustau D (2005) Agrometeorological research and applications needed to prepare agriculture and forestry to 21st century climate change. Clim Chang 70(1):319–340

Pletterbauer F, Melcher A, Graf W (2018) Climate change impacts in riverine ecosystems. In: Riverine ecosystem management. Springer, Cham, pp 203–223. https://doi.org/10.1007/978-3-319-73250-3_11

Prakash A, Rao J, Mukherjee AK et al (2014) Climate Change: Impact on Crop Pests. Applied Zoologists Research Association (AZRA), Central Rice Research

Radonic M, Lopez AV, Oka M et al (2005) Effect of the incubation temperature on the embryonic development and hatching time of eggs of the red porgy Pagrus pagrus (Linne, 1758) (Pisces: Sparidae). Rev Biol Mar Oceanogr 40:91–99

Rajmohan KS, Chandrasekaran R, Varjani S (2020) A review on occurrence of pesticides in environment and current technologies for their remediation and management. Indian J Microbiol 60(2):125–138

Reichenberger S, Bach M, Skitschak A et al (2007) Mitigation strategies to reduce pesticide inputs into the ground- and surface water and their effectiveness; a review. Sci Total Environ 384(1-3):1–35

Ritter L, Solomon K, Sibley P et al (2002) Sources, pathways, and relative risks of contaminants in surface water and groundwater: a perspective prepared for the Walkerton inquiry. J Toxicol Environ Heal Part A 65(1):1–142. https://doi.org/10.1080/152873902753338572

Robinet C, Roques A (2010) Direct impacts of recent climate warming on insect populations. Integr Zool 5(2):132–142

Rosenzweig C, Iglesius A, Yang XB, Epstein PR, Chivian E (2001) Climate change and extreme weather events-Implications for food production, plant diseases, and pests. Glob Chang Hum Health 2(2):90–104

Ruomeng B, Meihao O, Siru Z et al (2023) Degradation strategies of pesticide residue: from chemicals to synthetic biology. Synth Syst Biotechnol 8(2):302–313

Ruszkiewicz JA, Tinkov AA, Skalny AV et al (2019) Brain diseases in changing climate. Environ Res 177:108637

Sabarwal A, Kumar K, Singh RP (2018) Hazardous effects of chemical pesticides on human health–cancer and other associated disorders. Environ Toxicol Pharmacol 63:103–114

Scherm H (2004) Climate change: can we predict the impacts on plant pathology and pest management? Can J Plant Pathol 26(3):267–273. https://doi.org/10.1080/07060660409507143

Schiedek D, Sundelin B, Readman JW et al (2007) Interactions between climate change and contaminants. Mar Pollut Bull 54(12):1845–1856

Seebacher F (2009) Responses to temperature variation: integration of thermoregulation and metabolism in vertebrates. J Exp Biol 212(18):2885–2891

Serrano R, Barreda M, Blanes MA (2008) Investigating the presence of organochlorine pesticides and polychlorinated biphenyls in wild and farmed gilthead sea bream (Sparus aurata) from the western mediterranean sea. Mar Pollut Bull 56:963–972

Shivanna KR (2022) Climate change and its impact on biodiversity and human welfare. Proc Indian Natl Sci Acad 88(2):160–171. https://doi.org/10.1007/s43538-022-00073-6

Silva E, Cerejeira MJ (2015) Concentration addition-based approach for aquatic risk assessment of realistic pesticide mixtures in Portuguese river basins. Environ Sci Pollut Res 22:6756–6765

Skendžić S, Zovko M, Živković IP, Lešić V, Lemić D (2021) The impact of climate change on agricultural insect pests. Insects 12(5):440. https://doi.org/10.3390/insects12050440

Sobek A, McLachlan MS, Borgå K et al (2010) A comparison of PCB bioaccumulation factors between an arctic and a temperate marine food web. Sci Total Environ 408:2753–2760

Song SB, Xu Y, Zhou BS (2006) Effects of hexachloro-benzene on antioxidant status of liver and brain of the common carp (Cyprinus carpio). Chemosphere 65:699–706

Song Y, Kirkwood N, Maksimović C et al (2019) Nature-based solutions for contaminated land remediation and brownfield redevelopment in cities: a review. Sci Total Environ 663:568–579

Sorte CJB (2016) Invasive species, evolution and. In: Encyclopedia of evolutionary biology. Elsevier Inc, pp 282–287. https://doi.org/10.1016/B978-0-12-800049-6.00303-6

Subedi B, Poudel A, Aryal S (2023) The impact of climate change on insect pest biology and ecology: Implications for pest management strategies, crop production, and food security. J Agric Food Res 14:100733

Sulaiman NS, Rovina K, Joseph VM (2019) Classification, extraction and current analytical approaches for detection of pesticides in various food products. J Consum Prot Food Saf 14:209–221

Thredgold L, Gaskin S, Quy C et al (2019) Exposure of agriculture workers to pesticides: the effect of heat on protective glove performance and skin exposure to Dichlorvos. Int J Environ Res Public Health 16(23):4798

Tong S, Ebi K (2019) Preventing and mitigating health risks of climate change. Environ Res 174:9–13

Tudi M, Ruan HD, Wang L et al (2021) Agriculture development, pesticide application and its impact on the environment. Int J Environ Res Public Health 18(3):1–24. 1112. https://doi.org/10.3390/IJERPH18031112

Tudi M, Yang L, Wang L et al (2023) Environmental and human health hazards from Chlorpyrifos, Pymetrozine and Avermectin application in China under a climate change scenario: a comprehensive review. Agriculture 13:1683

UNEP/AMAP (2011) Climate change and POPs: predicting the impacts. Report of the UNEP/AMAP expert group, Secretariat of the Stockholm Convention, Geneva

US EPA (2021) Types of pesticide ingredients | United States Environmental Protection Agency. Ingredients used in pesticide products. Published 2021. https://www.epa.gov/ingredients-used-pesticide-products/types-pesticide-ingredients. Accessed 13 Mar 2024

USEPA (2011) Ecological Risk Assessment Ecological Risk Assessment | Superfund | US EPA. Risk Manag. Published online 2011:28–30. https://www.epa.gov/risk/ecological-risk-assessment. Accessed 19 Mar 2024

USEPA (2017) Climate impacts on water resources | Climate Change Impacts | US EPA. Published 2017. https://19january2017snapshot.epa.gov/climate-impacts/climate-impacts-ecosystems_.html. Accessed 18 Mar 2024

USEPA (2023) Climate change impacts on freshwater resources | US EPA. El cambio climático puede alterar la estabilidad de los recursos hídricos a escala local, regional y nacional. Published 2023. https://www.epa.gov/climateimpacts/climate-change-impacts-freshwater-resources. Accessed 19 Mar 2024

Ventakeswarlu B, Shanker AK (2009) Climate change and agriculture: adaptation and mitigation strategies. Indian J Agri 54:226–230

Verchot LV, Noordwijk MV, Kandji S et al (2007) Climate change: linking adaptation and mitigation through agroforestry. Mitig Adapt Strateg Glob Chang 12:901–918
Vieira P, Marques SM, Neto JM et al (2018) Brain as a target organ of climate events: Environmentally induced biochemical changes in three marine fish species. Ecol Indic 95:815–824
Visser ME, Both C (2005) Shifts in phenology due to global climate change: the need for a yardstick. Proc R Soc B Biol Sci 272(1581):2561–2569. https://doi.org/10.1098/rspb.2005.3356
Wang L, Li S, Ahmad IM et al (2023) Global face mask pollution: threats to the environment and wildlife, and potential solutions. Sci Total Environ 887:164055
Webster L, Walsham P, Russell M et al (2011) Halogenated persistent organic pollutants in deep water fish from waters to the west of Scotland. Chemosphere 83:839–850
Weiskopf SR, Rubenstein MA, Crozier LG et al (2020) Climate change effects on biodiversity, ecosystems, ecosystem services, and natural resource management in the United States. Sci Total Environ 733:137782
Werner I, Schneeweiss A, Segner H et al (2021) Environmental risk of pesticides for fish in small- and medium-sized streams of Switzerland. Toxics 9(4):79
Woodward A (2019) Climate change: disruption, risk and opportunity. Global Transit 1:44–49
Xu R, Yu P, Abramson MJ et al (2020) Wildfires, global climate change, and human health. N Engl J Med 383(22):2173–2181
Xu R, Yu P, Liu Y et al (2023) Climate change, environmental extremes, and human health in Australia: challenges, adaptation strategies, and policy gaps. Lancet Reg Health West Pac 40:100936
Yadav IC, Devi NL, Syed JH et al (2015) Current status of persistent organic pesticides residues in air, water, and soil, and their possible effect on neighboring countries: a comprehensive review of India. Sci Total Environ 511:123–137. https://doi.org/10.1016/j.scitotenv.2014.12.041
Zarrineh N, Abbaspour KC, Holzkämper A (2020) Integrated assessment of climate change impacts on multiple ecosystem services in Western Switzerland. Sci Total Environ 708:135212
Zhang W (2018) Global pesticide use: profile, trend, cost/benefit and more. Proc Int Acad Ecol Environ Sci 8(1):1
Zinyemba C, Archer E, Rother H-A (2021) Climate change, pesticides and health: considering the risks and opportunities of adaptation for Zimbabwean smallholder cotton growers. Int J Environ Res Public Health 18:121

Climate Change Impacts on Pharmaceutics, and Its Health Effects

8

Hanan Salah Eldin El-Abhar, Omneya Osama Galal,
Nesrine Salah El Dine El Sayed, Nadia Gamal Eldin Zaki,
and Sameh Mohamed Mostafa El Nabtity

Abstract

Climate change and pharmaceutical companies pose mutual impacts on each other to ultimately affect human health. Pharmaceutical operations alter climate through emissions and waste disposal, while transformed climate events impact human health through distorted disease patterns and mental well-being. Rising

H. S. E. El-Abhar (✉)
Department of Pharmacology, Toxicology, and Biochemistry, Faculty of Pharmacy, Future University in Egypt (FUE), Cairo, Egypt

National Committee of Toxicology, Academy of Scientific Research and Technology (ASRT), Ministry of Higher Education and Scientific Research, Cairo, Egypt
e-mail: Hanan.Elabhar@fue.edu.eg; Hanan.Elabhar@pharma.cu.edu.eg

O. O. Galal
National Committee of Toxicology, Academy of Scientific Research and Technology (ASRT), Ministry of Higher Education and Scientific Research, Cairo, Egypt

Department of Pharmacology and Toxicology, Faculty of Pharmacy, Ahram Canadian University, Giza, Egypt

N. S. E. D. El Sayed
Department of Pharmacology and Toxicology, Faculty of Pharmacy, Cairo University, Cairo, Egypt
e-mail: Nesrine.salah@pharma.cu.edu.eg

N. G. E. Zaki
National Committee of Toxicology, Academy of Scientific Research and Technology (ASRT), Ministry of Higher Education and Scientific Research, Cairo, Egypt

Department of Narcotic Research, The National Center for Social and Criminological Research, Cairo, Egypt

S. M. M. El Nabtity
Department of Pharmacology, Faculty of Veterrinary Medicine, Zagazig University, Zagazig, Egypt

A. Saad-Hussein et al. (eds.), *Climate Change Impacts on Toxins and Health Effects*,
https://doi.org/10.1007/978-981-96-3416-3_8

temperatures worsen air quality, exacerbating respiratory conditions, and extreme weather events can harm mental health. Additionally, changes in temperature affect the efficacy of anesthetic drugs during surgery, while their production and use participate in changing climate to close the loop. Furthermore, environmental alterations play a crucial role in the emergence and dissemination of infectious diseases among not only humans but also animals, leading to the augmentation of antibiotic consumption and resistance and provoking a significant global health threat. Climate change not only affects humans but also the aquaculture and livestock industries by affecting the growth, health, and productivity of species. Changes in temperature, precipitation patterns, and ocean acidity disrupt ecosystems and increase susceptibility to diseases in both sectors. Another player in the antibiotic resistance exacerbation is the employment of antibiotics to promote growth in livestock, while increased drug residues in animal tissues further compound the issue.

This chapter stresses the pressing necessity for joint efforts to address the interrelated issues of climate change, pharmaceutical practices, and health outcomes. It underscores that tackling planetary warming is crucial for environmental sustainability and the well-being of both humans and animals. Pharmaceutical companies can contribute by adopting sustainable practices and developing new treatments and vaccines. Adaptation strategies are crucial for mitigating climate change's adverse effects on aquaculture and livestock. Ultimately, collaboration among stakeholders is vital for effectively addressing these complex issues.

Keywords

Climate change · Pharmaceutical companies · Respiratory diseases · Metered dose inhalers · Seizure · Anesthesia · Antibiotic resistance · Livestock · Aquaculture · Drug residues

Abbreviations

ARB	Antibiotic Resistance Bacteria
ARGs	Antibiotic Resistance Genes
ASMs	Antiseizure Medications
CDC	Centers for Disease Control and Prevention
CFCs	Chlorofluorocarbons
CNS	Central Nervous System
CO_2	Carbon Dioxide
COPD	Chronic Obstructive Pulmonary Disease
DPIs	Dry-powder Inhalers
FGF	Fresh Gas Flow
GHGs	Greenhouse Gases
GW	Global Warming
GWP	Global Warming Potential

HFAs	Hydrofluoroalkanes
HFCs	Hydrofluorocarbons
HGT	Horizontal Gene Transfer
ICS	Inhaled Corticosteroids
IPCC	Intergovernmental Panel on Climate Change
kts	Kilotonnes
MDIs	Metered Dose Inhalers
MDR	Multidrug Resistance
NHS	National Health Service
NO_2	Nitrogen Dioxide
N_2O	Nitrous oxide
ODP	Ozone Depletion Potential
PM	Particulate Matter
pMDIs	Pressurized Metered Dose Inhalers
SABA	Short-acting β-agonist
SDGs	Sustainable Development Goals
SO_2	Sulfur Dioxide
SPC	Summary of Product Characteristics
TA	Thunderstorm-related Asthma
VBDs	Vector-Borne Diseases
WHO	World Health Organization
WOAH	World Organization for Animal Health

8.1 Interconnectedness of Climate Change and Pharmaceutical Practices

Introduction

As evident across nations, climate change represents the paramount crisis of the twenty-first century, unfolding at a pace exceeding previous projections. All parts of the world are at risk of the disturbing consequences of climate change. The gradual increase in temperature is fueling environmental degradation, natural disasters, food and water insecurity, and economic instability (UN 2019).

Society, organizations, and individuals must work together to tackle the fundamental causes of climate change, and the healthcare sector, including the pharmaceutical industry, is no exception.

The pharmaceutical industry plays a vital role in global healthcare, providing essential medications to billions of people worldwide. Through research, development, and innovation, pharmaceutical companies create life-saving medications, vaccines, and therapies that alleviate suffering, cure diseases, and improve the quality of life worldwide (McGuire et al. 2007).

The role of pharmaceutical companies is not limited to fighting infectious diseases or managing chronic conditions; the impact of pharmaceutical interventions is rather profound and far-reaching. The industry fosters economic growth, invests hugely in research, and generates employment opportunities. However, the pharmaceutical industry is often accused of being guilty for its negative impact on global climate change through the process of medicine production, transportation, and waste disposal, which contribute significantly to environmental degradation (Adeyeye et al. 2022). Conversely, the effects of climate change, such as extreme weather conditions, cause slow shifting in disease dynamics, arising new challenges to healthcare systems for medication stability and efficacy (Salas 2020).

Since the pharmaceutical industry is both a victim and contributor to environmental modification, this section seeks to explore the intricate connection between the pharmaceutical industry and climate change, illustrating both its impacts and potential adaptations that could possibly mitigate environmental harm and safeguard public health.

8.1.1 The Impact of the Pharmaceutical Industry on Climate Change

8.1.1.1 Greenhouse Gas Emissions

The production of pharmaceuticals involves a complex supply chain that involves various stages, from the extraction of raw materials to the point of market availability. A study by the United Nations Environment Program (UNEP) indicates that the pharmaceutical industry contributes significantly to global industrial greenhouse gas emissions (GHGs) (UNEP 2020).

The manufacturing development involves energy-intensive processes, including chemical synthesis, purification, and packaging. These activities release significant quantities of greenhouse gases, including carbon dioxide (CO_2), nitric oxide (NO), and methane (CH_4), into the atmosphere, contributing to 4–5% of national GHG emissions linked to climate change (Pichler et al. 2019). The share of pharmaceuticals on GHG emissions within the health systems represents a considerable percentage ranging from 10% to 55% (Wu 2019).

After a major oil spill off the California coast in 1969, industrial companies were under significant pressure to expand their social responsibilities to encompass environmental concerns. This led them to establish climate change targets and commit to regular reporting of greenhouse gas emissions (Latapí Agudelo et al. 2019).

In the 1980s and 1990s, a few significant international climate-related events took place, including the establishment of the Intergovernmental Panel on Climate Change (IPCC), a UN body created to evaluate climate science and recommend measures for reducing greenhouse gas emissions. Other notable events included the adoption of the Montreal Protocol, the UN Framework Convention on Climate Change, and the Kyoto Protocol. Together, these developments increased global awareness of environmental issues.

The 2000s marked the launch of the UN Global Compact initiative, which helps companies align their strategies with environmental principles, alongside the adoption of the Millennium Development Goals. In 2015, these were succeeded by the Sustainable Development Goals (SDGs) and the signing of the Paris Agreement, an international treaty aimed at limiting the increase in average global temperatures.

(Rogelj et al. 2017). In addition to the aforementioned treaties, several voluntary non-profit initiatives assisted corporations in setting goals to monitor and lower their emissions (Booth et al. 2023).

8.1.1.2 Land Deforestation

In addition to the contributions of pharmaceutical companies to GHG emissions, the pharmaceutical industry is found responsible for deforestation and habitat destruction, particularly in regions where raw materials for drug production are sourced. This is especially prevalent as many pharmaceutical products heavily rely on plant-based ingredients. (Cámara-Leret et al. 2020). Clearing land for agriculture or extracting plant-based ingredients can lead to the loss of biodiversity, disruption of ecosystems, global temperature increase, and release of deposited carbon into the atmosphere, ultimately contributing to global warming (Griscom et al. 2020).

8.1.1.3 Waste Disposal

Improper waste disposal tends to be one of the world's biggest environmental challenges. Pharmaceutical wastes obtained from the pharmaceutical factories or from the consumer himself cause pollution of the air and contamination of groundwater and soil resources, leading to environmental degradation and further impacting climate change (Küster & Adler 2014).

In this regard, unused or expired medications and their packaging materials, disposed of by flushing, burning, or landfilling, release toxic substances and emissions of GHG into the air (Daughton and Ruhoy 2013). Moreover, pharmaceutical residues can reach the aquatic life, affecting the marine life and disturbing the ecosystem (Kayode-Afolayan et al. 2022).

Additionally, the growing presence of pharmaceutically active compounds in surface water presents major challenges because they are resistant to traditional wastewater treatment methods (Lentz et al. 2024). Such substances are found difficult to remove and exceed water quality standards, affecting both ecosystems and drinking water quality (Deo 2014). The constant release of pharmaceuticals into surface waters is fueled by their large-scale production and widespread use (Tang et al. 2021).

More than 3000 different pharmaceuticals are used for human medical purposes in the European Union alone (Patel et al. 2019). Globally, over 713 pharmaceutical active compounds and 127 transformation products have been found in surface waters, with high concentrations exceeding detection limits. Noteworthy compounds such as diclofenac, carbamazepine, and ibuprofen have been found in water sources across the globe. The issue worsens during droughts, heightening concerns about water contamination (Aus der Beek et al. 2016).

All these varieties highlight the great responsibility of the pharmaceutical industry toward waste disposal management. The disposal of pharmaceutical waste often requires energy-intensive processes, such as incineration or transportation to specialized waste treatment facilities, and again, this energy consumption contributes directly to climate change by increasing carbon emissions (Al-Hazmi et al. 2024).

8.1.1.4 Transportation

The global nature of the pharmaceutical industry requires a wide transportation network to deliver products to markets worldwide. The use of trucks, airplanes, and ships for the distribution of medications generates GHG emissions that contribute to worldwide warming and climate change, besides environmental harm. The extent of emissions depends on factors such as the distance traveled, mode of transportation, and fuel type used. Additionally, the great dependence on refrigerated transport for temperature-sensitive medications greatly increases energy consumption and further contributes to GHG emissions (Kang and Yang 2020).

8.1.1.5 Antibiotic Resistance

In the past decade, the world has seen a sharp rise in resistant bacterial pathogens. Multidrug resistance (MDR) has been flagged as a critical issue by organizations like the US Centers for Disease Control and Prevention (CDC), the European Center for Disease Prevention and Control (ECDC), and the World Health Organization (WHO). MDR is now widely acknowledged as a growing global health threat and a major public health challenge. The causes of antibiotic resistance are complex and involve human behavior at various levels. For example, the overuse of antibiotics not only causes serious risks to global health but also impacts climate change. Furthermore, improper disposal of antibiotics worsens the problem by promoting the spread of antibiotic-resistant strains, eventually reducing the effectiveness of treatments and increasing mortality rates (Roca et al. 2015).

8.1.2 The Effect of Climate Change on the Pharmaceutical Industry

Climate change affected and will continue to affect human health in multiple ways, potentially leading to significant consequences for the prevalence, distribution, and transmission of infectious diseases across both human and animal populations (Watts et al. 2021). This also led to increasing heat-related death rates, contributing to food insecurity and decreased crop yields, increasing the risk of infectious disease transmission, causing rising sea levels, leading to severe wildfires, and resulting in life-threatening weather events (Burnham 2021).

8.1.2.1 Diseases and Infections

Climate changes have profoundly shaped human history. It has caused the rise and fall of civilizations and enabled the overthrow of new territories. As the global temperature slowly rises, the atmosphere's capacity to hold water increases exponentially, resulting in more intense weather, increased precipitation, and higher humidity. This can cause disruptions, especially in areas with inadequate sanitation infrastructure. Increased precipitation can result in flooding and related diseases, such as waterborne infections caused by the overflow of contaminated water from sewage systems, livestock, or nitrogen fertilizers.

While climate change itself does not directly cause antibiotic resistance, its impact on environmental conditions and human activities is what causes the appearance and spread of bacteria resistant to antibiotics. Another cause for increased antibiotic resistance is using nitrogen fertilizers; therefore, flooding with nitrogen-polluted water can lead to the spread of resistant pathogens and the development of antibiotic resistance genes, resulting in a vicious cycle of antibiotic resistance and its proliferation (Burnham 2021).

Additionally, water resources are greatly affected by the changes happening in global temperature and precipitation, which affect the quality and availability of these sources, leading to the proliferation of waterborne pathogens such as *Vibrio cholerae*, *Cryptosporidium*, and *Giardia*. Extreme weather will help exacerbate contamination of water supplies, increasing the incidence of diarrheal diseases and other waterborne infections (Levy et al. 2016).

Furthermore, global climate change significantly impacts the geographic range and seasonal activity of vector-borne diseases (VBDs) such as mosquitoes, ticks, and sandflies, which are responsible for the transmission of illnesses like malaria, dengue fever, Lyme disease, and leishmaniasis. Rising temperatures create favorable conditions for the breeding and survival of these vectors, altering disease transmission patterns and growing outbreaks.

Increased temperature can also influence the transmission dynamics of airborne pathogens such as the influenza virus, Mycobacterium tuberculosis, and fungal spores. Moreover, increased humidity and air pollution affect the survival and viability of these pathogens, causing potential increases in the incidence and severity of respiratory infections and allergies (Patz and Kovats 2002).

As a fact, climate changes are seen to influence the dynamics of diseases by altering environmental conditions, which leaves the health sector and pharmaceutical industry in continuous worry about the ecological disturbances and disruption of natural ecosystems that can lead to the emergence of novel infectious diseases triggering epidemics or pandemics (Jones et al. 2008).

8.1.2.2 Vaccine Stability

Vaccines are among the most successful and cost-effective interventions in the history of medicine; however, they are among the most fragile medicines temperature-wise (Kartoglu and Milstien 2014). The effect of climate change on immunization and vaccination programs cannot be overlooked. Temperature fluctuation causes significant challenges, from disruption in the raw material supply chain to compromise of vaccine production and stability, accelerating the degradation of vaccine components, and threatening global efforts to control infectious diseases (Sheffield and Landrigan 2011).

Furthermore, temperature fluctuations during transportation and storage can compromise vaccine stability and availability, particularly in regions with poor infrastructure for cold-chain management (Pambudi et al. 2022). The slightest temperature alteration in the environment can lead to a change in storage temperature (Vangroenweghe 2017), which means both medical and finance product losses (Bell et al. 2001). The impact of temperature on vaccine stability is most seen in developing countries that face a new challenge in the vaccine supply and are the most affected by the temperature increase because of global warming (Šklebar et al. 2022).

8.1.2.3 Supply Chain Disruption

Climate change can disrupt pharmaceutical supply chains through severe weather conditions such as hurricanes, floods, and wildfires. These events can damage manufacturing facilities, warehouses, and transportation networks, leading to delays in the production and distribution of medications (Godde et al. 2021).

8.1.2.4 Temperature Control Challenges

Maintaining temperature-controlled storage and transportation of pharmaceutical products is achieved through high-intense energy-consuming processes essential to ensure product quality and efficacy. However, heatwaves associated with climate change jeopardize the integrity of temperature-sensitive medications, leading to reduced potency and efficacy (Šklebar et al. 2022).

8.1.2.5 Raw Material Sourcing

Extreme weather conditions induced by climate change affect water and raw material availability. Water scarcity can influence pharmaceutical manufacturing processes, such as drug synthesis, purification, and formulation. Additionally, disruptions in water supply chains can impact the sourcing of raw materials and the availability of active pharmaceutical ingredients (Fernández-López et al. 2020).

8.1.3 Research and Development Priorities

- The climate change and challenges the world faces recently influenced research and development priorities within the pharmaceutical industry. There may be a growing emphasis on developing medications to address climate-related health problems, such as treatments for VBDs, respiratory ailments, and mental health disorders associated with extreme weather events.
- Given varying disease profiles, the specific pharmaceutical needs will change in terms of type and quality, irrespective of the implementation of control or mitigation measures. Pharmaceuticals could be fundamental to such initiatives, especially when a strategic prophylactic approach is implemented (Costello et al. 2011). Researches examining the environmental transport, fate, and ecotoxicology of drugs in ecosystems need to be prepared to reevaluate which pharmaceuticals are prioritized for research.
- Therefore, research and innovation play important roles in developing climate-stable vaccines and advanced storage techniques by using thermostable vaccine formulations that can withstand a wider range of temperatures without losing their efficacy.

8.2 Interactions Between Climate Change and Pharmaceutical Products: A Respiratory Perspective

The upcoming section aims to highlight the mutual impact of pharmaceutical products on climate change and vice versa regarding the respiratory system, addressing the pathways leading to ill health and strengthening health systems.

8.2.1 Impact of Metered Dose Inhalers on Climate Change: A Critical Environmental Concern

8.2.1.1 Chlorofluorocarbons: Agents of Ozone Layer Depletion

Introduced in the 1950s, metered dose inhalers (MDIs) revolutionized how medication was delivered directly to the lungs in patients with asthma and chronic obstructive pulmonary disease (COPD). Indeed, CFCs (chlorofluorocarbons) have been used as propellants since their development to release medication from the canister into the lungs (Forte and Dibble 1999). These compounds earned their value from the specific characteristics of being chemically stable, non-metabolized, and quickly released back into the environment when exhaled by the patient. Additionally, CFCs are flame-retardant under normal atmospheric conditions, their toxicity is very low, and they are cost-effective to produce (Kontny et al. 1999).

However, being minimally degraded in the atmosphere, as evidenced by their extended atmospheric lifespans (ranging from 60 to 400 years), these substances remain intact until they reach the stratosphere, where the chlorine resulting from their decomposition reacts with and depletes ozone molecules, which represent the Earth's natural shield against ultraviolet radiation (Kontny et al. 1999). Therefore, from the dawn of the 1970s, scientific studies have amassed considerable data linking the thinning and erosion of the stratospheric ozone layer to chemicals containing chlorine, as well as CFCs, with the emergence of notable adverse health consequences like cataracts, skin cancer, and diminished immune response following the penetration of higher levels of ultraviolet radiation (Forte and Dibble 1999).

Observed reductions in stratospheric ozone levels, particularly in the Antarctic region, confirm predictions of ozone depletion. The Antarctic ozone "hole" has worsened over the years, now covering an area twice the size of the continental United States. Similarly, the ozone layer in the northern mid-latitudes has steadily declined, with annual losses averaging around 5% over the past two decades. These declines are attributed to increased atmospheric chlorofluorocarbon concentrations, highlighting the significant role of CFCs in ozone layer depletion (Forte and Dibble 1999).

8.2.1.2 Hydrofluorocarbons: Agents That Emit Greenhouse Gases

Following the confirmed association between CFCs and ozone exhaustion, the Montreal Protocol on Substances that Deplete the Ozone Layer was established in response to public apprehension regarding this matter. It mandates the phase-out of

CFCs, leading to the need for alternative delivery methods for asthma and COPD treatment (United Nations 2018). This prompted the use of HFCs (hydrofluorocarbons) as propellants in the newly developed CFC-free MDIs, known as dry-powder inhalers (DPIs) (Janson et al. 2020).

Consequently, pharmaceutical companies have undertaken efforts to reformulate their MDIs to eliminate CFCs. However, these reformulations have encountered significant technical challenges where introducing a new MDI requires massive toxicological examinations for each constituent. At the beginning of the 1990s, MDI producers launched two distinct collaborative research projects to expedite and streamline the conclusion of toxicology evaluations on the propellant hydrofluoroalkane (HFA). These initiatives were comparable in magnitude to evaluating a new drug, with the industry having endowed over $1 billion in investigation and development trials thus far. As of 1995, both secure and efficient substitute propellants, HFA-134a and HFA-227ea, were formally acknowledged for application in MDIs in the European Union, and the utilization of HFA-134a in inhalers was authorized by the FDA before 1996 (Forte and Dibble 1999).

Despite their overlooked impact on depleting the ozone layer, pressurized metered dose inhalers (pMDIs) have been embroiled in another environmental setback, being identified as significant contributors to climate shifts through the production process of their HFC propellants. So, from an environmental standpoint, HFAs are favored over CFCs in MDIs since they do not participate in the loss of stratospheric ozone. Nonetheless, similar to other chlorofluorocarbon alternatives, HFAs are not entirely ecologically benign. They have the aptitude to capture heat in the atmosphere, thereby exacerbating the issue of global warming (GW) through their participation in increasing the emission of GHGs, which are crucial factors in environmental alteration due to their high global warming potential (GWP) (Janson et al. 2020).

Recently, it was reported that the carbon footprint generated by a 290-kilometer car journey can be comparably produced by the 200 doses of one pMDI. Although HFCs are also utilized in various other applications, such as air conditioning, cooling systems, refrigerants, and heat pump devices (80%), including foams, blowing agents (11%), cleaning agents, and fire suppressants (5%) (Climate and Clean Air Coalition 2019), their impact in MDIs cannot be overruled.

Globally, the release of HFCs from MDIs in 2014 amounted to approximately 0.013 gigatonnes of CO_2 equivalent, representing roughly 3% of the overall worldwide weighted CO_2 equivalent output of HFCs (Montreal protocol on substances that deplete the ozone 2014). Additionally, the propellants represent around 0.03% of annual worldwide GHG output, and medications for pMDIs make up roughly 3% of the overall healthcare-related outputs from the National Health Service of the United Kingdom (Fidler et al. 2022).

However, the prescription patterns across Europe have revealed significant disparities among countries regarding the selection of inhalation devices. For instance, a study published in 2011 (Lavorini et al. 2011) indicated that most of the devices of inhaled corticosteroids (ICS) (80%) utilized in the UK were MDIs, while in Sweden the majority (90%). This discrepancy was not clearly justified but could be attributed to biases among prescribers and patients, as well as marketing strategies (Lavorini et al. 2011). In the same context, it was recounted that MDIs contribute one-quarter of the carbon footprint from pharmaceutical procurement, which accounts for 16% of the National Health Service (NHS) carbon footprint in England, totaling approximately 23 megatonnes of CO_2 equivalent (NHS England sustainable development unit 2019).

In this regard, Janson et al. (2020) in their research reported that adopting the Sweden protocol and DPIs and switching from Evohaler MDIs to Ellipta and Accuhaler DPIs led to an annual reduction in carbon emissions equal to 422 kilograms of CO_2 per patient.

When comparing the proportions of MDI use across different classes, Janson et al. (2020) documented the potential reduction in kilotons (kts) of CO_2 in England that could be achieved by aligning this proportion with that of Sweden (Table 8.1).

From the NHS in England (https://digital.nhs.uk/), the community prescribing data in 2017 were analyzed, estimating DPIs at 1 kg and the MDIs carbon footprint at 20 kg CO_2e. Due to the lack of a DPI alternative, the analysis did not include SAMA. The possible per annum inhibition indicates the anticipated carbon cuts if England embraced Sweden's prescribing patterns of MDI. SABA, β2 agonist (short-acting); LABA, β2 agonist (long-acting); SAMA, muscarinic antagonist (short-acting); LAMA, muscarinic antagonist (long-acting); ICS; MDI; DPI.

8.2.1.3 Environment-Friendly Inhalers

In comparison to pMDIs, alternative inhaler devices with reduced carbon footprints are available, demonstrating comparable effectiveness, and are frequently favored by patients (Brocklebank et al. 2001; Welch et al. 2004; Dekhuijzen et al. 2016). These inhalers involve soft mist inhalers and dry powders that are utilized for both rapid relief and maintenance therapies in asthma and COPD management. These inhalers result in considerably lower CO_2 equivalence (<20 g CO_2e/inhalation)

Table 8.1 Potential CO_2e reduction in kilotons if MDI use in England matches Sweden's levels across different classes (Janson et al. 2020)

	England		Sweden		England	
	Inhalers (MDI)		Inhalers (MDI)			
	Per year	Percent	Per year	Percent	CO_2e (kt)/year	Possible inhibition of CO_2e (kt) per annum
SABA	21,931,511	94	1,477,692	10	414.00	350.0
LABA	700,195	65	377,415	2	9.30	8.4
SAMA	421,191	100	No data	100	8.40	0
ICS	6,733,445	94	765,796	15	127.00	101.0
ICS + LABA	14,075,067	47	1,719,428	13	140.00	91.0
LAMA and LAMA + LABA	6,549,448	0	428,732	0	6.55	0
LAMA + LABA + ICS	5211	99	2626	100	−0.10	0
Total	49,994,877	70	4,771,689	13	705.0	550.0

compared to pMDIs, which emit approximately 100 g CO_2e/inhalation (Montreal protocol 2015).

The guideline from the Canadian Thoracic Society underscores the use of a budesonide-formoterol DPI for reliever therapy as a suitable option instead of SABA pMDIs in certain asthma patients, especially those having milder disease (Yang et al. 2021).

8.2.2 Environmental Shift Threatens Human Respiratory Health

As the discussion pivots from the environmental impact of MDIs to the wider effects of global environmental shifts on respiratory health, it becomes evident that the relationship between these two domains is multifaceted. Since every coin has two sides, not only MDIs are implicated in contributing to climate change due to their GHG emissions, but the evolving climate also exerts significant influences on respiratory diseases. Changes in temperature, air quality, and the prevalence of allergens and pollutants can exacerbate respiratory conditions such as asthma and COPD, leading to increased morbidity and mortality rates among affected populations. Therefore, addressing both the environmental footprint of MDIs and the effect of global environmental shifts on respiratory health is essential for comprehensive disease management and sustainable healthcare practices.

8.2.2.1 Impacts of Climate Alteration and Increased Exposure to Pollutants on Human Health and Wellness: Role of Increased Carbon Dioxide Levels

The increased human activity has become a significant concern due to its impact on pollution. Fossil fuel consumption, including nitrous oxides (N_2O), methane, carbon dioxide (CO_2), and other fluorinated gases, has resulted in higher levels of GHGs. These gases, by trapping heat, raise GW and intensify thermal pollution, acting as the primary driver of climate change. As mentioned earlier, the propellants in MDIs contribute to CO_2 emissions, creating health risks in a harmful cycle. Extended exposure to concentrations of CO_2 as low as 1000 ppm has been reported to impose a serious health burden (Agache et al. 2022).

This threshold is already surpassed in certain indoor settings and is projected to be reached in urban outdoor air concentrations before the year 2100. Prolonged exposure to elevated CO_2 levels in the atmosphere has been linked to a broad spectrum of harmful health effects, ranging from bone demineralization and inflammation to respiratory acidosis and calcification of the kidney (Jacobson et al. 2019). Additionally, it has been observed that increasing atmospheric CO_2 levels lead to a decrease in the concentration of essential antioxidants for human wellness, such as plant carotenoids (Loladze et al. 2019). Furthermore, heightened CO_2 environments tend to increase pollen production, stimulate plant photosynthesis, and promote the proliferation of mold spores, worsening respiratory allergies and asthma, thereby compounding the impact on human well-being (D'Amato et al. 2020; Rauer et al. 2021).

8.2.2.2 Increased Levels of Air Pollutants Exacerbate Aggravate Human Health

In 2019, Campbell-Lendrum and Prüss-Ustün recounted that one of the foremost environmental hazards is air pollution that contributes, according to the WHO, to around 7 million deaths annually, with the main cause being respiratory illnesses, including stroke, ischemic heart disease, lung cancer, and COPD. This organization has identified sulfur dioxide (SO_2), nitrogen dioxide (NO_2), ozone, as well as particulate matter (PM) to be the crucial air pollutants. Other hazardous substances are involved such as inhalable particles (mercury, cadmium, chromium, lead, polycyclic aromatic hydrocarbons), liquid aerosols (perchloroethylene, methylene chloride), and gases (benzene, toluene, and xylenes).

As noted by Marchini et al. (2020), the effect of PM pollution varies based on particle size and composition, whether they originate from natural sources (like soil, dust, and bioaerosols) or human activities (such as ammonia, organic compounds, carbon, sulfates, nitrates, or lead). Particles measuring 10 μm or smaller (PM10) pose significant health risks upon inhalation, with finer particles (≤2.5 μm; PM2.5) having a more pronounced effect. These fine particles can penetrate the alveoli and bloodstream, leading to diffuse inflammation and other adverse effects, and prolonged exposure to the fine PM has been linked to hospitalizations for COVID-19. Additionally, extended and even acute exposure to these particles and others such as ozone, NO_2, and SO_2 have been associated with increased allergic rhinitis risk and asthma exacerbations, causing increased visits to emergency rooms and hospital admissions (Haikerwal et al. 2016). Moreover, Berger et al. (2020) stated that increased ozone level aggravates the severity of symptoms in those who suffer from pollen allergy during the pollen season.

8.2.2.3 Wildfires Blend with Particulate Matter to Harm Respiratory System

Wildfires are increasingly frequent and intense, amplifying the impact of climate change. The prominent origin of fine and ultrafine (PM2.5 & PM0.1, respectively) PM is the smoke from wildfires, characterized by a dynamic mixture of gases and particles. In addition to the naturally occurring hazardous smoke, wildfires can also incinerate structures and vehicles, emitting toxic chemicals from burned plastics and other materials (Agache et al. 2022). Whether exposed for long- or short-term periods to wildfire smoke, researchers reported that it can jeopardize health status, spanning from increased exacerbation of chronic respiratory (Black et al. 2017) and cardiovascular conditions (Stowell et al. 2019) to increasing the likelihood of acute cardiovascular events (Chen et al. 2021), metabolic disorders, dermatologic issues (Fadadu et al. 2021), and mental health disorders (Kilian and Kitazawa 2018). Additionally, vulnerability to respiratory infections (SARS-CoV-2, influenza) is heightened, along with dementia, infant developmental disorders, sperm damage, adverse pregnancy outcomes, and immune dysfunction. Studies have also linked wildfires to augmenting attacks of asthma and hospitalizations due to asthma (Agache et al. 2022).

8.2.2.4 Reductions in Green Spaces and Biodiversity Jeopardize Respiratory Function

Exposure to natural greenery and biodiverse environments maintains human well-being, including bolstering immune function, whereas their reduction is associated with changes in immune responsiveness that may reflect environmental conditions (Prescott 2020). Similarly, residing close to green spaces at birth has been linked with declined rates of asthma and allergic diseases by age 7 (Cavaleiro et al. 2021). Moreover, in regions with reduced ecosystem diversity, ragweed, as an example of an invasive species, can present new allergens to inhabitants, thereby raising the prevalence of asthma and allergies (Jones et al. 2019). Moreover, Islam et al. (2020) studied the impact of living in natural spaces during early childhood development and reported a direct correlation between greater exposure to green environments during childhood and enhanced physical activity levels and diminished incidence of wheezing, bronchitis, and other ailments, such as obesity and neurodevelopmental disorders, but with an emphasis on the type of plant species since some types heightened asthmatic symptoms in childhood. In the same milieu, Andersen et al. (2021) also accentuated the link between exposure to nature and its anti-inflammatory benefits and alleviation of asthma symptoms.

8.2.2.5 Thunderstorm Asthma and Climate Change

While individual thunderstorms are complex weather phenomena influenced by various factors, environmental change can impact the frequency, intensity, and spreading of thunderstorms by altering atmospheric conditions. Due to the increased GW, more moisture is present in the air, leading to increased evaporation and precipitation rates (Ralls 2023).

Intense thunderstorms not only affect weather patterns, leading to heavier rainfall that exacerbates flooding risks and more frequent severe weather events, but they also impact human health. For example, thunderstorm-related asthma (TA), a type of asthma triggered by the occurrences of thunderstorms, highlights this health concern. The asthmatic attack that initiates or intensifies following a thunderstorm is called TA, which predominantly affects individuals with seasonal allergic rhinitis, widely termed hay fever or allergies. This type is characterized by symptoms like a runny nose, sneezing, itchy eyes, and seasonal allergies that typically peak in spring, summer, or early fall (Lee et al. 2017).

While rain typically decreases pollen levels and alleviates asthma symptoms triggered by allergies, thunderstorms can exacerbate asthma due to a distinct chain of events starting with cold downdrafts that concentrate airborne particles, such as pollen and mold, which are then lifted into high-humidity clouds. Within the clouds, pollen undergoes osmotic rupture during thunderstorms, and wind and humidity fragment these particles, releasing hundreds of tiny (<5 μm) allergenic starch granules easily inhaled into the nose, sinuses, and lungs. Subsequently, wind gusts concentrate these smaller particles, leading to a substantial inhalation volume and exacerbating asthma symptoms (Agache et al. 2022; Rorie and Poole 2021).

Since the 1980s, there have been more than 23 documented occurrences of TA globally, with the most severe event taking place in November 2016, in Australia, concurring with the peak grass pollen season (D'Amato et al. 2020; Rorie and Poole 2021). During this catastrophic event, allergic rhinitis was identified as the primary risk factor for hospitalization, accounting for 87% of cases (Hew et al. 2019).

Although the connections between TA and climate change are still inexact, the influence of pollen is evident, potentially leading to an increased frequency of TA events. In this regard, a notable escalation in the number of children and adolescents visits to the asthma emergency department proportional to the increase of 10 grass pollen grains/m^3 was documented in a meta-analysis and systematic review (Erbas et al. 2018), proving the crosstalk between climate warming and the elongation of the pollen seasons, their higher concentration and load, as well as allergenicity (Rorie and Poole 2021). One study specifically linked around 50% of the prolonged pollen season to the shift in climate in North America. Furthermore, pollen exposure compromises innate immunity against respiratory viruses, as proven by the positive correlation between elevated concentrations of pollen and higher infection rates of SARS-CoV-2, as observed in 31 countries worldwide (Gilles et al. 2020; Damialis et al. 2021).

8.2.2.6 During Its Life Span, Plastic and Climate Change Endanger Human Health and Respiratory Function

Added to the previous climate change-related harmful events and their impact on health status, plastic pollution during its life span from manufacturing, consumption, and disposal adds to the human threat, especially with the global accelerating rise in its production and its imperishable fate (Wiesinger et al. 2021).

Exposure to plastic chemicals, which leach out into the environment, contributes significantly to environmental pollution that poses a marked hazard to human well-being. These illnesses arise from the thousands of chemicals present in plastics, which include known neurotoxicants, persistent organic pollutants, endocrine disruptors, and human carcinogens, among others. Therefore, any endeavors aimed at mitigating the hazards associated with plastics must address the dangers posed by the chemicals linked to plastic (Landrigan et al. 2023).

Additionally, workers involved in coal mining, oil extraction, and fracking to extract its raw materials; employees in plastic production; and individuals living in communities located near plastic manufacturing facilities endure increased risks of lung diseases, including asthma and COPD; cardiovascular ailments; and cancer, among others (O'Callaghan-Gordo et al. 2016; Symeonides et al. 2021). Furthermore, adults and children exposed to volatile organic compounds and airborne PM in the vicinity of a petrochemical site exhibit heightened rates of asthma, lung and respiratory infections, and cardiovascular problems (Rovira et al. 2014; Chang et al. 2020). Moreover, in a 2014 study, Shu et al. found that over a 10-year follow-up period in Sweden, children residing in polyvinylchloride flooring houses exhibited a high incidence of childhood asthma, despite plastics in the building and construction sector being one of the least conspicuous markets.

Furthermore, airborne pollutants generated during the extraction of coal, natural gas, and oil comprise PM, nitrogen oxides (NOx), and sulfur oxides (SOx), which impose health risks, where besides the harmful effect of PM on the respiratory system, they act as respiratory irritants (CCOHS 2022). Besides, ozone, which is produced in the atmosphere near gas and oil extraction sites, serves as a respiratory irritant resulting in asthma and COPD in children, the elderly, and outdoor aficionados (Colborn et al. 2014).

Contact with airborne microfibers among employees engaged in manufacturing "flock," which involves producing fleece and velvet-like fabrics from reclaimed or ground nylon, polypropylene, polyester, and polyethylene, creates a high possibility of developing flock workers' lung disease, which has symptoms that resemble those of fevers, asthma, chronic bronchitis, lower airway lesions, interstitial fibrosis, pneumothorax, chronic pneumonia, extrinsic allergic alveolitis, impaired lung function, and lung cancer (Kern et al. 2011).

8.2.3 Research Gap Between Climate Change, Global Warming, Respiratory Diseases, and Metered Dose Inhalers

8.2.3.1 Understanding Climate Change Impact on Respiratory Diseases

While there is substantial evidence linking climate change and GW to various environmental and health impacts, there remains a gap in fully understanding the specific mechanisms by which these changes exacerbate respiratory diseases. Additional studies are needed to elucidate the multifaceted interactions between climate variables, allergens, air pollution, and respiratory health outcomes.

8.2.3.2 Vulnerable Populations

Current research has identified the elderly, children, and patients with pre-existing respiratory abnormalities as susceptible persons to the adverse effects of climate change on respiratory health. However, there is a need for further investigation into the differential impacts across various demographics and regions, including urban and rural areas.

8.2.3.3 Mitigation Strategies and Adaptation Measures

While efforts to mitigate climate change are crucial for reducing future health impacts, there is a gap in research focusing on adaptation measures specifically targeted at minimizing the burden of respiratory diseases. Investigating and implementing adaptive strategies, such as improved air quality monitoring, urban planning, and healthcare infrastructure, could help mitigate the health consequences of climate change.

8.2.3.4 Impact of MDIs on Climate Change

Although MDIs are essential for managing respiratory diseases such as asthma and COPD, their propellant gases, typically HFAs, contribute to GHG emissions. Nevertheless, few research is available that quantifies the precise environmental influence of MDIs and explores potential alternatives or mitigation strategies to reduce their carbon footprint while ensuring effective treatment for patients. These approaches may encompass waste reduction, recycling, and the adoption of sustainable inhalers. Moreover, occasions exist to enhance patient care and simultaneously reduce emissions, such as optimizing inhaler device selection based on patient preferences. For example, maintaining reliever therapy for patients with asthma while minimizing redundant ICSs or steroid MDIs for the management of COPD patients. Developing green inhalers, particularly through transitioning to HFA152a propellant, is crucial for enhancing MDI efficiency, usability, and recyclability (Wilkinson and Anderson 2020).

8.2.3.5 Balancing Health Benefits and Environmental Concerns

There is a need for research focusing on finding a balance between the health benefits of MDIs for respiratory patients and the environmental concerns associated with their usage. This includes investigating alternative propellants, inhaler designs, recycling programs, and patient education initiatives to minimize the environmental impact of respiratory medication delivery systems without compromising patient care. Moreover, physicians are encouraged to engage in shared decision-making with patients, discussing the merits and drawbacks of different MDIs and involving their environmental effects regarding GW. Materials for facilitating modifications in inhaler prescriptions, such as posters, patient correspondence, cost comparison tables, electronic health record tools, and criteria for selecting appropriate patients, are readily accessible to the public (Fidler et al. 2022).

Additionally, the proper diagnosis of asthmatic patients can help minimize GHG emissions where one-third of patients initially diagnosed with asthma were found to lack objective evidence of the condition (Gagné et al. 2023). Implementing behavior change strategies to ensure consistent delivery of evidence-based real-world care is intricate, yet the prospect of reducing carbon footprints serves as a significant incentive for both providers and patients. Hence, avoiding unnecessary inhaler prescriptions could serve as a compelling catalyst for change in healthcare systems and can be utilized in future behavior modification initiatives. Another solution is to recycle inhalers through pharmacies instead of disposing of them in landfills, so it is possible to reuse plastic or aluminum components, consequently reducing CO_2 production (CoP Sustainable Inhaler Initiative 2022; Fidler et al. 2022).

8.3 Climate Change and Its Impact on Mental Health

The impact of climate change-associated events on mental well-being is becoming evident and clear. Consequently, a global concern caused by the prevalence of mental health disorders arises. In this sense, the increased risk of neurological and psychiatric disorders is significantly increased by the growth in GHG emissions, air pollution, and rising temperatures (Gulcebi et al. 2021).

Despite these facts, there are some gaps concerning the influence of rising global temperatures on mental health, as there is limited awareness among public health leaders, mental health professionals, and policy experts regarding this problem (Cianconi et al. 2020).

8.3.1 Impact of Changed Temperature on Mental Health

Mental health can be affected by various mechanisms, including physiological alterations like changes in blood flow as well as the level of serotonin, cognitive changes that result from disturbed sleep at elevated temperature, and the impact of temperature on functional connectivity in the brain (Lõhmus 2018). Other factors also include societal changes like increased aggression and stress caused by agricultural and economic output and high alcohol intake (Berry et al. 2010). Furthermore, the physiological state of individuals with pre-existing mental illness could be differently affected by temperature, partly due to impaired thermoregulation caused by psychiatric medications (Chong and Castle 2004).

8.3.2 Climate Change and Epilepsy

Generally, neurocognitive and cerebrovascular disorders are six times more common in people diagnosed with epilepsy (Athanasios et al. 2004). Moreover, life-threatening weather incidents such as heat waves and extensive daily temperature variations can easily disrupt the stability of people with neurocognitive illnesses, even more so when seizures are comorbid (Gulcebi et al. 2021).

Moreover, intrinsic features, such as the etiology of epilepsy or individual physiological characteristics, and extrinsic factors including ambient temperature, humidity, exposure to sunlight, exertional heatstroke, or pharmacological interventions, can exacerbate existing seizures or elevate the incidence of developed epilepsy resulting from vector-borne disorders, infections, and trauma of the CNS (Bartolini and Sander 2019; Sisodiya et al. 2019). Furthermore, individuals with compromised CNS and prone to seizures among other diseases may experience intensified susceptibility compared to those with healthier brains at high temperatures (Peters et al. 2016).

Demonstrating this concept, hyperthermia in adult rats was reported to aggravate both seizures and hippocampal impairment caused by either neurotoxic or non-neurotoxic doses of kainic acid after the rise of body temperature to 42 °C (Liu et al. 1993). Additionally, on the tenth day after birth, exposure of rats to 30 min of hyperthermia (core and brain temperature at 39 °C) elevated the epileptogenicity of status epilepticus and the neuropathological consequences compared to some conditions wherein body temperature is maintained at 35 °C (Sarkisian et al. 1999).

Such effects may be attributed to the potential impairment of the mechanisms which usually protect against circuit variability in various neurological diseases. The adverse effects of these changes might remain concealed until exposed to temperature stress. Similarly, it can be theorized that individuals with posttraumatic stress disorders may be similarly susceptible to circuit disruptions, with symptoms manifesting only under environmental stress that unveils prior alterations in brain circuitry (Marder 2023).

In the same setting, the variations between indoor and outdoor temperatures and the level of humidity, along with their daily fluctuations, due to emerging forms of climate extremes, are possible to impact epilepsy. Moreover, the occurrence of many seizure triggers is likely to rise with climate variation. These promoters can exert not only direct consequences on human physiology but also indirect ones, such as disruption in socioeconomic status that contributes to stress, fatigue, and deprivation of sleep, all of which are common seizure triggers (Ferlisi and Shorvon 2014).

Moreover, in another study done in Brazil, Zhang et al. (2023) revealed that the temperature impact threshold in Brazil nationally was 26 °C, and each 1 °C rise beyond this threshold was associated with an elevation in the risk of hospitalization for seizures on the current day of admission and up to 7 days prior.

Furthermore, Sisodya reported in 2023 that people diagnosed with Dravet syndrome, which is one of the rare conditions known as developmental and epileptic encephalopathies, usually experience seizures that are resistant to medication and can be triggered by fever and high temperatures. Besides, heat waves affect these patients' lives by disrupting their sleeping patterns. In addition, individuals struggle to employ effective physiological and behavioral heat dissipation mechanisms due to thermoregulatory dysfunction, intellectual disability, and treatment with medications like topiramate (which can reduce sweating), combined with limited mobility.

In addition to temperature, epilepsy is influenced by other climate change factors. Research conducted in a highly polluted city in China reveals temporary spikes in air pollutants like NO_2 and SO_2 are correlated with heightened hospital visits for epilepsy. Similarly, these findings are consistent with research conducted in Chile, which has detected an elevated risk of epilepsy-related hospitalizations with increased air pollution levels (Cakmak et al. 2010; Sisodiya et al. 2019).

Additionally, a recent study revealed a potential correlation between higher relative air humidity and an elevated likelihood of admission for epileptic seizures, while higher ambient temperatures were linked with a reduced hazard. Although these studies are preliminary, they indicate complex relationships that necessitate further investigation with additional data (Sisodiya et al. 2019).

8.3.3 Climate Change and Antiseizure Drugs

Few published research studies are available on the potential variations in the way antiseizure medications (ASMs) work in different climate circumstances or whether their stability is influenced by humidity and/or temperature. However, the efficacy of ASMs was found to be affected by seasonal variation, and ambient temperature was identified to be a possible reason. Additionally, serum levels of certain ASMs may be influenced by increased temperature resulting in elevated body sweat levels. In a prior study carried out on some people diagnosed with epilepsy and receiving chronic ASM treatment, phenobarbitone sweat concentration increased with increasing sweat flow while phenytoin sweat concentration was independent of sweat flow (Parnas et al. 1978).

Moreover, a reduction in serum levels of diphenylhydantoin was observed toward the end of summer due to an elevation in perspiration (Cohn 1982). Later, the results of a subsequent study performed on patients receiving valproic acid or carbamazepine revealed that serum concentrations of both drugs were considerably decreased in spring versus autumn (Avakian et al. 2013).

Additionally, it has been proposed that sunlight can be a potential palliative management in epilepsy. In this context, endogenous circadian rhythms were suggested to contribute to seizures, with ASM chronotherapy proposed as a tool to improve seizure control in selected individuals with epilepsy (Ramgopal et al. 2013).

Besides, storage conditions for ASMs may regulate product shelf life, which may be mainly crucial in formulations that are sensitive to humidity, temperature, and sunlight, such as certain blister pack preparations, syrups, and injections. Thus, storing carbamazepine and phenytoin formulations in hot and/or humid conditions damaged the stability of pharmaceutical forms and decreased bioavailability by up to 50% (Cloyd 1991). Furthermore, the stability of enteric-coated tablets may change when removed from their original packaging and repackaged into dosette boxes (or dose administration aids) because of the hygroscopic nature of the antiepileptic drug valproate; thus, valproate in a dosette should be refrigerated in hot and humid conditions. In the same milieu, the summary of product features of phenobarbital determines that it should be stored below 25 °C in a dry place, while no such recommendations are found for other ASMs (Redmayne et al. 2015).

8.3.4 Research Priorities

- Referring to the limited available data, there is a serious necessity for additional information to improve our understanding of the potential impacts of climate change on epilepsy care. Therefore, it is important to conduct comprehensive epidemiological studies to assess the relationship between climate change and epilepsy, including long-term observational studies that track seizure frequency and severity relative to changing environmental circumstances.
- Implement public health initiatives to raise awareness among healthcare providers and epilepsy patients about the potential impact of climate change on seizure management and treatment outcomes, providing guidance on adapting seizure management strategies during extreme weather events or environmental emergencies.
- Pharmaceutical companies should invest in manufacturing stable medications that can tolerate and adapt to climate changes such as increased heat and humidity. Additionally, the medical health system should invest in the development and deployment of telemedicine and remote monitoring technologies to ensure continued access to epilepsy care and medication management during climate-related disruptions to healthcare services, facilitating remote consultations and medication delivery to affected individuals.

8.3.5 Anesthetic Practices and Their Climate Footprint: Exploring the Environmental Impact

The involvement of the healthcare sector in climate change is a pivotal factor. One notable issue is the release of GHG from hospitals, accounting for approximately half of these emissions (Malik et al. 2018). Additionally, operating rooms and intensive care units, in particular, contribute to nearly one-third of hospital waste, with a daily rate of 13.6 kg per patient and an overall carbon intensity of 160 kg CO_2 per operation (McGain et al. 2009; MacNeill et al. 2017; Martin et al. 2017).

Mainly, anesthesia, which is a fundamental component of medical practices, possesses an environmental impact that is frequently snubbed. The environmental footprint of perioperative services is among the largest in all of health care (Seidman and Parker 1998; Sherman and Ryan 2010; Kwakye et al. 2011). The emission of volatile anesthetics can lead to significant atmospheric consequences, including their involvement in ozone depletion within the stratosphere and their contribution to greenhouse warming within the troposphere (Langbein et al. 1999).

Previous publications reported the role of inhaled anesthetics on GWP, reproducing a growing concern for the effect of anesthesia practice on the environment (Ryan and Nielsen 2010). These investigations scrutinized the immediate climatic repercussions of inhaled anesthetics. While their influence appears modest when juxtaposed with the collective impact of other GHGs like CO (Sulbaek Andersen et al. 2010), findings underscored that desflurane, noted for its superior heat-trapping capacity, and N_2O, acknowledged for its significant emissions, exert the most pronounced effects compared to sevoflurane and isoflurane. By the same token, the CO_2 emitted from a car journeying distances of 783 km, 667 km, and 3924 km can be likened to the amount emitted from 2% sevoflurane, 1.2% isoflurane, and 6% desflurane, respectively, used during a seven-hour surgical procedure employing inhalational anesthetics with a fresh gas flow (FGF) rate set at 0.5 L/min (Özelsel et al. 2019). When clinically viable, anesthesiologists ought to prioritize intravenous, regional, or neuraxial anesthesia. Should inhalation anesthetics be deemed necessary, sevoflurane or isoflurane are favored over desflurane, as the latter demands higher gas concentrations for general anesthesia. Moreover, the minimization of the FGF rate, even during induction, stands as a crucial strategy (Gordon 2020).

The problem of the inhaled anesthetic gases can be explained by undergoing minimal in vivo metabolism, predominantly being exhaled either into the anesthesia apparatus and subsequently vented into the atmosphere as waste anesthetic gas or directly into the indoor environment by patients. In the absence of gas-capturing mechanisms, all emissions of unmetabolized drugs are presumed to be released into the atmosphere (Yasny and White 2012). The impact of the released gases involves ozone depletion, which is influenced by substances like isoflurane and methoxyflurane, both containing chlorine, as well as N_2O. The ozone depletion potential (ODP) compares their effectiveness in depleting ozone relative to CFC-11 (CCl3F). However, due to their relatively short atmospheric lifetimes, especially for methoxyflurane, their impact on ozone depletion is anticipated to be minimal. The ODP of N_2O varies depending on the concentration of other GHGs and the extent of halogen loading in the stratosphere. While the ODP was calculated at 0.015 for the year 2000, it is projected to increase in the future across all conceivable emission scenarios (Sulbaek Andersen et al. 2023).

An alternative approach for gas capture technologies impact is propofol, which is rapidly and highly metabolized, with only minimal levels excreted unchanged. Therefore, it was assumed that no unmetabolized propofol entered the environment directly (Dawidowicz et al. 2002).

Technologies on the near horizon involve photochemical air purification. This approach can theoretically destroy all waste anesthetic gases. Instead, existing volatile waste anesthesia gas-capturing systems can reclaim volatile gases for reuse rather than discharge waste into the atmosphere. Hence, what practical steps can anesthesiologists take to effectively mitigate the environmental impact of functioning rooms and intensive care units (ICUs)? First, adopting the principle of "Reduce, Reuse, Recycle" is imperative. In pre-operative evaluations, anesthesiologists can play a significant role in minimizing healthcare's carbon footprint by avoiding excessive or unwarranted tests, often conducted for defensive medical practices. Initiatives such as "Choosing Wisely," aimed at reducing healthcare waste and unnecessary procedures, offer a pathway to mitigate patient risks and cut costs simultaneously (Onuoha et al. 2014; Corrente et al. 2024).

Hence, it is advantageous for anesthesiologists to gain a deeper comprehension of the environmental repercussions of clinical practice and leverage this knowledge to mitigate adverse effects while upholding stringent standards for patient safety.

8.4 Impacts of Climate Change on Antibiotic-Resistant Bacteria

Introduction

With the escalation of the climate crisis, millions globally face health impacts from extreme weather events like droughts, floods, and heatwaves. A significant result is the rising occurrence and transmission of bacterial, viral, parasitic, and fungal infections, and VBD in living organisms, along with the increased potential for pathogens to evolve and develop resistance to treatments.

This results in the emergence and dissemination of drug-resistant infections, which are challenging or even untreatable using antimicrobial drugs (antibiotics, antivirals, antifungal, antiparasitics), thereby endangering updated medications for the treatment of different diseases (Dogan 2022). Thus, as a consequence of environmental shifts driven by climate change, there are observable modifications in the prevalence and global dissemination of human diseases, involving infectious ones and accordingly, it is expected that the fluctuating disease rates attributed to environmental change will provoke the utilization of a wide array of medications (Dogan 2022).

8.4.1 Climate Change Aggravates Human Pathogenic Diseases

A wide taxonomic range of pathogenic human ailments, including bacteria, viruses, fungi, and protozoa, as well as various transmission methods such as VB, airborne, and direct contact, may be influenced by ongoing climate changes. The interaction of multiple climatic hazards with various pathogens could significantly worsen human diseases. Research shows that over half of known human pathogenic diseases may be exacerbated by climate change.

A systematic investigation was conducted to explore the influence of ten environmental threats vulnerable to GHG outputs on all identified human pathogenic illnesses (Mora et al. 2022). It was discovered that 58% of infectious diseases globally have been exacerbated by climatic hazards at some point, while 16% were occasionally diminished. Empirical evidence identified 1006 distinct signalings through which environmental risks, via various transmission methods, contributed to pathogenic ailments, which makes comprehensive societal adaptation impractical, emphasizing the urgent need to address the root cause, which is reducing GHG emissions (Mora et al. 2022).

8.4.2 Effect of Climate Change on Infectious Diseases

Over the next decades, and because of the changing climate, it is anticipated that there will be an upsurge in the occurrence of infectious ailments, such as cryptococcal infection. Moreover, significant expansion is expected in both human and animal diarrheal diseases, driven by the facilitated transmission of pathogens like *Cryptosporidium*, *Vibrio* spp., *Escherichia coli*, *Salmonella*, and others. Bacterial infections, encompassing both primary infections like scabies and conjunctivitis and secondary infections such as impetigo, *Staphylococcus*, and *Streptococcus*, are anticipated to proliferate in response to climate change (Redshaw et al. 2013).

Environmental change plays a pivotal role in shaping the dynamics of infectious diseases, impacting them both overtly and covertly. The overt impact includes influencing a subset of diseases such as VBDs, cholera as a waterborne disease, and soil- and foodborne pathogens, as illustrated by McIntyre et al. (2017). Moreover, its indirect effects, intertwined with socioeconomic factors, are equally significant. For instance, climate-related events such as flooding can disrupt established disease control measures, including those aimed at vector control (Boyce et al. 2016).

8.4.2.1 Vector-Borne Diseases

VBDs stand out as the subset of infectious diseases most sensitive to climate, exhibiting vulnerability to a wide array of climate initiators (McIntyre et al. 2017). These diseases, predominantly transmitted by arthropod vectors, are particularly responsive to climate fluctuations for various reasons. Arthropods, being ectothermic, rely on external environmental conditions to regulate their internal temperature. Additionally, their larval development typically hinges on specific humidity conditions and the presence of water bodies. Noteworthy examples of emerging VBDs impacting not only animals but humans as well include malaria, caused by five Plasmodium parasite species and spread by female Anopheles mosquitoes, and arboviruses such as dengue, chikungunya, yellow fever, Zika, and West Nile, which are transmitted by *Aedes* mosquito species like *Aedes albopictus* and *Aedes aegypti*. In addition, some diseases are tick-borne, such as those spread by *Ixodes* ticks that are capable of transmitting bacteria like *Borrelia* and *Flaviviridae* viruses, such as tick-borne encephalitis. Moreover, midge-borne diseases, such as the Bluetongue virus and helminths, along with other parasites, illustrate the influence of weather on the parasitic worm dynamics (Caminade et al. 2019; WHO 2020).

8.4.2.2 Waterborne Diseases

The bacterium responsible for cholera, *Vibrio cholerae*, exists inherently in nature. However, the occurrence of cholera outbreaks is facilitated by the conjunction of floods and unusually elevated sea surface temperatures, as observed by Caminade et al. (2019).

8.4.2.3 Soil-Borne Diseases

To comprehend the implications of atmospheric alteration on soil-borne diseases, it is crucial to recognize the extensive alterations expected in land-use practices, such as shifts in crop species, cultivation methods, and environmental conditions like moisture and temperature. These changes are anticipated to fundamentally reshape the complicated interactions between plants and microorganisms within soil ecosystems, particularly affecting soil-borne pathogens, including fungi, viruses, oomycetes, plant pathogenic bacteria, and nematodes (Lamichhane et al. 2024).

These pathogens often form complex disease syndromes in soil, resulting in chronic declines in both yield and quality of crops. Within this dynamic context, root exudates emerge as a pivotal mediator of plant-microbe interactions, being a complex group of organic material in the rhizosphere with the potential to impact other organisms. Hence, their composition and function are profoundly influenced by ongoing climate change. By understanding the mechanisms underlying these interactions, there emerges the potential to efficiently control plant-related microbes, including disease-causing agents, as elucidated by Lamichhane et al. (2024).

8.4.2.4 Foodborne Diseases

As the climate crisis progresses, the risk of foodborne illnesses is on the rise. According to the WHO, roughly 600 million occurrences of foodborne maladies take place each year, leading to 420,000 deaths. Many of the pathogens responsible for these infections are particularly sensitive to temperature variations, where warm and moist weather conditions serve as catalysts for bacterial growth (Naumova 2022).

In this context, rising temperatures contribute to the proliferation of various types of foodborne infections. Among these, intestinal infections are caused by *Cyclospora*, a minute parasite, which is transmitted through the consumption of feces-contaminated food or water. Also, *Campylobacter*, a bacterium frequently associated with undercooked meat, is a common cause of diarrhea, whereas *Vibrio*, associated with the eating of raw or insufficiently cooked shellfish, is another pathogen exacerbated by heat. *Salmonella,* which induces diarrhea and is often associated with animal feces, also thrives in warmer conditions. Additionally, the incidence of STEC, a prevalent strain of *E. coli,* as well as the parasite *Cryptosporidium,* germ *Listeria,* and the bacterium *Shigella*, all escalate with rising temperatures (Naumova 2022).

Human infections caused by the yeast species *Cryptococcus neoformans* and *C. gattii* usually lead to cryptococcal meningitis or pneumonia. Additionally, fungal spore inhalation can lead to concurrent cryptococcosis affecting the muscle tissues, brain, and the lung. Hence, climate change plays a significant role in fostering conditions conducive to the survival and proliferation of fungal spores, thus contributing to the emergence of *C. gattii* in temperate regions or facilitating its adaptation to new climatic niches (Redshaw et al. 2013).

Climate change gives rise to indirect health consequences, such as an increase in diarrheal diseases like cryptosporidiosis and cholera. These illnesses are commonly connected to sewage overflows, effluent discharges from wastewater treatment facilities, and the pollution of water sources caused by stormwater runoff. Furthermore, *Salmonella* outbreaks exhibit a strong correlation with rising temperatures, as elevated temperatures facilitate the rapid replication of the bacteria, thus exacerbating the incidence of infections (Redshaw et al. 2013).

8.4.2.5 Plastic Pollution and Infectious Diseases

During the closing years of the twentieth century, plastic became extensively utilized owing to its remarkable versatility and durability. However, its extensive use has led to environmental pollution, primarily due to its poor degradability and inefficient waste management practices, resulting in prolonged persistence in the environment. Approximately 55% of plastic debris ultimately ends up in landfills or natural habitats, where it disrupts ecosystems and contributes to the spread of infectious diseases, adversely affecting both human and livestock health (Maquart et al. 2022).

Plastic waste interacts with the environment in numerous ways, exacerbating the propagation of infectious diseases. The accumulation of plastic debris, particularly those capable of retaining water, provides breeding grounds for arthropod vectors and shelters for medically significant species, thus fostering local vector populations and heightening disease transmission risks. Moreover, as stagnant water reservoirs, plastic waste allows the proliferation of pathogenic bacteria like leptospirosis and harmful algae. These microorganisms can form biofilms on plastic parts, facilitating their colonization of new water bodies and perpetuating disease dissemination (Maquart et al. 2022).

As previously mentioned, *Aedes aegypti* and *Aedes albopictus* are among the common mosquitoes that transmit most of the prominent examples of arthropod-borne diseases. Given the widespread distribution of these highly human-associated vector species, over half of the global population is at risk of contracting viruses transmitted by Aedes mosquitoes. (Brady and Hey 2020; Fontenille and Powell 2020). Moreover, plastic wastes are used by the vectors of the Japanese encephalitis virus in their larval stages in aquatic environments. Additionally, discarded plastics serve as cover for other blood-feeding arthropods (e.g., triatomine bugs), which transmit Chagas disease (Boyer et al. 2021). In addition, stagnant water ponds formed by plastic debris can foster the spread of waterborne diseases. These include trematodiasis, dracunculiasis (Guinea worm disease), schistosomiasis (bilharzia), lymphatic filariasis, and onchocerciasis (Fenwick 2006; Maquart et al. 2022). The influence of environmental changes on these microorganisms is summarized in Table 8.2 (Magnano San Lio et al. 2023).

8.4.3 Antibiotics

Infections in humans and animals are commonly treated by antibiotics, but these drugs are also widely employed globally to promote animal growth, regardless of bacterial infections (Stocker et al. 2013). This widespread usage has led to a notable rise in global consumption of these drugs, with a 35% rise in the first decade of the twenty-first century (Field et al. 2014; Van Boeckel et al. 2014).

Accordingly, animals have a sharp buildup of antibiotic-resistant bacteria and antibiotic deposits, which are subsequently excreted in their feces (Sarmah et al. 2006; Watts et al. 2018; Jian et al. 2021). Livestock manure across various regions was found to harbor numerous antibiotic resistance genes (ARGs), playing a role in the diminished efficacy of treatment options for bacterial infections on a global

Table 8.2 The correlation between climate change and infections (Magnano San Lio et al. 2023)

Microorganisms	Role of climate change	Disease
Campylobacter spp. and *Salmonella* spp.	Rising temperatures in the water system contribute to better survival of these microorganisms	Waterborne and foodborne diseases
Vibrio cholerae	Rising temperatures led to natural disasters, determining better conditions for microorganism survival	Waterborne diseases (Cholera)
Candida auris	Gained thermotolerance and salinity tolerance in the wetland ecosystem	Fungal infection (Candidiasis)
Plasmodium falciparum	Rising temperatures and humidity contribute to increased transmissibility	Vector-borne disease (Malaria)
Zika, Chikungunya and Dengue viruses, *Tripanosoma cruzi*	Warmer temperatures led to rising spread of vectors, even in winter months	VBDs (Zika, Chikungunya, Dengue and Chagas diseases)
Pseudomonas aeruginosa, *Klebsiella pneumoniae*, *Escherichia coli*, and *Staphylococcus aureus*	Warm-season changes in temperature contribute to their optimal growth conditions at 32–36 °C	Gram-negative infections (especially in healthcare settings)
SARS-CoV-2	Increased aridity and prolonged droughts led to bats migration and increased viral transmission	

scale, an issue that is primarily attributed to pathogens-induced antibiotic resistance. For instance, in 2008, Souli and colleagues reported certain strains of *Klebsiella pneumoniae, Acinetobacter baumannii,* and *Pseudomonas aeruginosa* developing opposition to all known antibiotics (Souli et al. 2008).

In the European Union, infections resistant to antibiotics are estimated to result in over 30,000 fatalities each year, with Italy and Greece having the highest incidence. Moreover, in low- and middle-income nations in Asia, Africa, and South America, MDR infections exhibit even greater rates of illness and death. Globally, deaths from antibiotic-resistant infections are expected to increase from 700,000 in 2014 to 10 million annually by 2050 (O'Neill 2014; Rodríguez-Verdugo et al. 2020). This resistance contributes to more than 200,000 deaths worldwide each year, including over 2000 from extensively drug-resistant tuberculosis (XDR-TB; WHO 2023).

8.4.3.1 Mechanism of Antibiotic Resistance

Three primary mechanisms have evolved in bacteria to thrive and multiply in the presence of antibiotics (Brauner et al. 2016; Balaban et al. 2019). First, by developing **tolerance**, bacteria can transiently reduce their growth to endure antibiotics (Handwerger and Tomasz 1985; Kester and Fortune 2014). Secondly, the same effect can be obtained by reducing or even preventing the growth of a subpopulation of their cells in a phenomenon known as **persistence** (Balaban et al. 2013; Wakamoto et al. 2013). Finally, by **resistance,** which means modifying their genes, bacteria can withstand higher concentrations of antibiotics for longer periods (Rodríguez-Verdugo et al. 2020).

8.4.3.2 Antibiotic Resistance Genes

ARGs are a global concern because they allow pathogens to resist treatment. They are the main cause of bacterial resistance, with pathogenic bacteria acquiring them through plasmid exchange. Plasmids, integrons (In), and transposons (Tn) can transfer horizontally among various strains and species. Even after resistant strains die, DNA containing ARGs persists in the environment for long periods, shielded by deoxynucleotide enzymes (Bertolla et al. 2000; Dantas et al. 2008).

Jian et al. (2021) reviewed the occurrence and release of ARGs in animal farms, their environmental transmission, and potential routes of contact with human pathogens. They also analyzed the mechanisms of plasmids and phages in horizontal gene transfer (HGT) and summarized the transmission pathways of ARGs.

8.4.3.3 Classification of Resistance Genes

One of the recognized emerging contaminants in the environment is the ARGs, which include both intra- and extra-cellular ARGs (iARGs and eARGs). In 2021, Zarei-Baygi and Smith argued about the horizontal transfer and the persistence of the two types in different environments, and they concluded that in aquatic environments, the eARGs are widely spread, whereas in environments that are rich in nutrients, the chief constituents are the iARGs. The latter type was reported to endure for a short period, whereas the eARGs prevail longer by evading degradation by DNase following their adsorption to soil and sediment particles. Together, these traits highlight the pivotal importance of eARGs in driving the dissemination of resistance to antibiotics throughout the ecosystem (Zarei-Baygi and Smith 2021).

ARGs are categorized according to the classes of antibiotics they confer resistance to, and these involve tetracyclines (tet), sulfonamides (sul), β-lactams (bla), macrolides (erm), aminoglycosides (aac), fluoroquinolone (fca), colistin (mcr), glycopeptide/vancomycin (van), and multidrug resistant to all β-lactam antibiotics (mdr/mec). Recently, Zhuang et al. (2021) conducted an extensive review of the resistome by searching online databases and articles published from 1990 to 2022, providing a comprehensive overview of ARG-related literature. These authors concluded that from hospitals, the most common ARGs are mec, mecA, a genetic element found in MRSA strains, vanA, vanB, and bla, whereas those that are frequently present in farms, soil, wastewater treatment plants (WWTPs), and water were the sul and tet ARGs.

Moreover, the first 50 ARG subtypes in the habitats of air, WWTPs, farms, cities, and soil included eight antibiotic families, namely, bla, sul, tet, aac, mdr, van, amphenicol (flo), and trimethopolyl (dfr). Among them, mecA was the most conventional one worldwide, followed by bla, vanA, vanB, sul1, sul2, and tetM. The top 10 ARGs reported from Asia included $bla_{NDM\text{-}1}$, $bla_{CTX\text{-}M\text{-}15}$, mdrA, $bla_{TEM\text{-}1}$, sul1, vanA, $bla_{KPC\text{-}2}$, sul2, $bla_{CTX\text{-}M\text{-}14}$, and $bla_{OXA\text{-}48}$, which are responsible for β-lactam, MDR, sulfonamide, and glycopeptide antibiotics (Jian et al. 2021).

8.4.3.4 Distribution of Antibiotic Resistance Genes

In various investigations, researchers have explored the occurrence and spread of ARGs in rivers across the globe. Concurrently, additional studies have delved into the fate, associated risks, and origins of the antibiotic resistome, alongside its attenuation dynamics within river water-sediment systems. These inquiries have been the focus of field and microcosm studies, as highlighted by Liu et al. (2024).

The utilization of antibiotics in aquaculture precipitates the emergence of antibiotic resistance in surrounding bacteria within the water column, sediment, and fish-associated bacterial strains. The dissemination of ARGs is also facilitated by HGT among aquatic microbial communities, potentially transporting harmful pathogenic bacteria to humans (Pepi and Focardi 2021).

The emergence and dissemination of ARGs in soil from animal waste and organic matter pose significant threats to ecosystems and human health. An earlier study was conducted to carefully typify ARGs and mobile genetic elements in the soil from three land-use types in China. High-throughput quantitative polymerase chain reaction and 16S rRNA amplicon sequencing demonstrated that ARGs with high diversity and abundance were detected in the agricultural soil, followed by tea plantations and forested areas. The study revealed that in all samples, a total of 124 distinct ARGs were identified, with overlapping subtypes suggesting a common source or high transmission rates. Furthermore, notable regional variations in ARG distribution were observed, with the highest concentration located in southern China (Wu et al. 2023).

Additionally, wetlands serve as another site for the distribution of ARGs, where previous research has documented the levels of ARGs in natural wetlands at high altitudes. An inclusive ARGs profiling across various altitudes and environmental factors in wetlands across the Qinghai-Tibetan Plateau was provided by Yang et al. (2019). The study also identified chief factors that form the geographic distribution and diversity of ARGs, shedding light on the impact of environmental change and elevated human actions on ARG dissemination (Yang et al. 2019). The impacts of constructed wetlands and nanotechnology on ARB and ARG were abridged in the review written by Sharma et al. (2016).

Moreover, ARGs and the recently detected last-resort ARGs (LARGs) have been detected in groundwater across the Tibetan Plateau, along with bacterial communities. Considering that groundwater is a vital drinking water supply for Tibetans, it is plausible that naturally occurring pathogenic and non-pathogenic bacteria carry multiple LARGs. Furthermore, disparities in ARG distribution and bacterial community composition have been observed between samples from low- and high-altitude groundwater (Yang et al. 2023).

In a separate investigation, a recent study explored ARGs in aircraft sewage collected from onboard tanks, revealing a high diversity of ARGs compared to conventional sewage sources. Despite the small amount of bacterial DNA analyzed, numerous distinct ARGs were identified. Notably, blaCARB-4, the β-lactamase gene, was prevalent in aircraft wastewater but absent in municipal wastewater samples unrelated to airports. Additionally, the relative abundance of ARGs in the effluent of aircraft exceeded that in untreated municipal effluent, highlighting the significant role of contaminated water transported by aircraft in facilitating the rapid and global distribution of antibiotic resistance (Jian et al. 2021).

Sharma et al. (2016) conducted a review examining the presence of ARB and ARGs across various environmental systems, including natural environments and drinking water resources. Their findings revealed the detection of ARGs or ARBs exhibiting resistance to antibiotics such as ciprofloxacin, sulfamethoxazole, trimethoprim, quinolone, vancomycin, and tetracycline. Notable examples included tet(A), tet(B), tet(C), tet(G), tet(O), tet(M), tet(W), as well as sul I and sul II. The authors highlighted that resistance development could occur through intrinsic mechanisms, spontaneous mutations (de novo), or HGT from donor bacteria, phages, or free DNA to recipient bacteria (Sharma et al. 2016).

8.4.4 Climate Crisis and Antibiotic Resistance

Altered climate patterns and resistance to antibiotics are intertwined global challenges with complex interactions. Environmental change can exacerbate the dissemination of antibiotic resistance through various mechanisms. For instance, rising temperatures can directly impact bacterial physiology as this rise mimics the effects of antibiotics on bacterial cell physiology, exacerbating antibiotic resistance in aquaculture and heightening threats to human health (Pepi and Focardi 2021).

Additionally, tremendous weather events such as storms and floods can lead to the uncleanness of water sources with antibiotic residues and resistant bacteria, further propagating resistance. Changes in environmental conditions driven by climate disruption, such as alterations in humidity and precipitation patterns, can also influence the survival and transmission of pathogens and their associated resistance genes. Furthermore, environmental change-induced transformations in ecosystems may affect the distribution and abundance of antibiotic-producing organisms and their natural competitors, potentially altering environmental antibiotic resistance dynamics. Addressing the complex interplay between climate change and antibiotic resistance requires integrated strategies that prioritize both environmental sustainability and public health.

8.4.4.1 Heat and Antibiotic Resistance

Temperature serves as a critical factor influencing both bacterial survival in the occurrence of antimicrobial agents and an environmental factor experiencing significant shifts due to climate change. Recent insights highlight temperature's pivotal role across cellular, physiological, ecological, and evolutionary processes impacting bacterial survival. These responses unfold at various biological levels and spatiotemporal scales, encompassing transient physiological changes altering cellular behavior, genetic adaptations linked to thermal stress, and large-scale impacts on antibiotic resistance dynamics (Rodríguez-Verdugo et al. 2020).

Researches examining the interplay between antibiotics and temperature reveal potential thermal adaptations in microbes, potentially leading to elevated antibiotic resistance levels. Although the majority of microbes may not be pathogens, they constitute a reservoir of antibiotic-resistant genes, underscoring the importance of investigating resistance emergence and persistence in both clinical pathogens and environmental bacteria (Rodríguez-Verdugo et al. 2020). Additionally, research examining the influence of climate factors on antibiotic resistance distribution has demonstrated associations between increasing local temperature and rising antibiotic resistance rates across common pathogens (MacFadden et al. 2018). Studies conducted in Europe further corroborate these findings, indicating a prolonged impact of atmospheric temperature on the acceleration of antibiotic resistance rates, with warmer temperatures correlating with swifter surges in resistance levels over time. For instance, the antibiotic resistance of common pathogens such as *Escherichia coli, Klebsiella pneumoniae, and Staphylococcus aureus* was recorded to reach 4.2%, 2.2%, and 2.7%, respectively, when the temperature increases 10 °C. These insights emphasize the urgent need to tackle the intricate relations between climate change, temperature, and antibiotic resistance to mitigate the escalating global threat posed by antimicrobial resistance (MacFadden et al. 2018).

In 28 European countries, another study carried out by McGough et al. (2020) assessed the impact of temperature on antibiotic resistance during the period from 2000 to 2016. Analyzing bacterial pathogens with four antibiotic classes showed a sustained impact of the minimum ambient temperature on the rise in antibiotic resistance rates. In European nations experiencing a 10 °C rise in ambient temperatures, *E. coli* and *K. pneumoniae* exhibited notably swifter elevation in the levels of antibiotic resistance over the study period. The observed rises in antibiotic resistance ranged between 0.33% and 1.2% annually across all antibiotic classes and pathogens.

8.4.4.2 Genetic Adaptations to Temperature and Antibiotic Stressors

Temperature influences antibiotic susceptibility by altering patterns of gene expression and bacterial growth. These impacts are temporary, affecting individual bacteria through plastic changes.

Indeed, genetic modifications can induce the development of antibiotic resistance, known as long-term adaptations. Resistance may arise from spontaneous mutations or HGT, where genetic material is exchanged between bacteria via plasmids or bacteriophages. Studies have shown that HGT is more prevalent at higher temperatures and is increased at 30° relative to 25 °C (Walsh et al. 2011).

Additionally, spontaneous mutations can confer resistance to antibiotics, providing a selective advantage to resistant individuals in antibiotic-rich environments. Temperature, acting on similar cellular functions as antibiotics, can also be selected for resistance mutations. As a result, mutations providing resistance to temperature stress may also grant resistance to antibiotics, and vice versa (Rodríguez-Verdugo et al. 2016), a phenomenon referred to as collateral resistance or cross-protection. This interplay between temperature and antibiotics underscores the intricacy of antibiotic resistance development (Rodríguez-Verdugo et al. 2016, 2020; Dragosits et al. 2013).

8.4.4.3 Other Factors Contributing to Increased Antibiotic Resistance

While antibiotic use has traditionally been regarded as the primary driver of antibiotic resistance escalation in populations, it is essential to recognize that additional factors may contribute to the regional dissemination of antibiotic resistance (Wellington et al. 2013; Holmes et al. 2016).

- *Rising Regional Temperature and Population Concentration*

The correlation between temperature and population density applies to common pathogens like *E. coli*, *K. pneumoniae*, and *S. aureus* (MacFadden et al. 2018). As infection rates rise and antibiotic-resistant strains emerge, the intensification of climate change is expected to lead to an increase in antibiotic-resistant pathogens.

- *Disasters and Infections*

Heat-related flooding can worsen the transmission of antibiotic resistance by enabling the dispersal of waterborne pathogens and contaminants, including antibiotic-resistant bacteria. As floodwaters overflow, they can carry antibiotic-resistant bacteria from sewage systems, livestock facilities, and other sources into the environment, increasing the risk of contamination of water sources and further dissemination of resistance genes. Additionally, flooding can disrupt sanitation infrastructure, leading to poor hygiene conditions that promote the transmission of resistant pathogens among humans and animals. Overall, heat-associated flooding poses significant challenges for managing antibiotic resistance and controlling the dissemination of resistant bacteria (Vieira et al. 2016; Blakiston and Freeman 2020).

– *Pollution and Antibiotic Resistance*

Pollutants have been shown to trigger the activation of ARGs and bacterial mutations. Elevated levels of agricultural runoff, such as eutrophication stemming from fertilizer usage, are anticipated to spur bacterial proliferation in aquatic ecosystems. This heightened bacterial presence amplifies the likelihood of horizontal transfer of ARGs. Moreover, pollutants like heavy metals, stemming from industrial activities, can infiltrate the environment during flooding events, which are exacerbated by climate change. This influx of contaminants is projected to elevate the prevalence of ARGs within soil environments (Dogan 2022).

8.4.5 Research Gaps and Needs

- Given these challenges, there is a pressing demand for research and development efforts to combat emerging infectious diseases. These diseases spread more rapidly and extensively in a warming world, where climate changes significantly affect illness patterns and consequently impact pharmaceutical usage.
- It is urgent to advance the discovery of new antimicrobial treatments as environmental change exacerbates the ongoing risk of drug-resistant infections.
- Promoting equitable access to care for drug-resistant infections and reinforcing the global health system (Dogan 2022).
- Increase funds and investments in discovering new diagnostic methods to determine drug-resistant infections, the route these infections are transported, and inventing new therapeutics that can treat and prevent these infections. For instance, the Antimicrobial Resistance (AMR) Action Fund was raised by the Wellcome.org (2021) to invent new antibiotics to treat drug-resistant infections.

8.5 Climate Changes and Veterinary Medicine

Introduction

The projected effects of climate change on livestock's health are primarily negative. Climate change involves several variations that can impact animal health, such as droughts, pests and diseases, floods, and humidity, with the heat waves being considered the primary threat. Studies have shown conflicting results regarding the effects of climate change on animal health and welfare. On one hand, there is evidence that higher temperatures may lead to fewer deaths and better living conditions for humans and animals in regions with extremely cold winters. On the other hand, other studies have shown that changes in air temperature, precipitation patterns, and the frequency of extreme weather events can have a negative impact on animal welfare. These effects can be both direct and indirect. The direct consequences encompass alterations in the accessibility of the suitable besides heat stress and dryness. Meanwhile, the shift in both hosts and vectors, dissemination and inhabitants, along with the modifications to the host-pathogen dynamics and alterations in the duration and severity of infectious diseases, construct the indirect effects (Ezzati et al. 2004; Altizer et al. 2013).

8.5.1 Heat Stress, Livestock Health, and Aquaculture

The primary causal factor of climate change-induced immediate impact is elevated temperature and the heightened occurrence and severity of heat waves. Heat stress hinders metabolic processes, causes oxidative stress, and suppresses the immune system, all of which can harm cattle and increase their risk of infection and death (Das et al. 2016; Lacetera 2019). In addition, the availability and quality of water and feed sources, as well as the persistence and spread of disease-causing pathogens and their vectors, are all significantly affected by climate change. Indeed, increased temperature acts as a double-edged sword; although it has a negative impact on animals' appetite centers, leading to reduced feed intake with decreased body weight and body condition scores, this reduction in feed intake helps decrease heat production, as digestion generates heat, particularly in ruminants (Baile and Forbes 1974).

In this regard, the decline in food intake was reported to start around 25–26 °C for lactating cows and accelerate as temperatures rise further. At extreme temperatures like 40 °C, feed intake can drop significantly, by up to 40% in cows, 22–35% in dairy goats, or 8–10% in buffalo heifers (Das et al. 2016).

Along with reducing productivity and fertility in animals (Forman et al. 2008), increased temperature also weakens the immunological and endocrine systems, making animals more susceptible to diseases (Das et al. 2016). Therefore, the decreased weight and food intake, along with the compromised immunity, can jeopardize the animal's health and leave it vulnerable to the occurrence and recurrence of infectious illnesses. Several factors can participate, such as the altered interaction between hosts and pathogens, as well as changes in the distribution and quantity of vectors and hosts. Thus, understanding the connection between climate change and animal diseases is crucial for formulating efficient approaches to prevent and manage diseases (Chang et al. 2023).

In the same pattern, rising temperatures in marine and aquatic environments pose risks to fisheries and aquaculture. As water temperatures increase, the quality of the water declines, leading to more frequent occurrences of hazardous algae blooms, reduced levels of oxygen, and slower rates of feeding and growth. These factors contribute to an increase in disease incidence (Handisyde et al. 2014).

The lessons learned from the human health sector highlight the need to address heat-related difficulties. This emphasizes the need to include methods for adapting to and reducing rising temperatures in Veterinary Services' climate change agenda (Stephen et al. 2019).

8.5.2 Adapting Veterinary Services to Combat Climate Change Challenges

Following significant environmental changes, animals will encounter the effects of climate change through various interconnected mechanisms, including alterations in infectious disease patterns, increased vulnerability to heat, pollutants, and extreme weather events; changes in access to essential natural resources for survival; and shifts in animal ecology, sociobiology, and population dynamics.

Veterinarians specializing in livestock care possess the expertise to identify and implement early interventions that safeguard animals from heat stress, as well as to provide services that facilitate recovery from heat-related incidents. Additionally, they may improve the overall well-being of populations, thereby increasing their capacity to address future climate change risks (resilience) and offering recommendations on dietary and other approaches to reduce emissions from animals (Kiran et al. 2022).

To address the requirements for action in the veterinary field, veterinary services must concentrate on several critical areas to alleviate the impacts of climate change. The areas encompass: (a) the provision of essential services; (b) the reduction of population vulnerability; (c) the enhancement of population resilience; and (d) the mitigation of the underlying causes of climate change risks. The objectives of disease prevention, productivity maintenance, and the preservation of healthy systems will remain, yet veterinary services will necessitate structural and extensive modifications.

Effective resource allocation requires the integration of health information that combines hazard monitoring with an assessment of population susceptibility. Initiatives should not only prioritize risk management but also integrate capacity building for robust and resilient animal populations and health systems.

Climate change will impact the determinants of animal health in various direct and indirect manners across multiple levels of influence. Veterinary services must implement substantial modifications to address the interrelated challenges of sustainable development, climate change, and biodiversity loss effectively. Effective collaborations and governance structures are essential for acquiring and integrating knowledge regarding changes in global and regional socioecological systems (Stephen and Soos 2021).

8.5.3 Climate Changes and Veterinary Drugs and Vaccines

Numerous veterinary professionals routinely transport prescription medications in their vehicles as they conduct their daily duties, which often involve visiting farms or other locations where they administer care to animals.

Veterinary drugs include substances or combinations used for the prevention, treatment, or diagnosis of animal diseases, as well as for the intentional regulation of animal physiological functions (Wang et al. 2021a, b). In recent decades, a variety of veterinary drugs have been used in animals, encompassing numerous categories aimed at disease control, infection management, and growth promotion. These include antibacterial agents, parasiticides, hormone drugs, anti-inflammatory agents, analgesics, and vaccines (Wang et al. 2021a, b). All drugs possess a limited shelf life and are sensitive to temperature variations, necessitating storage under specific thermal conditions as outlined in their respective "Summary of Product Characteristics (SPC)."

The recommended storage temperatures are divided into two categories: controlled room temperature (cool) and cold storage circumstances. Controlled room temperature ranges from 2 to 25 °C, while cold storage conditions range from 2 to 8 °C (Taylor 2001; Summerhays 2010). Preventing freezing is critical for the majority of vaccines and many liquid products. Nevertheless, temperatures within a vehicle can vary, potentially diverging from the ideal storage temperatures indicated in the SPC. Exposure to unsuitable temperatures can weaken the quality of drugs, resulting in thermal degradation and reduced therapeutic effectiveness (Johansen et al. 1993; Grant et al. 1994; Church et al. 1994; Ballereau et al. 1997). Therefore, Küpper et al. (2006) recommended the replacement of temperature-sensitive medications exposed to temperatures beyond the manufacturer's specified limits. Additionally, they recommended replacing medications at least once annually.

Regarding livestock vaccinations, the primary goal of this practice is to enhance animal well-being and minimize the financial losses linked to disease outbreaks (Meeusen et al. 2007). Hence, vaccines are essential for maintaining international trade in animal goods, as they effectively limit the transmission of infection and minimize the need for widespread slaughter of vulnerable animals (Paton et al. 2006). The World Organization for Animal Health (WOAH) has developed vaccines for 53% (63/117) of the diseases listed in their Manual of Diagnostic Tests and Vaccines for Terrestrial Animals, 2021. However, members of the Office International des Epizooties (OIE) have historically reported the production of vaccines for 68 diseases. It is important to acknowledge that the data provided to the WOAH on vaccine production only includes information from laboratories under the National Veterinary Services. The data do not include vaccines produced by private firms that have not reported to the WOAH (2023).

From the production process until their administration, vaccines often require consistent storage at temperatures ranging from 2 to 8 °C, which necessitates the use of a cold-chain system for their transit (Brandau et al. 2003). It is strongly recommended to implement vaccination efforts in areas where diseases like foot-and-mouth disease and rabies are prevalent. These campaigns strive to avert disease outbreaks and mitigate the economic damage these diseases inflict. These countries are typically characterized as low- and middle-income nations that lack widespread access to a reliable electrical supply or an efficient cold-chain infrastructure for vaccines. Given this, the heat stability of most existing vaccinations against internationally significant animal diseases is a crucial problem (WOAH 2023).

It is remarkable that even in 2021, most vaccinations for both human and animal diseases will still rely on cold-chain systems to maintain their effectiveness during production, transportation, storage, and application. Vaccines that are impervious to heat and freezing could yield significant economic and health advantages in both human and animal healthcare. Immunizations that are stable under both heat and freezing conditions could potentially decrease vaccine waste and mitigate the risks associated with administering ineffective immunizations (Chen and Kristensen 2009). The World Health Organization's Global Vaccine Action Plan 2011–2020 (WHO 2013) has designated thermostable vaccines as the top research focus due to these reasons.

8.5.4 Climate Changes and Veterinary Drug Residues

The administration of veterinary pharmaceuticals to animals raised for food might lead to the existence of remnants in animal-derived commodities such as meat, milk, eggs, fish, and honey, which could potentially endanger the health of consumers. While these veterinary drugs have effectively treated diseases in farm animals, their prolonged usage has led to drug resistance. Additionally, their administration in aquaculture can pollute water and jeopardize the safety of drinking water (Gehring et al. 2006; Boobis et al. 2017). Additionally, the excessive and illicit use of certain veterinary drugs poses significant risks to human health, causing environmental pollution and adversely affecting animals, plants, and microorganisms (Wang et al. 2021a, b). Antimicrobial resistance, hypersensitivity reactions, changes to the natural intestinal flora, as well as carcinogenic, mutagenic, and teratogenic effects are just some of the health problems that drug residues can cause.

Certain residues possess the capacity to accumulate and biomagnify within the food chain. This scenario begins with the administration of these drugs to animals through various routes, leading to metabolism and resulting in residues within the animal's body or excretion into the environment. Furthermore, the introduction of veterinary drugs into aquatic organisms results in their presence in aquatic products and water bodies. These drugs are absorbed by plants and trees through water and soil and subsequently reintroduced into the bodies of animals and humans upon consumption, thereby posing a significant risk to human health (Fig. 8.1; Wang et al. 2021a, b).

To protect human health, governments and related agencies have established regulations, standards, and protocols for monitoring veterinary drug residues in the edible tissues of food-producing animals (Pratiwi et al. 2023). The issue stems from the recurrent improper administration of drugs and the failure to comply with withdrawal periods necessary for the removal of these residues from animal tissues. Although the quantities of these residues are typically minimal and not inherently toxic, there remains a deficiency in comprehensive global data regarding their prevalence. To mitigate the presence of veterinary drug residues, it is essential for all animal health professionals to understand the pharmacokinetics, pharmacodynamics, and toxicological effects of these drugs in order to avert potential adverse impacts on human health (Beyene 2016). The FDA establishes regulatory thresholds for approved veterinary medications in Title 21, Part 556, detailing tolerances for residues of new animal drugs in food (FDA 2023). Similarly, in the European Union (EU), Commission Regulation (EU) No. 37/2010 regulates the maximum

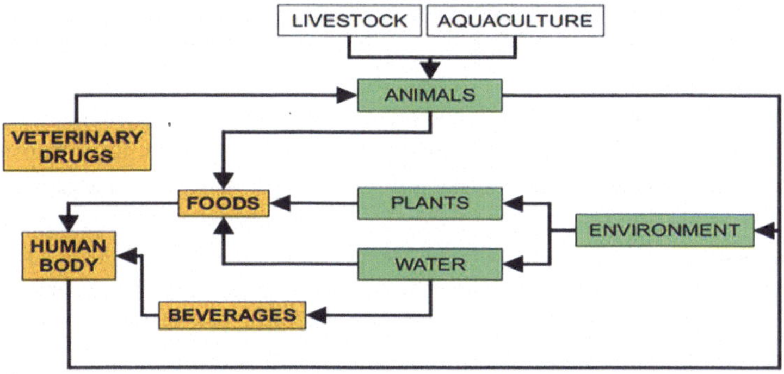

Fig. 8.1 A series of processes involving veterinary drug residues in the human body. (Wang et al. 2021a, b)

residue limits of pharmacologically active substances in food derived from animals (European Commission Regulation 2010).

In 2009, Gale et al. reported findings that demonstrate a significant impact of climate change on the occurrence and prevalence of animal diseases in Great Britain. The insular nature may provide some defense against climate-induced alterations in animal diseases; however, this could merely delay essential adaptations due to the persistent trend of GW. The island of Ireland has consistently encountered substantial introductions of emerging diseases via the importation of infected animals.

Cooper et al. (2015) examined the potential impact of twenty-first-century climate change on veterinary drug residues in food, highlighting that environmental changes may elevate disease rates in farm animals, thereby increasing the need for drug administration. Multiple published reviews indicate that the utilization of veterinary medicines is expected to rise in the forthcoming decades due to heightened disease challenges, although some researchers have expressed skepticism regarding this prediction (Cooper et al. 2015).

With the exception of parasitic diseases, published evidence on the subject is limited; however, it is likely that the alterations will not be significant. Cooper and colleagues anticipate an increase in the use of veterinary medicines in food animals, despite the difficulty in isolating the impacts of climate change from other contributing factors. Advancing drug resistance may necessitate increased medication dosages or the inappropriate application of alternative therapies. There is a potential for an increased presence of veterinary drug residues in locally produced foods over an extended duration (Cooper et al. 2015).

8.5.5 Climate Changes and Prevalence of Antibiotic-Resistant Bacteria in Aquaculture and Livestock Health

Aquaculture emerges as a pivotal endeavor in addressing millennium challenges, meeting the demand for protein while upholding environmental stewardship (Maite et al. 2021; Pepi and Focardi 2021). Particularly significant in the Mediterranean region, aquaculture nevertheless raises concerns due to its reliance on antibiotics for treatment and prevention. This practice triggers antibiotic resistance in bacteria within the water column, sediment, and among bacterial strains associated with fish, posing consequential impacts. Additionally, this practice can entail the transmission of this resistance to humans, rendering antibiotic usage ineffective for human health (Rigos and Troisi 2005).

Does climate change contribute to bacterial resistance in aquaculture, was a question raised by Pepi and Focardi (2021) who posited that rising temperatures, akin to the effect of antibiotics, could influence bacterial cell physiology, potentially fostering antibiotic resistance. This scenario is highly noticed in the Mediterranean Sea, that emerges as a climate change "hot spot," amplifying aspects of antibiotic resistance in aquaculture and heightening threats to human health. Moreover, changes in climate and rising temperatures result in the bacteria adapting, leading to a substantial increase in antibiotic resistance.

Additionally, it is crucial to recognize that climate change is increasingly leading to more frequent interactions between humans and animals, resulting in the emergence of zoonotic and VBDs with pandemic potential. Furthermore, there is widespread acknowledgment that the utilization of antimicrobials across human, animal, and environmental spheres serves as a principal driver for antimicrobial resistance.

Strategies are imperative to mitigate adverse impacts on human health, with a pivotal focus on reducing antibiotic usage. Concurrently, addressing climate change through the reduction of anthropogenic CO_2 emissions is essential. Globally, the Mediterranean climate areas can be considered the world's most vulnerable ecosystems due to their high susceptibility to climate change. Hence, to ensure the effective and beneficial growth of aquaculture resources, it is imperative to adhere to the principles of sustainability as outlined in the 2030 Agenda and Global Development Goals recommendations.

This agenda involves prioritizing the well-being of humans and animals, as well as maintaining harmony with the environment. To accomplish these goals, it is crucial to work through different hinges, such as decreasing the utilization of antibiotics in aquaculture and addressing the declining effectiveness of antibiotics by implementing rigorous hygiene practices and maintaining clean fish farms. Furthermore, researchers should address current concerns surrounding antibiotic usage in human therapy by conducting additional investigations to explore novel antibiotics derived from marine microorganisms for potential use in aquaculture, besides introducing innovative vaccines to reduce the incidence of fish infections. Meanwhile, it is vital to significantly reduce the factors contributing to climate change and address the issue through a comprehensive global approach, considering the environmental, medical, and social perspectives.

Therefore, adopting a One Health approach, which involves interdisciplinary collaboration across veterinary, ecological, and medical domains, aligning with sustainability principles, is imperative to confront these significant challenges to human and animal health and environmental integrity in the Mediterranean region.

The Mediterranean Sea region boasts unique local attributes due to differences among the surrounding countries. Consequently, reducing antibiotic usage and addressing the repercussions of increasing temperatures and climate change demand involvement from local authorities and residents. Their profound and well-founded comprehension of regional conditions is indispensable for effective action. Moreover, it is important to take measures to mitigate climate change, such as reducing CO_2 emissions as the primary step to control temperature rise, which could otherwise escalate by 3–4 °C by 2030 (Pepi and Focardi 2021).

References

Adeyeye E, New BJM, Chen F et al (2022) Sustainable medicines use in clinical practice: a clinical pharmacological view on eco-pharmaco-stewardship. Br J Clin Pharmacol 88:3023–3029. https://doi.org/10.1111/bcp.15140

Agache I, Sampath V, Aguilera J et al (2022) Climate change and global health: a call to more research and more action. Allergy 77:1389–1407. https://doi.org/10.1111/all.15229

Al-Hazmi HE, Hassan GK, Kurniawan TA et al (2024) Technological solutions to landfill management: towards recovery of biomethane and carbon neutrality. J Environ Manag 354:120414. https://doi.org/10.1016/j.jenvman.2024.120414

Altizer S, Ostfeld RS, Johnson PT et al (2013) Climate change and infectious diseases: from evidence to a predictive framework. Science 341:514–519. https://doi.org/10.1126/science.1239401

Andersen L, Corazon SS, Stigsdotter UK (2021) Nature exposure and its effects on immune system functioning: a systematic review. Int J Environ Res Public Health 18:1416. https://doi.org/10.3390/ijerph18041416

Athanasios G, Carroll K, Majeed A et al (2004) The epidemiology of the comorbidity of epilepsy in the general population. Epilepsia 45:1613–1622. https://doi.org/10.1111/j.0013-9580.2004.17504

Aus der Beek T, Weber F-A, Bergmann A et al (2016) Pharmaceuticals in the environment: global occurrence and perspectives. Environ Toxicol Chem 35:823–835. https://doi.org/10.1002/etc.3339

Avakian GN, Oleĭnikova OM, Lagutin IV et al (2013) Season fluctuations of carbamazepine and valproate concentration in patients with epilepsy. Zh Nevrol Psikhiatr Im S S Korsakova 113:34–39. PMID: 24429946

Baile CA, Forbes JM (1974) Control of feed intake and regulation of energy balance in ruminants. Physiol Rev 54:160–214. https://doi.org/10.1152/physrev.1974.54.1.160

Balaban NQ, Gerdes K, Lewis K et al (2013) A problem of persistence: still more questions than answers? Nat Rev Microbiol 11:587–591. https://doi.org/10.1038/nrmicro3076

Balaban NQ, Helaine S, Lewis K et al (2019) Definitions and guidelines for research on antibiotic persistence. Nat Rev Microbiol 17:441–448. https://doi.org/10.1038/s41579-019-0196-3

Ballereau F, Prazuck T, Schrive I et al (1997) Stability of essential drugs in the field: results of a study conducted over a two-year period in Burkina Faso. Am J Trop Med Hyg 57:31–36. https://doi.org/10.4269/ajtmh.1997.57.31

Bartolini E, Sander JW (2019) Dealing with the storm: an overview of seizure precipitants and spontaneous seizure worsening in drug-resistant epilepsy. Epilepsy Behav 97:212–218. https://doi.org/10.1016/j.yebeh.2019.05.036

Bell KN, Hogue CJR, Manning C et al (2001) Risk factors for improper vaccine storage and handling in private provider offices. Pediatrics 107:E100. https://doi.org/10.1542/peds.107.6.e100

Berger M, Bastl K, Bastl M et al (2020) Impact of air pollution on symptom severity during the birch, grass and ragweed pollen period in Vienna, Austria: importance of O3 in 2010–2018. Environ Pollut 263(Pt A):114526. https://doi.org/10.1016/j.envpol.2020.114526

Berry HL, Bowen K, Kjellstrom T (2010) Climate change and mental health: a causal pathways framework. Int J Public Health 55:123–132. https://doi.org/10.1007/s00038-009-0112-0

Bertolla F, Kay E, Simonet P (2000) Potential dissemination of antibiotic resistance genes from transgenic plants to microorganisms. Infect Control Hosp Epidemiol 21:390–393. https://doi.org/10.1086/501779

Beyene T (2016) Veterinary drug residues in food-animal products: its risk factors and potential effects on public health. J Vet Sci Technol 7:285. https://doi.org/10.4172/2157-7579.1000285

Black C, Gerriets JE, Fontaine JH et al (2017) Early life wildfire smoke exposure is associated with immune dysregulation and lung function decrements in adolescence. Am J Respir Cell Mol Biol 56:657–666. https://doi.org/10.1165/rcmb.2016-0380OC

Blakiston MR, Freeman JT (2020) Population-level exposures associated with MRSA and ESBL-E. coli infection across district health boards in Aotearoa New Zealand: an ecological study. N Z Med J 133:62–69

Boobis A, Cerniglia C, Chicoine A et al (2017) Characterizing chronic and acute health risks of residues of veterinary drugs in food: latest methodological developments by the joint FAO/WHO expert committee on food additives. Crit Rev Toxicol 47:885–899. https://doi.org/10.1080/10408444.2017.1340259

Booth A, Jager A, Faulkner SD et al (2023) Pharmaceutical company targets and strategies to address climate change: content analysis of public reports from 20 pharmaceutical companies. Int J Environ Res Public Health 20:3206. https://doi.org/10.3390/ijerph20043206

Boyce RR, Reyes M, Matte M et al (2016) Severe flooding and malaria transmission in the western Ugandan Highlands: implications for disease control in an era of global climate change. J Infect Dis 214:1403–1410. https://doi.org/10.1093/infdis/jiw363

Boyer S, Durand B, Yean S et al (2021) Host-feeding preference and diel activity of mosquito vectors of the Japanese encephalitis virus in rural Cambodia. Pathogens 10:376. https://doi.org/10.3390/pathogens10030376

Brady OJ, Hay SI (2020) The global expansion of dengue: how Aedes aegypti mosquitoes enabled the first pandemic arbovirus. Annu Rev Entomol 65:191–208. https://doi.org/10.1146/annurev-ento-011019-024918

Brandau DT, Jones LS, Wiethoff CM et al (2003) Thermal stability of vaccines. J Pharm Sci 92:218–231. https://doi.org/10.1002/jps.10296

Brauner A, Fridman O, Gefen O et al (2016) Distinguishing between resistance, tolerance and persistence to antibiotic treatment. Nat Rev Microbiol 14:320–330. https://doi.org/10.1038/nrmicro.2016.34

Brocklebank D, Ram F, Wright J et al (2001) Comparison of the effectiveness of inhaler devices in asthma and chronic obstructive airways disease: a systematic review of the literature. Health Technol Assess 5:1–149. https://doi.org/10.3310/hta5260

Burnham JP (2021) Climate change and antibiotic resistance: a deadly combination. Ther Adv Infect Dis 8:2049936121991374. https://doi.org/10.1177/2049936121991374

Cakmak S, Dales RE, Vidal CB (2010) Air pollution and hospitalization for epilepsy in Chile. Environ Int 36:501–505. https://doi.org/10.1016/j.envint.2010.03.008

Cámara-Leret R, Frodin DG, Adema F et al (2020) New Guinea has the world's richest Island flora. Nature 584:579–583. https://doi.org/10.1038/s41586-020-2549-5

Caminade C, McIntyre KM, Jones AE (2019) Impact of recent and future climate change on vector-borne diseases 2019. Ann N Y Acad Sci 1436:157–173. https://doi.org/10.1111/nyas.13950

Campbell-Lendrum D, Prüss-Ustün A (2019) Climate change, air pollution and noncommunicable diseases. Bull World Health Organ 97:160–161. https://doi.org/10.2471/BLT.18.224295

Canadian Centre for Occupational Health and Safety (CCOHS) (2022) OSH answers fact sheets: sulfur dioxide. Published November 2, 2022. Accessed November 3, 2022. https://www.ccohs.ca/oshanswers/chemicals/chem_profiles/sulfurdi.html

Cavaleiro RJ, Paciência I, Hoffimann E et al (2021) The neighbourhood natural environment is associated with asthma in children: a birth cohort study. Allergy 76:348–358. https://doi.org/10.1111/all.14493

Climate and Clean Air Coalition (2019) Hydrofluorocarbons (HFC). https://www.ccacoalition.org/fr/slcps/hydrofluorocarbons-hfc. Accessed 19 Aug 2019.

Chang Q, Zhou H, Khan N et al (2023) Can climate change increase the spread of animal diseases? Evidence from 278 villages in China. Atmos 14:1581. https://doi.org/10.3390/atmos14101581

Chang WW, Boonhat H, Lin RT (2020) Incidence of respiratory symptoms for residents living near a petrochemical industrial complex: a meta-analysis. Int J Environ Res Public Health 17:2474. https://doi.org/10.3390/ijerph17072474

Chen D, Kristensen D (2009) Opportunities and challenges of developing thermostable vaccines. Expert Rev Vaccines 8:547–557. https://doi.org/10.1586/erv.09.20

Chen H, Samet JM, Bromberg PA et al (2021) Cardiovascular health impacts of wildfire smoke exposure. Part Fibre Toxicol 18:2. https://doi.org/10.1186/s12989-020-00394-8

Chong TW, Castle DJ (2004) Layer upon layer: thermoregulation in schizophrenia. Schizophr Res 69:149–157. https://doi.org/10.1016/s0920964(03)00222-6

Church WH, Hu SS, Henry AJ (1994) Thermal degradation of injectable epinephrine. Am J Emerg Med 12:306–309. https://doi.org/10.1016/0735-6757(94)90145-7

Cianconi P, Betrò S, Janiri L (2020) The impact of climate change on mental health: a systematic descriptive review. Front Psych 11:74. https://doi.org/10.3389/fpsyt.2020.00074

Cloyd J (1991) Pharmacokinetic pitfalls of present antiepileptic medications. Epilepsia 32:S53–S65. Review. PMID: 1743171

Cohn DF (1982) Is there a climatic influence on the blood level of anticonvulsive drugs? Med Hypotheses 8:427–430. https://doi.org/10.1016/0306-9877(82)90037-8

Colborn T, Schultz K, Herrick L et al (2014) An exploratory study of air quality near natural gas operations. Hum Ecol Risk Assess Int J 20:86–105. https://doi.org/10.1080/10807039.2012.749447

Cooper KM, McMahon C, Fairweather I et al (2015) Potential impacts of climate change on veterinary medicinal residues in livestock produce: an Island of Ireland perspective. Trends Food Sci Technol 44:21–35. https://doi.org/10.1016/j.tifs.2014.03.007

CoP Sustainable Inhaler Initiative (2022) Toronto: Centre for Sustainable Health Systems. https://www.sustainablehealthsystems.ca/copsustainableinhalerinitiative. Accessed 15 Feb 2022

Corrente A, Pace MC, Fiore M (2024) Climate change and human health: last call to arms for us. World J Clin Cases 12:1870–1874. https://doi.org/10.12998/wjcc.v12.i11.1870

Costello EJ, Copeland W, Angold A (2011) Trends in psychopathology across the adolescent years: what changes when children become adolescents, and when adolescents become adults? J Child Psychol Psychiatry Allied Discip 52:1015–1025. https://doi.org/10.1111/j.1469-7610.2011.02446.x

D'Amato G, Chong-Neto HJ, Monge Ortega OP et al (2020) The effects of climate change on respiratory allergy and asthma induced by pollen and mold allergens. Allergy 75:2219–2228. https://doi.org/10.1111/all.14476

Damialis A, Gilles S, Sofiev M et al (2021) Higher airborne pollen concentrations correlated with increased SARS-CoV-2 infection rates, as evidenced from 31 countries across the globe. Proc Natl Acad Sci USA 118(12):e2019034118. https://doi.org/10.1073/pnas.2019034118

Dantas G, Sommer MO, Oluwasegun RD et al (2008) Bacteria subsisting on antibiotic. Science 320:100–103. https://doi.org/10.1126/science.1155157

Das R, Sailo L, Verma N et al (2016) Impact of heat stress on health and performance of dairy animals: a review. Vet World 9:260–268. https://doi.org/10.14202/vetworld.2016.260-268

Daughton CG, Ruhoy IS (2013) Lower-dose prescribing: minimizing "side effects" of pharmaceuticals on society and the environment. Sci Total Environ 443:324–327. https://doi.org/10.1016/j.scitotenv.2012.10.092

Dawidowicz AL, Kalityński R, Trocewicz J et al (2002) Investigation of propofol renal elimination by HPLC using supported liquid membrane procedure for sample preparation. Biomed Chromatogr 16:455–458. https://doi.org/10.1002/bmc.183

Dekhuijzen PNR, Lavorini F, Usmani OS (2016) Patients' perspectives and preferences in the choice of inhalers: the case for the Respimat or HandiHaler. Patient Prefer Adherence 10:1561–1572. https://doi.org/10.2147/PPA.S82857

Deo RP (2014) Pharmaceuticals in the Surface Water of the USA: a review. Curr Environ Health Rep 1:113–122. https://doi.org/10.1007/s40572-014-0015-y

Dogan SE (2022) Will climate change lead to more drug-resistant infections?. https://www.preventionweb.net/news/will-climate-change-lead-more-drug-resistant-infections (10 May 2024)

Dragosits M, Mozhayskiy V, Quinones-Soto S et al (2013) Evolutionary potential, cross-stress behavior and the genetic basis of acquired stress resistance in *Escherichia coli*. Mol Syst Biol 9:643. https://doi.org/10.1038/msb.2012.76

Erbas B, Jazayeri M, Lambert KA et al (2018) Outdoor pollen is a trigger of child and adolescent asthma emergency department presentations: a systematic review and meta-analysis. Allergy 73:1632–1641. https://doi.org/10.1111/all.13407

European Commission Regulation (EU) No 37/2010 of 22 December 2009 on Pharmacologically Active Substances and Their Classification Regarding Maximum Residue Limits in Foodstuffs of Animal Origin. 2010. [(accessed on 5 July 2023)]. Off J Eur Union. Available online: https://eur-lex.europa.eu/LexUriServ/LexUriServ.do?uri=OJ:L:2010:015:0001:0072:en:PDF

Ezzati M, Lopez AD, Rodgers AA et al (2004) Comparative quantification of health risks: global and regional burden of disease attributable to selected major risk factors. World Health Organization, Geneva, p 2258

Fadadu RP, Grimes B, Jewell NP et al (2021) Association of wildfire air pollution and health care use for atopic dermatitis and itch. JAMA Dermatol 157:658–666. https://doi.org/10.1001/jamadermatol.2021.0179

Fenwick A (2006) Waterborne infectious diseases—could they be consigned to history? Science 313:1077–1081. https://doi.org/10.1126/science.1127184

Ferlisi M, Shorvon S (2014) Seizure precipitants (triggering factors) in patients with epilepsy. Epilepsy Behav 33:101–105. https://doi.org/10.1016/j.yebeh.2014.02.019

Fernández-López J, Botella-Martínez C, de Vera CNR et al (2020) Vegetable soups and creams: raw materials, processing, health benefits, and innovation trends. Plan Theory 9:1769. https://doi.org/10.3390/plants9121769

Fidler L, Green S, Wintemute K (2022) Pressurized metered-dose inhalers and their impact on climate change. Can Med Assoc J 194:E460. https://doi.org/10.1503/cmaj.211747

Field CB, Barros VR, Dokken DJ et al (2014) Climate Change 2014: impacts, adaptation, and vulnerability. Part A: Global and sectoral aspects. Working Group II Contribution to the Fifth Assessment Report of the Intergovernmental Panel on Climate Change. Cambridge University Press, Cambridge/New York

Food and Drug Administration (2023) CFR-code of Federal Regulations Title 21. PART 556: tolerances for residues of new animal drugs in food. Food and Drug Administration, Silver Spring

Forman S, Hungerford N, Yamakawa M et al (2008) Climate change impacts and risks for animal health in Asia. Revue Scientifique et Technique (International Office of Epizootics) 27:581–597. https://doi.org/10.20506/rst.27.2.1814

Forte R Jr, Dibble C (1999) The role of international environmental agreements in metered-dose inhaler technology changes. J Allergy Clin Immunol 104:S217–S220. https://doi.org/10.1016/s0091-6749(99)70033-3

Fontenille D, Powell JR (2020) From anonymous to public enemy: how does a mosquito become a feared arbovirus vector? Pathogens 9:265. https://doi.org/10.3390/pathogens9040265

Gagné M, Karanikas A, Green S et al (2023) Reductions in inhaler greenhouse gas emissions by addressing care gaps in asthma and chronic obstructive pulmonary disease: an analysis. BMJ Open Respir Res 10:e001716. https://doi.org/10.1136/bmjresp-2023-001716

Gale P, Drew T, Phipps LP et al (2009) The effect of climate change on the occurrence and prevalence of livestock diseases in Great Britain. J Appl Microbiol 106:1409–1423. https://doi.org/10.1111/j.1365-2672.2008.04036.x

Gehring R, Baynes RE, Riviere JE (2006) Application of risk assessment and management principles to the extralabel use of drugs in food-producing animals. J Vet Pharmacol Ther 29:5–14. https://doi.org/10.1111/j.1365-2885.2006.00707.x

Gilles S, Blume C, Wimmer M et al (2020) Pollen exposure weakens innate defense against respiratory viruses. Allergy 75:576–587. https://doi.org/10.1111/all.14047

Godde CM, Mason-D'Croz D, Mayberry DE et al (2021) Impacts of climate change on the livestock food supply chain; a review of the evidence. Global. Food Secur 28:100488. https://doi.org/10.1016/j.gfs.2020.100488

Gordon D (2020) Sustainability in the operating room: reducing our impact on the planet. Anesthesiol Clin 38:679–692. https://doi.org/10.1016/j.anclin.2020.06.006

Grant TA, Carroll RG, Church WH et al (1994) Environmental temperature variations cause degradations in epinephrine concentration and biological activity. Am J Emerg Med 12:319–322. https://doi.org/10.1016/0735-6757(94)90148-1

Griscom BW, Busch J, Cook-Patton SC et al (2020) National mitigation potential from natural climate solutions in the tropics. Philos Trans R Soc B Biol Sci 375:20190126. https://doi.org/10.1098/rstb.2019.0126

Gulcebi MI, Bartolini E, Lee O et al (2021) Climate change and epilepsy: insights from clinical and basic science studies. Epilepsy Behav 116:107791. https://doi.org/10.1016/j.yebeh.2021.107791

Haikerwal A, Akram M, Sim MR et al (2016) Fine particulate matter (PM 2.5) exposure during a prolonged wildfire period and emergency department visits for asthma. Respirology 21:88–94. https://doi.org/10.1111/resp.12613

Handisyde NT, Ross LG, Badjeck M-C et al (2014) The effects of climate change on world aquaculture: a global perspective. Technical Report. https://www.researchgate.net/publication/265497647_The_Effects_of_Climate_Change_on_World_Aquaculture_A_Global_Perspective

Handwerger S, Tomasz A (1985) Antibiotic tolerance among clinical isolates of bacteria. Annu Rev Pharmacol Toxicol 25:349–380. https://doi.org/10.1146/annurev.pa.25.040185.002025

Hew M, Lee J, Susanto NH et al (2019) The 2016 Melbourne thunderstorm asthma epidemic: risk factors for severe attacks requiring hospital admission. Allergy 74:122–130. https://doi.org/10.1111/all.13609

Holmes AH, Moore LS, Sundsfjord A et al (2016) Understanding the mechanisms and drivers of antimicrobial resistance. Lancet 387:176–187. https://doi.org/10.1016/S0140-6736(15)00473-0

Islam MZ, Johnston J, Sly PD (2020) Green space and early childhood development: a systematic review. Rev Environ Health 35:189–200. https://doi.org/10.1515/reveh-2019-0046

Jacobson TA, Kler JS, Hernke MT et al (2019) Direct human health risks of increased atmospheric carbon dioxide. Nat Sustain 2:691–701

Janson C, Henderson R, Löfdahl M et al (2020) Carbon footprint impact of the choice of inhalers for asthma and COPD. Thorax 75:82–84. https://doi.org/10.1136/thoraxjnl-2019-213744

Jian Z, Zeng L, Xu T et al (2021) Antibiotic resistance genes in bacteria: occurrence, spread, and control. J Basic Microbiol 61:1049–1070. https://doi.org/10.1002/jobm.202100201

Johansen RB, Schafer NC, Brown PI (1993) Effect of extreme temperatures on drugs for prehospital ACLS. Am J Emerg Med 11:450–452. https://doi.org/10.1016/0735-6757(93)90080-u

Jones KE, Patel NG, Levy MA et al (2008) Global trends in emerging infectious diseases. Nature 451:990–993. https://doi.org/10.1038/nature06536

Jones NR, Agnew M, Banic I et al (2019) Ragweed pollen and allergic symptoms in children: results from a three-year longitudinal study. Sci Total Environ 683:240–248. https://doi.org/10.1016/j.scitotenv.2019.05.284

Kang J, Yang Y (2020) Energy carbon emission structure and reduction potential focused on the supply side and demand-side. PLoS One 15:e0239634. https://doi.org/10.1371/journal.pone.0239634

Kartoglu U, Milstien J (2014) Tools and approaches to ensure quality of vaccines throughout the cold chain. Expert Rev Vaccines 13:843–854. https://doi.org/10.1586/14760584.2014.923761

Kayode-Afolayan SD, Ahuekwe EF, Nwinyi OC (2022) Impacts of pharmaceutical effluents on aquatic ecosystems. Sci Afr 17:e01288. https://doi.org/10.1016/J.SCIAF.2022.E01288

Kern DG, Kern E, Crausman RS et al (2011) A retrospective cohort study of lung cancer incidence in nylon flock workers, 1998–2008. Int J Occup Environ Health 17:345–351. https://doi.org/10.1179/oeh.2011.17.4.345

Kester JC, Fortune SM (2014) Persisters and beyond: mechanisms of phenotypic drug resistance and drug tolerance in bacteria. Crit Rev Biochem Mol Biol 49:91–101. https://doi.org/10.3109/10409238.2013.869543

Kilian J, Kitazawa M (2018) The emerging risk of exposure to air pollution on cognitive decline and Alzheimer's disease – evidence from epidemiological and animal studies. Biom J 41:141–162. https://doi.org/10.1016/j.bj.2018.06.001

Kiran D, Sander WE, Duncan C (2022) Empowering veterinarians to be planetary health stewards through policy and practice. Front Vet Sci 9:775411. https://doi.org/10.3389/fvets.2022.77541

Kontny MJ, Destefano G, Jager PD et al (1999) Issues surrounding mdi formulation development with non-cfc propellants. J Aerosol Med 4:181–187. https://doi.org/10.1089/jam.1991.4.181

Küpper TEAH, Schraut B, Rieke B et al (2006) Drugs and drug administration in extreme environments. J Travel Med 13:35–47. https://doi.org/10.1111/j.1708-8305.2006.00007.x

Küster A, Adler N (2014) Pharmaceuticals in the environment: scientific evidence of risks and its regulation. Philos Trans R Soc B Biol Sci 369:20130587. https://doi.org/10.1098/rstb.2013.0587

Kwakye G, Brat GA, Makary MA (2011) Green surgical practices for health care. Arch Surg 146:131–136. https://doi.org/10.1001/archsurg.2010.343

Lacetera N (2019) Impact of climate changes on animal health and welfare. Anim Front 9:26–31. https://doi.org/10.1093/af/vfy030

Lamichhane JR, Barbetti MJ, Chilvers MI et al (2024) Exploiting root exudates to manage soil-borne disease complexes in a changing climate. Trends in Microbiology 32:27–37. https://doi.org/10.1016/j.tim.2023.07.011

Landrigan PJ, Raps H, Cropper M et al (2023) The Minderoo-Monaco commission on plastics and human health. Ann Glob Health 89:23. https://doi.org/10.5334/aogh.4056

Langbein T, Sonntag H, Trapp D et al (1999) Volatile anaesthetics and the atmosphere: atmospheric lifetimes and atmospheric effects of halothane, enflurane, isoflurane, desflurane and sevoflurane. Br J Anaesth 82:66–73. https://doi.org/10.1093/bja/82.1.66

Latapí Agudelo MA, Jóhannsdóttir L, Davídsdóttir B (2019) A literature review of the history and evolution of corporate social responsibility. Int J Corp Soc Responsib 4:1. https://doi.org/10.1186/s40991-018-0039-y

Lavorini F, Corrigan CJ, Barnes PJ et al (2011) Retail sales of inhalation devices in European countries: so much for a global policy. Respir Med 105:1099–10103. https://doi.org/10.1016/j.rmed.2011.03.012

Lee J, Kronborg C, O'Hehir RE, Hew M (2017) Who's at risk of thunderstorm asthma? The ryegrass pollen trifecta and lessons learnt from the Melbourne thunderstorm epidemic. 132:146–148. https://doi.org/10.1016/j.rmed.2017.10.012. Epub 2017 Oct 18.

Lentz MP, Graham DJ, van Vliet MTH (2024) Drought impact on pharmaceuticals in surface waters in Europe: case study for the Rhine and Elbe basins. Sci Total Environ 922:171186. https://doi.org/10.1016/j.scitotenv.2024.171186

Levy K, Woster AP, Goldstein RS et al (2016) Untangling the impacts of climate change on waterborne diseases: a systematic review of relationships between diarrheal diseases and temperature, rainfall, flooding, and drought. Environ Sci Technol 50:4905–4922. https://doi.org/10.1021/acs.est.5b06186

Lõhmus M (2018) Possible biological mechanisms linking mental health and heat-a contemplative review. Int J Environ Res Public Health 15:1515. https://doi.org/10.3390/ijerph15071515

Loladze I, Nolan JM, Ziska LH et al (2019) Rising atmospheric CO_2 lowers concentrations of plant carotenoids essential to human health: a meta-analysis. Mol Nutr Food Res 63:1801047. https://doi.org/10.1002/mnfr.201801047

Liu C, Shan X, Jinping CJ et al (2024) Fate, risk and sources of antibiotic resistome and its attenuation dynamics in the river water-sediment system: field and microcosm study. Environ Pollut 340(Pt 2):122853. https://doi.org/10.1016/j.envpol.2023.122853

Liu Z, Gatt A, Mikati M et al (1993) Effect of temperature on kainic acid induced seizures. Brain Res 631:51–58. https://doi.org/10.1016/0006-8993(93)91185-u

MacFadden D, McGough F, Fisman D et al (2018) Antibiotic resistance increases with local temperature. Nat Clim Chang 8:510–514. https://doi.org/10.1038/s41558-018-0161-6

MacNeill AJ, Lillywhite R, Brown CJ (2017) The impact of surgery on global climate: a carbon footprinting study of operating theatres in three health systems. Lancet Planet Health 1:e381–e388. https://doi.org/10.1016/S2542-5196(17)30162-6

Magnano San Lio R, Favara G, Maugeri A et al (2023) How antimicrobial resistance is linked to climate change: an overview of two intertwined global challenges. Int J Environ Res Public Health 20:1681. https://doi.org/10.3390/ijerph20031681

Maite C, Peral IP, Ramos S, Basurco B, López-Francos MA, Cavallo M, Perez J, Aguilera C, Fu-rones D, Reverté C et al (2021) Deliverable 1.2 of the Horizon 2020 Project MedAID (GA number 727315) [(accessed on 5 January 2021)]; Available online: https://archimer.ifremer.fr/doc/00515/62630

Malik A, Lenzen M, McAlister S et al (2018) The carbon footprint of Australian health care. Lancet Planet Health 2:e27–e35. https://doi.org/10.1016/S2542-5196(17)30180-8

Maquart PO, Froehlich Y, Boyer S (2022) Plastic pollution and infectious diseases. Lancet Planet Health 6:e842–e845. https://doi.org/10.1016/S2542-5196(22)00198-X

Marchini T, Zirlik A, Wolf D (2020) Pathogenic role of air pollution particulate matter in cardiometabolic disease: evidence from mice and humans. Antioxid Redox Signal 33:263–279. https://doi.org/10.1089/ars.2020.8096

Marder E (2023) Individual variability, statistics, and the resilience of nervous systems of crabs and humans to temperature and other perturbations. eNeuro 10:ENEURO.0425-23.2023. https://doi.org/10.1523/ENEURO.0425-23.2023

Martin DM, Yanez ND, Treggiari MM (2017) An initiative to optimize waste streams in the operating room: RECycling in the Operating Room (RECOR) Project. AANA J 85:108–112

McGain F, Story D, Hendel S (2009) An audit of intensive care unit recyclable waste. Anaesthesia 64:1299–1302. https://doi.org/10.1111/j.1365-2044.2009.06102

McGough SF, MacFadden DR, Hattab MW et al (2020) Rates of increase of antibiotic resistance and ambient temperature in Europe: a cross-national analysis of 28 countries between 2000 and 2016. Euro Surveill 25:1900414. https://doi.org/10.2807/1560-7917.ES.2020.25.45.1900414

McGuire JL, Hasskarl H, Bode G et al (2007) Pharmaceuticals, general survey. In: Ullmann's encyclopedia of industrial chemistry. VCH, Weinheim. https://doi.org/10.1002/14356007.a19_273.pub2

McIntyre KM, Setzkorn C, Hepworth PG et al (2017) Systematic assessment of the climate sensitivity of important human and domestic animal's pathogens in Europe. Sci Rep 7:7134. https://doi.org/10.1038/s41598-017-06948-9

Meeusen ENT, Walker J, Peters A et al (2007) Current status of veterinary vaccines. Clin Microbiol Rev 20:489–510. https://doi.org/10.1128/CMR.00005-07

Montreal protocol on substances that deplete the ozone layer. Report of the UNEP medical technical options Committee, 2014 assessment. http://conf.montreal-protocol.org/meeting/oewg/oewg-6/presession/Background%20Documents%20are%20available%20in%20English%20only/MTOC-Assessment Report-2014.pdf. Accessed 11 Jan 2019

Montreal protocol on substances that deplete the ozone layer: UNEP 2014 report of the Medical Technical Options Committee. Nairobi [Kenya]: United Nations Environment Programme; 2015

Mora C, McKenzie T, Gaw IM et al (2022) Over half of known human pathogenic diseases can be aggravated by climate change. Nat Clim Chang 12:869–875. https://doi.org/10.1038/s41558-022-01426-1

Naumova EN (2022) Climate change and food safety risk, Friedman School of Nutrition Science and Policy (tufts.edu), June 29, 2022. https://nutrition.tufts.edu/news/climate-change-and-food-safety-risk

NHS England sustainable development unit (2019) Reducing the use of natural resources in health and social care 2018 report. https://www.sduhealth.org.uk/policy-strategy/reporting/natural-resource-footprint-2018.aspx. Accessed 20 Jun 2019

O'Callaghan-Gordo C, Orta-Martínez M, Kogevinas M (2016) Health effects of non-occupational exposure to oil extraction. Environ Health 15:56, s12940-016-0140-0141. https://doi.org/10.1186/s12940-016-0140-1

O'Neill J (2014) Antimicrobial resistance: tackling a crisis for the health and wealth of nations:1–16. https://www.who.int/news/item/29-04-2019-new-report-calls-for-urgent-action-to-avert-antimicrobial-resistance-crisis

Onuoha OC, Arkoosh VA, Fleisher LA (2014) Choosing wisely in anesthesiology: the gap between evidence and practice. JAMA Int Med 174:1391–1395. https://doi.org/10.1001/jamainternmed.2014.2309

Özelsel TJ, Sondekoppam RV, Buro K (2019) The future is now-it's time to rethink the application of the global warming potential to anesthesia. Can J Anaesth 66:1291–1295. https://doi.org/10.1007/s12630-019-01385-w

Pambudi NA, Sarifudin A, Gandidi IM et al (2022) Vaccine cold chain management and cold storage technology to address the challenges of vaccination programs. Energy Rep 8:955–972. https://doi.org/10.1016/j.egyr.2021.12.039

Parnas J, Flachs H, Gram L et al (1978) Excretion of antiepileptic drugs in sweat. Acta Neurol Scand 58:197–204. https://doi.org/10.1111/j.1600-0404.1978.tb02879

Patel M, Kumar R, Kishor K et al (2019) Pharmaceuticals of emerging concern in aquatic systems: chemistry, occurrence, effects, and removal methods. Chem Rev 119:3510–3673. https://doi.org/10.1021/acs.chemrev.8b00299

Paton DJ, de Clercq K, Greiner M et al (2006) Application of non-structural protein antibody tests in substantiating freedom from foot-and-mouth disease virus infection after emergency vaccination of cattle. Vaccine 24:6503–6512. https://doi.org/10.1016/j.vaccine.2006.06.032

Patz JA, Kovats RS (2002) Hotspots in climate change and human health. Br Med J 325:1094–1098. https://doi.org/10.1136/bmj.325.7372.1094

Pepi M, Focardi S (2021) Antibiotic-resistant bacteria in aquaculture and climate change: a challenge for health in the Mediterranean area. Int J Environ Res Public Health 18(11):5723. https://doi.org/10.3390/ijerph18115723

Peters C, Rosch RE, Hughes E et al (2016) Temperature-dependent changes in neuronal dynamics in a patient with an SCN1A mutation and hyperthermia induced seizures. Sci Rep 1(6):31879. https://doi.org/10.1038/srep31879

Pichler PP, Jaccard IS, Weisz U et al (2019) International comparison of health care carbon footprints. Environ Res Lett 14:064004. https://doi.org/10.1088/1748-9326/ab19e1

Pratiwi R, Ramadhanti SP, Amatulloh A et al (2023) Recent advances in the determination of veterinary drug residues in food. Food Secur 12:3422. https://doi.org/10.3390/foods12183422

Prescott SL (2020) Allergy as a sentinel measure of planetary health and biodiversity loss. Allergy 75:2358–2360. https://doi.org/10.1111/all.14255

Ralls E (2023) Thunderstormes are becoming more intense due to climate change. https://www.earth.com/news/thunderstorms-are-becoming-more-intense-due-to-climate-change/

Ramgopal S, Thome-Souza S, Loddenkemper T (2013) Chronopharmacology of anti-convulsive therapy. Curr Neurol Neurosci Rep 13:339. https://doi.org/10.1007/s11910-013-0339-2

Rauer D, Gilles S, Wimmer M et al (2021) Ragweed plants grown under elevated CO_2 levels produce pollen which elicit stronger allergic lung inflammation. Allergy 76:1718–1730. https://doi.org/10.1111/all.14618

Redmayne N, Robertson S, Kockler J et al (2015) Repackaged sodium valproate tablets–Meeting quality and adherence to ensure seizure control. Seizure 31:108–111. https://doi.org/10.1016/j.seizure.2015.07.007

Redshaw CH, Stahl-Timmins WM, Lora E et al (2013) Potential changes in disease patterns and pharmaceutical use in response to climate change. J Toxicol Environ Health Part B Crit Rev 16:285–320. https://doi.org/10.1080/10937404.2013.802265

Roca I, Akova M, Baquero F et al (2015) The global threat of antimicrobial resistance: science for intervention. New Microbes New Infect 6:22–29. https://doi.org/10.1016/j.nmni.2015.02.007

Rodríguez-Verdugo A, Lozano-Huntelman L, Cruz-Loya M et al (2020) Compounding effects of climate warming and antibiotic resistance. iScience 23:101024. https://doi.org/10.1016/j.isci.2020.101024

Rodríguez-Verdugo A, Tenaillon O, Gaut BS (2016) First-step mutations during adaptation restore the expression of hundreds of genes. Mol Biol Evol 33:25–39. https://doi.org/10.1093/molbev/msv228

Rogelj J, Fricko O, Meinshausen M et al (2017) Understanding the origin of Paris Agreement emission uncertainties. Nat Commun 8:15748. https://doi.org/10.1038/ncomms15748

Rorie A, Poole JA (2021) The role of extreme weather and climate related events on asthma outcomes. Immunol Allergy Clin N Am 41:73–84. https://doi.org/10.1016/j.iac.2020.09.009

Rovira E, Cuadras A, Aguilar X et al (2014) Asthma, respiratory symptoms and lung function in children living near a petrochemical site. Environmental Research 133:156–163. https://doi.org/10.1016/j.envres.2014.05.0221026

Rigos G, Troisi GM (2005) Antibacterial agents in Mediterranean finfish farming: a synopsis of drug pharmacokinetics in important euryhaline fish species and possible environmental implications. Rev Fish Biol Fish 15:53–73. https://doi.org/10.1007/s11160-005-7850-8

Ryan SM, Nielsen CJ (2010) Global warming potential of inhaled anesthetics: application to clinical use. Anesth Analg 111:92–98. https://doi.org/10.1213/ANE.0b013e3181e058d7

Salas RN (2020) The climate crisis and clinical practice. N Engl J Med 382:589–591. https://doi.org/10.1056/nejmp2000331

Sarkisian MR, Holmes GL, Carmant L et al (1999) Effects of hyperthermia and continuous hippocampal stimulation on the immature and adult brain. Brain Dev 21:318–325. https://doi.org/10.1016/s0387-7604(99)00032-7

Sarmah AK, Meyer MT, Boxall ABA (2006) A global perspective on the use, sales, exposure pathways, occurrence, fate and effects of veterinary antibiotics (VAs) in the environment. Chemosphere 65:725–729. https://doi.org/10.1016/j.chemosphere.2006.03.026

Seidman PA, Parker BM (1998) Sharps disposal in the operating room: current clinical practices and costs. Anesth Analg 87:634–636. https://doi.org/10.1097/00000539-199809000-00027

Sharma VK, Johnson N, Cizmas L et al (2016) A review of the influence of treatment strategies on antibiotic resistant bacteria and antibiotic resistance genes. Chemosphere 150:702–714. https://doi.org/10.1016/j.chemosphere.2015.12.084

Sheffield PE, Landrigan PJ (2011) Global climate change and children's health: threats and strategies for prevention. Environ Health Perspect 119:291–298. https://doi.org/10.1289/ehp.1002233

Sherman JD, Ryan S (2010) Ecological responsibility in anesthesia practice. Int Anesthesiol Clin 48:139–151. https://doi.org/10.1097/AIA.0b013e3181ea7587

Shu H, Jönsson BA, Larsson M et al (2014) PVC flooring at home and development of asthma among young children in Sweden, a 10-year follow-up. Indoor Air 24:227–235. https://doi.org/10.1111/ina.12074

Šklebar T, Rudež KD, Rudež LK et al (2022) Global warming and prescribing: a review on medicines' effects and precautions. Psychiatr Danub 34:5–12. PMID: 36752238

United Nations (2018) Montreal protocol on substances that deplete the ozone layer. https://ozone.unep.org/treaties/montreal-protocol#:~:text=Treaty%20relevant%20annex-,The%20Montreal%20Protocol%20on%20Substances%20that%20Deplete%20the%20Ozone%20Layer,consumption%20of%20ozone%2Ddepleting%20substances

Sisodiya SM, Fowler HJ, Lake I et al (2019) Climate change and epilepsy: time to take action. Epilepsia Open 4:524–536. https://doi.org/10.1002/epi4.12359

Souli M, Galani I, Giamarellou H (2008) Emergence of extensively drug-resistant and pandrug-resistant Gram-negative bacilli in Europe. Euro Surveill 13:19045. PMID: 19021957

Summerhays GES (2010) Monitoring of temperature in cars with regard to the pharmaceutical precautions of medicine storage. Equine Vet Educ 12:307–311. https://doi.org/10.1111/j.2042-3292.2000.tb00066.x

Stephen C, Soos C (2021) The implications of climate change for Veterinary Services. Revue Scientifique et Technique (International Office of Epizootics) 40:421–430. https://doi.org/10.20506/rst.40.2.3234.2

Stephen C, Carron M, Stemshorn B (2019) Climate change and veterinary medicine: action is needed to retain social relevance. Can Vet J 60:1356–1358

Stocker TF, Qin D, Plattner G-K et al (2013) Climate change 2013: the physical science basis. Working Group I Contribution to the Fifth Assessment Report of the Intergovernmental Panel on Climate Change. Cambridge University Press, Cambridge/New York

Stowell JD, Geng G, Saikawa E et al (2019) Associations of wildfire smoke PM2.5 exposure with cardiorespiratory events in Colorado 2011–2014. Environment International 133(Pt A):105151. https://doi.org/10.1016/j.envint.2019.105151

Sulbaek Andersen MP, Sander SP, Nielsen OJ et al (2010) Inhalation anaesthetics and climate change. Br J Anaesth 105:760–766. https://doi.org/10.1093/bja/aeq259

Sulbaek Andersen MP, Nielsen OJ, Sherman JD (2023) Assessing the potential climate impact of anaesthetic gases. Lancet Planet Health 7:e622–e629. https://doi.org/10.1016/S2542-5196(23)00084-0

Symeonides C, Brunner M, Mulders Y et al (2021) Buy-now-pay-later: hazards to human and planetary health from plastics production, use and waste. J Paediatr Child Health 57:1795–1804. https://doi.org/10.1111/jpc.15777

Tang J, Sun J, Wang W et al (2021) Pharmaceuticals in two watersheds in Eastern China and their ecological risks. Environ Pollut 277:116773. https://doi.org/10.1016/j.envpol.2021.116773

Taylor J (2001) Recommendations on the control and monitoring of storage and transportation temperatures of medicinal products. Pharm J 267:128–131. https://www.researchgate.net/publication/228807303_Recommendations_on_the_control_and_monitoring_of_storage_and_transportation_temperatures_of_medicinal_products

United Nations Environment Programme (UNEP) and United Nations Environment Programme Technical University of Denmark Partnership. Emissions Gap Report 2020. 20.500.11822/34467

UN (2019) https://www.un.org/en/un75/climate-crisis-race-we-canwin#:~:text=Rising%20temperatures%20are%20fueling%20environmental,acidifying%2C%20and%20forests%20are%20burning.

Van Boeckel TP, Gandra S, Ashok A et al (2014) Global antibiotic consumption 2000 to 2010: an analysis of national pharmaceutical sales data. Lancet Infect Dis 14:742–750. https://doi.org/10.1016/S1473-3099(14)70780-7

Vangroenweghe F (2017) Good vaccination practice: it all starts with a good vaccine storage temperature. Porcine Health Manag 3:24. https://doi.org/10.1186/s40813-017-0071-4

Vieira MT, Marlow MA, Aguiar-Alves F et al (2016) Living conditions as a driving factor in persistent methicillin-resistant Staphylococcus aureus colonization among HIV-infected youth. Pediatr Infect Dis J 35:1126–1131. https://doi.org/10.1097/INF.0000000000001246

Wakamoto Y, Dhar N, Chait R et al (2013) Dynamic persistence of antibiotic-stressed mycobacteria. Science 339:91–95. https://doi.org/10.1126/science.1229858

Walsh TR, Weeks J, Livermore DM et al (2011) Dissemination of NDM-1 positive bacteria in the New Delhi environment and its implications for human health: an environmental point prevalence study. Lancet Infect Dis 11:355–362. https://doi.org/10.1016/S1473-3099(11)70059-7

Wang B, Xie K, Lee K (2021a) Veterinary drug residues in animal-derived foods: sample preparation and analytical methods. Food Secur 10:555. https://doi.org/10.3390/foods10030555

Wang F, Harindintwali JD, Yuan Z et al (2021b) Technologies and perspectives for achieving carbon neutrality. Innovation (Cambridge) 2:100180. https://doi.org/10.1016/j.xinn.2021.100180

Watts N, Amann M, Ayeb-Karlsson S et al (2018) The Lancet countdown on health and climate change: from 25 years of inaction to a global transformation for public health. Lancet 391:581–630. https://doi.org/10.1016/S0140-6736(17)32464-9

Watts N, Amann M, Arnell N et al (2021) The 2020 report of the Lancet Countdown on health and climate change: responding to converging crises. Lancet 397:129–170. https://doi.org/10.1016/S0140-6736(20)32290-X

Welch MJ, Nelson HS, Shapiro G et al (2004) Comparison of patient preference and ease of teaching inhaler technique for Pulmicort Turbuhaler versus pressurized metered-dose inhalers. J Aerosol Med 17:129–139. https://doi.org/10.1089/089426804l457174

Wellcome.org (2021). https://wellcome.org/news/were-backing-amr-action-fund-what-it-means-antibiotic-innovation

Wellington EMH, Boxall AB, Cross P et al (2013) The role of the natural environment in the emergence of antibiotic resistance in gram-negative bacteria. Lancet Infect Dis 13:155–165. https://doi.org/10.1016/S1473-3099(12)70317-1

WHO (2013) Global Vaccine Action Plan. WHO, Geneva. https://www.who.int/teams/immunization-vaccines-and-biologicals/strategies/global-vaccine-action-plan#:~:text=The%20Global%20Vaccine%20Action%20Plan%20(GVAP)%20%E2%80%95%20endorsed%20by%20the,for%20people%20in%20all%20communities

WHO (2020) Vector-borne diseases. https://www.who.int/news-room/fact-sheets/detail/vector-borne-diseases

WHO (2023) Global Tuberculosis Report 2023. https://www.who.int/teams/global-tuberculosis-programme/tb-reports/global-tuberculosis-report-2023/tb-disease-burden/1-3-drug-resistant-tb

Wiesinger H, Wang Z, Hellweg S (2021) Deep dive into plastic monomers, additives, and processing aids. Environ Sci Technol 55:9339–9351. https://doi.org/10.1021/acs.est.1c00976

Wilkinson AJK, Anderson G (2020) Sustainability in inhaled drug delivery. Pharm Med 34:191–199. https://doi.org/10.1007/s40290-020-00339-8

WOAH (2023) Manual of Diagnostic tests and Vaccines for Terrestrial Animals, 12th edition. https://www.woah.org/fileadmin/Home/eng/Health_standards/tahm/A_summry.htm

Wu J, Guo S, Lin H et al (2023) Uncovering the prevalence and drivers of antibiotic resistance genes in soils across different land-use types. J Environ Manag 344:118920. https://doi.org/10.1016/j.jenvman.2023.118920

Wu R (2019) The carbon footprint of the Chinese health-care system: an environmentally extended input–output and structural path analysis study. Lancet Planet Health 3:e413–e419. https://doi.org/10.1016/S2542-5196(19)30192-5

Yang CL, Hicks EA, Mitchel P et al (2021) 2021 Canadian thoracic society guideline - a focused update on the management of very mild and mild asthma. Can J Respir Crit Care Sleep Med 5:205–245. https://doi.org/10.1080/24745332.2021.1877043

Yang Y, Liu G, Ye C et al (2019) Bacterial community and climate change implication affected the diversity and abundance of antibiotic resistance genes in wetlands on the Qinghai-Tibetan Plateau. J Hazard Mater 361:283–293. https://doi.org/10.1016/j.jhazmat.2018.09.002

Yang Y, Zhou J, Shi D et al (2023) Landscape of antibiotic resistance genes and bacterial communities in groundwater on the Tibetan Plateau and distinguishing their difference with low-altitude counterparts. J Hazard Mater 459:132300. https://doi.org/10.1016/j.jhazmat.2023.132300

Yasny JS, White J (2012) Environmental implications of anesthetic gases. Anesth Prog 59:154–158. https://doi.org/10.2344/0003-3006-59.4.154

Zarei-Baygi A, Smith AL (2021) Intracellular versus extracellular antibiotic resistance genes in the environment: prevalence, horizontal transfer, and mitigation strategies. Bioresour Technol 319:124181. https://doi.org/10.1016/j.biortech.2020.124181

Zhang Y, Xu R, Ye T et al (2023) Heat exposure and hospitalization for epileptic seizures: a nationwide case-crossover study in Brazil. Urban Clim 49:101497. https://doi.org/10.1016/j.uclim.2023.101497

Zhuang M, Achmon Y, Cao Y et al (2021) Distribution of antibiotic resistance genes in the environment. Environ Pollut 285:117402. https://doi.org/10.1016/j.envpol.2021.117402

Climate Change Impacts on Freshwater Toxins, and Its Health Effects

9

Nahed Mohamed Mohamed Ismail,
Marwa Tamim Ahmed Abdel-Wareth,
and Sara Sayed Mahmoud Sayed

Abstract

The influence of the global warming resulting from climate change (CC) together with the presence of freshwater toxins has led to many risks for different water uses and human health. The present chapter discussed the potential risks associated with the effect of global warming on the diffusion of freshwater toxins into the production of drinking water and its impact on the transmission of waterborne diseases affecting human health and other aquatic organisms. Freshwater toxins were classified into algal toxins (cyanotoxins) and fungal toxins (mycotoxins). Cyanotoxins were classified into cyclic peptides (microcystins (MCs)) having different chemical compounds. These are attributed to the variability of their characteristic function. Cyanobacterial toxins may lead to instability of the health of aquatic ecosystems and should not be ignored with their specific harmful effects. Fungal toxins (mycotoxins) were also considered natural toxins coinciding with the prediction of climate change scenarios, which may affect human health and the ecosystem. The adverse effects of freshwater toxins on human health and the aquatic environment were illustrated. The harmful algal bloom (HAB) toxins have acute and chronic health consequences in mammals (including humans) and aquatic organisms. Mycotoxins negatively affect aquatic organisms and animal and human health. Regarding the detection methods of freshwater toxins, various traditional analytical methods were reported for the detection of cyanotoxins, including high-performance liquid chromatography (HPLC), protein phosphatase inhibition assay (PPIA), and enzyme-linked immunosorbent assay (ELISA). It is important to realize the natural toxins in an engagement in climate policy plans, planning, and action. Mitigation and adapta-

N. M. M. Ismail (✉) · M. T. A. Abdel-Wareth · S. S. M. Sayed
Department of Environmental Research, Theodor Bilharz Research Institute, Giza, Egypt

A. Saad-Hussein et al. (eds.), *Climate Change Impacts on Toxins and Health Effects*,
https://doi.org/10.1007/978-981-96-3416-3_9

tion efforts should be focused on surveillance schedules, more laws, and comprehension of the toxicity of the toxic substances.

Keywords

Climate change · Natural toxins · Cyanotoxins · Mycotoxins · Human health · Aquatic ecosystem · Aquatic organisms · Adaptations and Mitigations

Abbreviations

AChE	Acetylcholinesterase
AFB1	Aflatoxin B1
AFM1	Aflatoxin M1
ALT	Alanine transaminase
AST	Aspartate aminotransferase
ATXs	*Anatoxin-a*
BEA	*Beauvericin*
BMAA	*β-N-methylamino-L-alanine*
CC	Climate change
CYNs	*Cylindrospermopsin*
DON	*Deoxynivalenol*
ELISA	Enzyme-linked immunosorbent assay
ENNs	*Enniatins*
FL	Fluorescence
GC	Gas chromatography
HABs	Harmful algal blooms
HCC	Hepatocellular carcinoma
HPLC	High-performance liquid chromatography
HT2	HT-2 toxin
IPCC	Intergovernmental Panel on Climate Change
LC	Liquid chromatography
LDCs	Least developed countries
LODs	Limits of detection
MCs	*Microcystins*
MS	Mass spectrometry
NAPAs	National Adaptation Programmes of Action
NODs	*Nodularins*
OTA	Ochratoxin A
PAT	Patulin
PCR	Polymerase chain reaction
PKC	Protein kinase C
PPIA	Protein phosphatase inhibition assay
RAPD	Random amplified polymorphic DNA
STX	*Saxitoxin*

T2	T-2 toxin
UNFCCC	United Nations Framework Convention on Climate Change
UV	Ultraviolet
ZEN	*Zearalenone*

9.1 Introduction

According to the definition of the Sixth Assessment Report of the Intergovernmental Panel on Climate Change (IPCC): "Earth's climate is undergoing adverse global changes as an unequivocal result of anthropogenic activity. The planet's climate has been going through a period of significant changes since the mid-twentieth century, as demonstrated by the rise in global air and ocean temperatures, the increase in sea level, and the shrinking ice sheets" (Zingales et al. 2022; IPCC 2021).

The influence of climate change (CC) in the presence of freshwater toxins has led to many risks for different water uses. The potential risks associated with the proliferation of freshwater toxins on drinking water production and its impact on the transmission of water-borne diseases affecting human health were discussed in this chapter (Zamyadi 2014).

The fungal life cycle stages and their ability to survive, output toxins, and infect crops are mainly affected by climatic factors such as temperature, water, light quality/quantity, extreme drought, desertification, and fluctuations of humid/dry cycles (Paterson 2006). Thus, mycotoxigenic fungi may alter their geographical distribution and occurrence according to climate change (CC) (Zingales et al. 2022). In general, water is a significant resource for various economic sectors such as agriculture, industry, energy generation, and urban and recreational activities (Koop and Leeuwen 2017; Morseletto et al. 2022). Therefore, the water quality faces several significant challenges resulting from these resource-intensive activities and global environmental changes, which have different impacts on water quality and availability (Boretti 2019; UN-Water 2020). This is because natural toxins are created from microorganisms (fungal biotoxins or mycotoxins, algal and bacterial toxins), phytotoxins, and animal toxins (Picardo et al. 2019). These biotoxins can affect the aquatic environment, causing pollution in drinking water tanks, leading to risk on human health (Morabito et al. 2018), dissolved oxygen deficiency (Janssen 2019), and fish mortality (Sabart et al. 2015). Environmental factors have a significant role in enhancing water contamination by toxins leading to the subsequent formation of blue-green algal blooms (eutrophication processes) (Heisler et al. 2008). Global warming, which leads to increased nutrient enrichment (Wells et al. 2015), high temperatures, and excessive precipitation in conjunction with extended drought (Stocker et al. 2013), is the factor contributing to the increasingly frequent occurrence of hazardous blooms.

Cyanobacteria produce algal toxins (cyanotoxins) (Igwaran et al. 2024). Cyanotoxins are mainly divided into chemical groups: cyclic peptides (microcystins (MCs)), alkaloids, lipopolysaccharides polyketides, amino acids (β-aminomethyl-L-alanine), and lipoglycans (Picardo et al. 2019; Du et al. 2019). According to tissue targets, cyanotoxins are classified into three main groups: dermatoxins, hepatotoxins, and neurotoxins (Chorus and Bartram 1999). Bioassays, molecular, biochemical, and chemical assays were used to detect the natural toxins in aquatic ecosystem with high performance (Humbert et al. 2010; Moreira et al. 2014; Marie et al. 2016). Nevertheless, advanced detection techniques are needed because the current traditional methods have proven unable to identify all forms and kinds of cyanotoxins (Du et al. 2019).

Mycotoxins are harmful secondary compounds created by certain types of fungi. These mycotoxins are part of a wide range of fungal metabolites with low molecular weights that have been found to cause genetic damage and interfere with hormonal functions (Marin et al. 2013; Rocha et al. 2014; Ostry et al. 2017; Cimbalo et al. 2020). Researchers have also identified these mycotoxins as emerging contaminants in waterbodies (Gromadzka et al. 2009; Picardo et al. 2019). Mycotoxins encompass a variety of substances, including aflatoxins (AFs) like aflatoxin M1 (AFM1) and aflatoxin B1 (AFB1), T-2 toxin (T2) and HT-2 toxin (HT2), deoxynivalenol (DON), zearalenone (ZEN), ochratoxins, and fumonisins (Gruber-Dorninger et al. 2016). The majority of mycotoxins are produced by *Aspergillus*, *Fusarium*, *Penicillium*, and *Alternaria* spp. (Nicholson 2004).

The toxicity of mycotoxins on aquatic organisms was demonstrated in many studies. Chronic toxic effects such as reduced fecundity were observed for ZEN on zebrafish (Schwartz et al. 2010). Also, Persi et al. (2014) and Nomura et al. (2011) observed traces of aflatoxin B1 (AFB1) in the muscle of fish during laboratory tests. Meanwhile, Takadong et al. (2020) suggested that the mycotoxins found in fish could stem from water ponds being previously contaminated or from the buildup of mycotoxins in mud ponds. Furthermore, it was noted that the existence of these mycotoxins in water could significantly alter the way ecosystems operate (Schenzel et al. 2010).

As for the occurrence of mycotoxins in waterbodies, it was found that these harmful substances, which are produced by fungi in plants, can be released from the plant material into the water if the plant's surface touches the water via rain and during irrigation (Gautam and Dill-Macky 2012). In a similar vein, ZEN was found in the water that drains from fields where wheat and corn were infected with *Fusarium* (Bucheli et al. 2008). This suggests that mycotoxins can be carried into rivers and streams. However, these waterbodies can also contain fungi that are mycotoxin producers like *Aspergillus niger*, *Fusarium foetens*, and *Purpureocillium lilacinum* (Oliveira et al. 2018). Moreover, the breakdown of mycotoxins in animals that produce food can lead to these substances ending up in the water through runoff from fields that have been fertilized with manure from these animals (Biswas et al. 2017). Therefore, it is important to investigate how mycotoxins appear and behave in aquatic environments (Bucheli et al. 2005; Gautam and Dill-Macky 2012).

These mycotoxins have been identified in surface waters and drinking water tanks at concentrations reaching up to 35 ng L^{-1} (Oliveira et al. 2018). Additionally, the occurrence of fungal metabolites, including fumonisins, has been documented at the ng/L level in water samples (Waśkiewicz et al. 2015). Furthermore, various species of mycotoxins have been detected in the biofilms of drinking water distribution systems (Hurtado-McCormick et al. 2016). The detection of mycotoxins in bottled water in Portugal has also been reported, with aflatoxin B2 (AFB2) being the most commonly found, then with aflatoxin B1 (AFB1, aflatoxin G1 (AFG1), and OTA-A (Mata et al. 2015). Mycotoxins have also been found in wastewater (Gracia-Lor et al. 2020; Kolpin et al. 2014) and surface water (Bucheli et al. 2008; Emídio et al. 2020; Kolpin et al. 2014).

To determine mycotoxin concentrations, liquid chromatography together with tandem mass spectrometry (LC-MS/MS) and high-resolution mass spectrometry (LC-HRMS) are the most two methods, as they are characterized by their sensitivity and selectivity (Alshannaq and Yu 2017; Anfossi et al. 2016; Righetti et al. 2016; Stroka and Maragos 2016; Turner et al. 2015). It should be mentioned that most analytical methods usually focus on quantitative analytes such as deoxynivalenol (DON), zearalenone (ZEN), patulin (PAT), trichothecenes, and ochratoxin (OTA) (Bucheli et al. 2008; Gromadzka et al. 2009; Hoerger et al. 2009; Schenzel et al. 2010; Lucci et al. 2017).

9.2 Types of Freshwater Toxins

9.2.1 Algal Toxins (Cyanotoxins)

The algal bloom affects water resources, harming drinking water, fish farming, agriculture, and economic and industrial activities (Zhang et al. 2019a, b; Olson et al. 2020; Heil and Muni-Morgan 2021). Algal toxins may be released by certain algal species, including *Microcystis aeruginosa, Raphidiopsis mediterranea, Cylindrospermopsis raciborskii, Anabaena circinalis, Planktothrix rubescens, and Planktothrix agardhii* (Mantzouki et al. 2018). Cyanotoxins are secondary metabolites that can harm people via the food web (Skafi et al. 2021; Suo et al. 2021) and cause hepatotoxicity, nephrotoxicity, neurotoxicity, and reproductive toxicity (Chen et al. 2016; Li et al. 2017; Hinojosa et al. 2019a).

According to functional properties of cyanotoxins, they are divided into different groups: hepatotoxins, neurotoxins, and dermatotoxins produced by freshwater cyanobacteria and cytotoxins produced from marine cyanobacteria (Du et al. 2019; Nugumanova et al. 2023). Also, cytotoxins are classified based on their chemical structure to cyclic peptides microcystins (MCs), nodularin (NOD), and neurotoxins such as anatoxins (anatoxin-a, homoanatoxin-a, guanitoxin), ciguatoxins, saxitoxins, ß-N-methylamino-L-alanine (BMAA), and its isomers (2,4-diaminobutyric acid (2,4-DAB) and N-(2-aminoethyl)-glycine (AEG) (Huisman et al. 2018). Thus, cyanotoxin construction participates in the variety of their functional features (Du et al. 2019).

9.2.1.1 Hepatotoxins

Hepatotoxins are toxins that cause acute hepatotoxicosis in humans and animals, including a cyclic polypeptide chain, which is classified according to the chemical structure into cyanobacterial toxins microcystins (MCs and nodularins (NODs

(Mutoti et al. 2023), and alkaloids which are organic compounds containing nitrogen such as cylindrospermopsin (CYNs) (Martínez-Ruiz et al. 2020).

- Microcystins (MCs)

MCs are cyclic heptapeptides that can be harmful to liver. The first determination of MCs was in the 1990s using nuclear magnetic resonance (NMR) (Harada et al. 1990). Currently, highly sensitive mass spectrometry (MS) approaches can detect MCs in tiny amounts in cyanobacterial specimens (Du et al. 2019; Suo et al. 2021). MC-LR, MC-YR, and MC-RR congeners are common according to the L-forms of leucine (L), tyrosine (Y), or arginine (R) found in positions 2 and 4 (Bouaïcha et al. 2019; Díez-Quijada et al. 2019). The toxicity of MCs depends on variations in their chemical structure (Buratti et al. 2017; Massey et al. 2018), which affects different organs (Hu et al. 2016). Multiple neurotoxic effects of MC-LR were demonstrated using multiple biological models, including birds, fishes, and mammals (Pašková et al. 2008; Zhao et al. 2015; Tzima et al. 2017). MCs are distributed in freshwater habitats and generated by *Anabaena, Dolichospermum, Geitlerinema, Leptolyngbya, Microcystis, Nostoc, Phormidium*, and *Planktothrix* (Salmaso et al. 2017). In addition, MCs were produced by many marine cyanobacteria such as *Spirulina, Synechococcus*, and *Trichodesmium* (Tamele et al. 2019).

- Nodularins (NODs)

NODs are hepatotoxic cyclic pentapeptides. NODs are produced by cyanobacterium *Nodularia spumigena* (Mutoti et al. 2023). The earliest description of a NOD was recorded in 1988. Meriluoto et al. (2017) identified 10 congeners, the most common of which is NOD-R. NODs are widespread in temperate and subtropical climates and mainly found in coastal waters and freshwater. NODs can promote liver tumors (Chen et al. 2013), are related to animal and human death, and are more toxic than MCs (Du et al. 2019).

- Cylindrospermopsin (CYNs)

A CYN is one of the alkaloid hepatotoxins composed of a tricyclic guanidine moiety that bind to hydroxymethyl-uracil (Hinojosa et al. 2019b). The first chemical structure of a CYN was identified by Ohtani et al. (1992), which is distributed in freshwater and ocean in tropical, subtropical, and warm zones. CYNs are created by several cyanobacterium genera (Pearson et al. 2016). CYNs are most likely to cause liver failure and lead to death (Lee et al. 2017). In addition, CYNs showed more toxicity than MCs regarding neurotoxicity and may cause nervous system damage (Hinojosa et al. 2019a, b).

9.2.1.2 Neurotoxins

- Anatoxin-a (ATXs)

Anatoxins (ATXs) are neurotoxins produced by *Anabaena* sp., *Aphanizomenon* sp., *Arthrospira* sp., *Cuspidothrix* sp., *Cylindrospermum* sp., *Dolichospermum* sp.,

Oscillatoria sp., and *Phormidium* sp. (Jaja-Chimedza 2014; Du et al. 2019). The most prevalent toxins are alkaloid neurotoxins with excellent post-synaptic depolarizing neuromuscular blocking properties (McLaughlin 2013; Hameed 2013). However, these ATXs are unsteady, have a limited half-life, and may lose their toxicity, particularly under light and at high pH levels (Bruno et al. 2017).

- Saxitoxin (STX)

Saxitoxins (STXs) are produced by different genera of cyanobacteria. STXs are a group of carbamate alkaloid toxins and consist of two guanidinium moieties and a tetrahydropurine group (Duran-Riveroll and Cembella 2017). They are classified into nonsulfated, singly sulfated (gonyautoxins (GTX)), doubly sulfated (C-toxins), and decarbamylated analogs (Dittmann et al. 2013).

- β-N-methylamino-L-alanine (BMAA)

The BMAA, isolated from *Chroococcus*, *Merismopedia*, *Microcystis*, and *Nostoc* sp., was reported to be the cause of neurodegenerative diseases that affect a high number of people (Nunes-Costa et al. 2020). BMAA are vital bioactive compounds, with amino and carboxyl groups, without protein constituents, and are mainly present in microorganisms and plants. Ploux et al. (2017) declared that BMAA could be a cause of certain neurodegenerative diseases. Three relevant isomers of BMAA have been found (2,4-diaminobutyric acid (2,4-DAB), N-2-aminoethylglycine (AEG), and β-amino-N-methyl lanine (BMAA) (Banack and Murch 2018).

9.2.1.3 Dermatotoxins

- Lyngbyatoxins

Lyngbyatoxins are dermatoxic alkaloids found in coastal and estuarine waters in subtropical and tropical climates and are produced by *Lyngbya majuscule* (Tamele et al. 2019). Lyngbyatoxins have been found in the *Lyngbya*, *Oscillatoria*, and *Schizothrix* (Cullen et al. 2018; Rastogi et al. 2015) and may cause human skin irritation and marine turtle poisoning (Ito et al. 2002). Besides, they may induce protein kinase C (PKC) activity (Jiang et al. 2014).

- Aplysiatoxin

Aplysiatoxin is a dermatoxic alkaloid, also produced by *Lyngbya majuscule* (Sanseverino et al. 2017), which is composed of phenolic bis-lactones and, like lyngbyatoxins, is thought to be the primary cause of severe contact dermatitis. These poisons have high inflammatory potential, and they can promote the growth of tumors. Aplysiatoxin has five analogs: aplysiatoxin, debromoaplysiatoxin, anhydroaplysiatoxin, 3-methoxydebromoaplysiatoxin, oscillatoxin A, and 31-noroscillatoxin B (Ding et al. 2018). Also, they cause contact dermatitis by activating PKC

(Pearson et al. 2016). Furthermore, aplysiatoxin and debromoaplysiatoxin have potent tumor enhancers in the skin of mice (Du et al. 2019).

9.2.2 Fungal Toxins (Mycotoxins)

Mycotoxins are secondary metabolites with low molecular weight which grow on organic compounds. They are produced by some filamentous fungal species (Nicholson 2004). These toxic compounds pose a risk to human and animal health in addition to contaminating certain food products such as nuts, maize, grains, and fruits. Mycotoxins are produced by nearly one hundred filamentous fungi, and they have different structures, and hence different chemical and physical properties (Mata et al. 2015).

Most mycotoxins relevant to agriculture are produced by *Aspergillus, Penicillium, Fusarium,* and *Alternaria*, where the last two genera infect plants during cultivation, while *Aspergillus* and *Penicillium* affect crops when stored in unsuitable conditions (Nicholson 2004). The agriculturally relevant mycotoxins are aflatoxin B1 (AFB1), aflatoxin M1 (AFM1); trichothecenes which include HT2-toxin, T2-toxin, and deoxynivalenol (DON; zearalenone (ZEN; and ochratoxin A (OTA) and fumonisins (Gruber-Dorninger et al. 2016).

More than 20 fungal species produce aflatoxins, where the most common species are *Aspergillus flavus* and *Aspergillus parasiticus.* These aflatoxins are polyketide compounds that are considered very toxic. The major types of aflatoxins are aflatoxin B1 (AFB1), aflatoxin B2 (AFB2), aflatoxin G1 (AFG1), aflatoxin G2 (AFG2), aflatoxin M1 (AFM1), and aflatoxin M2 (AFM2). On the other hand, ochratoxins are produced by various fungal species such as *Aspergillus niger, A. carbonarius, A. ochraceus*, and *Penicillium verrucosum*, while zearalenone (ZEN) is a mycotoxin produced by *Fusarium* family; it was reported to have strong estrogenic activity (Bennett and Klich 2003; Zinedine et al. 2007). Fumonisins were reported in cereals as *Fusarium proliferatum* and *F. verticillioides* are the most common producers (Rheeder et al. 2002). Also, in grape, maize, and peanut, *Aspergillus Nigri* group can produce fumonisins (Astoreca et al. 2007; Mogensen et al. 2009; Kumar et al. 2017).

There is another group of mycotoxins which is known as modified mycotoxins. They represent mycotoxin metabolites and emerging mycotoxins that are not usually detected by conventional analytical methods due to their biotransformation (Rychlik et al. 2014). The data about the presence and toxic effect of these modified mycotoxins are limited. Plants, microorganisms, animals, and humans can biotransform mycotoxins by phase I metabolism, which includes oxidation, reduction, or hydrolysis of the parent compound, and phase II metabolism, which depends on conjugation reactions. Additionally, there is phase III metabolism, in which plant vacuoles are used for storing conjugates or they can be combined with plant cell walls (De Boevre et al. 2018; European Commission 2006).

Fusarium metabolites, beauvericin (BEA) and enniatins (ENNs), belong to emerging mycotoxins (Gruber-Dorninger et al. 2016; Khoshal et al. 2019).

Additionally, *Fusarium proliferatum* and *F. subglutinans*, which use cereals as a substrate, are the main producers of BEA (Gruber-Dorninger et al. 2016).

Recently, mycotoxins have been identified as emerging pollutants of aquatic environments (Picardo et al. 2019). It was found that mycotoxins from phytopathogens can reach water via irrigation or rains as they are filtrated out from the infected plant by its direct contact with water (Gautam and Dill-Macky 2012). DON and 15-ADON (< 0.02 to 3 μg L^{-1}) which are *Fusarium* mycotoxins were detected in water samples discharged from a wheat field (Gautam and Dill-Macky 2012). The same investigations were shown by Bucheli et al. (2008) where ZEN was reported in water discharged from corn fields infected with *Fusarium*.

On the other hand, aquatic fungi are also found in surface water. In Egypt, Abdel-Wareth and Sayed (2023) found that in Giza Governorate waterbodies, the most frequent fungal genus was *Aspergillus*, where its percentage reached 78% in spring, followed by 66% in winter, and *Fusarium* sp. showed a noticeable frequency rate in summer (10%). Gharbia Governorate surface water in the same study showed the highest frequency of *Aspergillus* sp. in autumn (52%), while *Fusarium* sp. was recorded with slightly high frequency (17%) in winter. Also, *Purpureocillium lilacinum, Fusarium foetens,* and *Aspergillus niger* were recorded in surface water by Oliveira et al. (2018); they proved via in vitro studies that these species were capable of producing various kinds of toxins such as ZEN, AFB1, AFB2, fumonisins, and OTA.

Moreover, as a result of bioconversion in animals, mycotoxin metabolites may be found in water via surface runaway from croplands filled up with fertilizers, as it was shown for ZEN and its metabolites (Biswas et al. 2017). It was reported that in natural waters, most mycotoxins are stable with 38 days (DON) and 180 days (beauvericin) half-lives (Schenzel et al. 2012), so it is important to study their presence in freshwater sources (Bucheli et al. 2005; Gautam and Dill-Macky 2012). Unfortunately, researchers are interested in mycotoxins detected in cereals and biological samples such as urine, feces, and blood (Nicholson 2004; Ajandouz et al. 2016; Meerpoel et al. 2018; Lauwers et al. 2019). Few researchers studied the detection of mycotoxins in aquatic samples from bottled water (Mata et al. 2015, Zhang et al. 2019a, b), wastewater (Kolpin et al. 2014; Gracia-Lor et al. 2020), and streams (Bucheli et al. 2008; Kolpin et al. 2014; Emídio et al. 2020).

9.3 Effect of Freshwater Toxins on Aquatic Ecosystem and Human Health

9.3.1 Adverse Effects of Cyanotoxins

Cyanobacterial blooms have significant social and environmental consequences, affecting drinking water, fisheries, agriculture, tourism, real estate, water quality, food web resilience and contributing to anoxia and fish kills (Carmichael and Boyer 2016), where algal blooms with high biomass buildup and degradation might decrease dissolved oxygen, cause light attenuation, and block fish gills. These impacts can cause thousands of fish and other aquatic organisms to die, resulting in ecological degradation (Sonak et al. 2018). In addition, the harmful algal bloom (HAB) toxins have acute and long-term health implications in all organisms. Toxins can directly kill fish and aquatic organisms, and also they can accumulate in fish and seafood and cause human illness (Griffith and Gobler 2020; Saleh et al. 2024).

9.3.1.1 Impact on Human Health

Cyanotoxins have different impacts depending on the exposure mode and the type of toxins (Taylor et al. 2014). According to the exposure mode, algal toxins may cause skin issues and eye discomfort, respiratory tract effects by inhalation of algal toxins during swimming, water skiing, and boating ride. Drinking water and eating food polluted with algae can produce gastroenteritis and headaches. On the other hand, hepatotoxicity, neurotoxicity, and genotoxicity may also occur according to the type of toxins (Shah et al. 2019; Soydan and Erbaş 2022).

Microcystin is the major hepatotoxins, which resist gastrointestinal-mediated digestion and concentrate in the liver with an active transport system (Soydan and Erbaş 2022). In acute poisoning, hemorrhagic shock occurs with liver damage and blood loss (Zhang et al. 2015a, b), where hepatotoxin enters the bloodstream after consumption by the ileum and then transfers to hepatocytes through the transport system. Furthermore, hepatotoxins degrade the arrangement and integrity of endothelial cells in the capillaries of the hepatic circulation, disrupt the cellular skeleton, and destroy of actin microfilament structures, blood loss, intrahepatic bleeding, and liver malfunction (Kalaitzidou et al. 2016). They also suppress protein phosphatase type 1 or type 2A, which is responsible for the regulation of liver cellular structure and function (Zanchett and Oliveira-Filho 2013).

Neurotoxins can cross the blood-brain barrier and enter the central nervous system (CNS) (Berry 2013). Although anatoxins are not among the most common neurotoxins in the world, they pose serious health risks to aquatic and terrestrial creatures and even cause death due to their high toxicity values (Metcalf and Codd 2009). There are two anatoxin-a variants, which are anatoxin-a and homoanatoxin-a (Bruno et al. 2017) and can bind to nicotinic receptors and work as a kind of mimic of acetylcholine results in the successive stimulation of neurons and their subsequent depolarization. And induced symptoms such as a decrease in locomotor activity, rapid loss of coordination, irregular breathing, change in gait, paralysis of the peripheral skeletal and respiratory muscles, tremors, and, even worse, contractions that appear before acute death are examples of the acute effects of anatoxins in vertebrates (Colas et al. 2020).

Also, β-N- methylamino-l-alanine (BMAA is known to cause mitochondrial dysfunction, which has a specific effect on the pro-inflammatory profile in neurodegenerative diseases (Nunes-Costa et al. 2020). Manolidi et al. (2019) declared that BMAA can be incorporated into the protein structures of nerve cells and cause protein misfolding, neuron damage, oxidative stress, and inhibition of certain specific enzymes that directly interfere with the folding of proteins.

Furthermore, the cylindrospermopsin (CYN) toxin is produced by *Cylindrospermopsis raciborskii*, which is the primary target of the liver. However,

CYN toxin was also described as pro-genotoxic (Puerto et al. 2018). Also, Bazin et al. (2010) suggested that CYN toxicity mediates cytochrome p450 (CYP) metabolites and may be linked to a possible carcinogenic effect. CYN may prevent chromatin condensation and inhibit the formation of metaphase chromosomes, affecting the chromatin-specific structure. Díez-Quijada et al. (2020) found that Cylindrospermopsin and microcystins mixture leads to an increase in micronucleus formation in the bone marrow of rats. Besides, histopathological alterations are shown in the liver, such as fatty degeneration and necrotic hepatocytes in centrilobular areas, as well as a light mononuclear inflammatory periportal infiltrate. Additionally, the stomach had flaking epithelium and mild necrosis of epithelial cells. Therefore, cyanobacterial toxins may lead to instability of the health of aquatic ecosystems and should not be ignored with their specific harmful effects.

9.3.1.2 Impact on Aquatic Ecosystem

Large amounts of organic matter associated with algal blooms can stimulate microbial respiration that depletes dissolved oxygen, produces CO_2, and promotes hypoxia (Breitburg et al. 2018) and acidification (Baumann et al. 2015). Increased pH can cause blindness, loss of signal recognition, sensitivity, and death.

Huisman et al. (2018) reported that blind spots were found in many freshwater lakes during summer. Although hypoxic and anaerobic circumstances still threaten sustainable commercial and recreational fishing around the world, some algal blooms provide an extra concern by producing toxic (Weirich and Miller 2014).

Consequently, cyanotoxins in drinking water and aquaculture can cause significant local and regional economic consequences. Furthermore, Paerl (2014) discovered that cyanotoxins created a significant impact on public health and influenced the zooplankton to ingest algae.

Although algae are considered a part of primary production in the food chain and act as a food organism for many fishes, *Microcystis, Cylindrospermopsis, Synechococcus, Anabaena, Lyngbya,* and *Oscillatoria* have been reported to cause toxic effects in the aquatic system (Aklakur et al. 2023). Drobac Backović et al. (2021) attributed the histopathological alterations observed in the gills and kidneys of fish (*Abramis brama* and *Carassius gibelio*) collected from the pond, including hypertrophy of the epithelium, hyperemia, hyperplasia, fusions of secondary lamellae, vacuolization of tubular epithelial cells, separation of the tubular basal lamina, lymphocyte infiltration, and karyolysis due to the presence of cyanobacteria blooming during April, May, July, and September.

9.3.2 Adverse Effects of Mycotoxins

9.3.2.1 Effect on Aquatic Organisms

Mycotoxins have negative effects on aquatic organisms. The chronic toxicity of ZEN was demonstrated in the decreased reproduction of zebrafish at 100 ng L^{-1} as the lowest observed effect concentration (LOEC21d) (Schwartz et al. 2010). Also, it was stated that the mycotoxins detected in the aquatic ecosystems might deleteriously affect their functioning (Kagami et al. 2014). Moreover, water hyacinth (*Eichhornia crassipes*), which is used as a fish diet, was found to be infected with the fungal phytopathogen *Alternaria alternata*; thus, when it is used as fish feed, it resulted in a reduction of aquatic fish production (Aneja et al. 2014).

Gareis and Wolff (2000) and Iqbal et al. (2014) highlighted the food safety risk resulted from contamination of edible tissues of fishes with mycotoxins such as aflatoxins (AFs), zearalenone (ZEN), and ochratoxin A (OTA). Zinedine et al. (2007) showed that ZEN and its derivatives were found to be the only endocrine-disrupting mycotoxins. Furthermore, Persi et al. (2014) and Nomura et al. (2011) detected AFB1 residues in fish muscles under laboratory conditions, while Takadong et al. (2020) suggested that detection of mycotoxins in fish tissues might be due to previous contamination of pond water or from mycotoxin accumulation in their muddy substrate. Also, Abdel-Wahaab et al. (2005) observed sterigmatocystin residues in muscles of Nile tilapia.

El-Sayed and Khalil (2009) observed various effects after 120 days of exposure to 0.2 mg/kg AFB1 in common carp, *Cyprinus carpio*. These effects include disturbances in blood circulation and renal and liver damages. In addition, Diab et al. (2018) investigated the impact of ochratoxicosis in Nile tilapia and observed that ochratoxin A (OTA) caused sluggish swimming and poor growth. Pathological lesions were also detected in spleen, kidneys, and liver. Alanine transaminase (ALT), aspartate aminotransferase (AST), creatinine, and urea significantly increased with reduced total protein, albumin, and globulin in the ochratoxicated fish group. Trichothecene toxic effects in animals include decreased plasma glucose, reduced blood cell and leukocyte count, weight loss, alimentary toxic aleukia, and pathological changes in the liver and stomach (Diab et al. 2018). The mechanism involved in T-2 and DON toxicity is generally via oxidative stress-mediated DNA damage and apoptosis (Wu and Munkvold 2008; Liew and Mohd-Redzwan 2018). Furthermore, T-2 and DON are inhibitors of protein synthesis, so they are responsible for trichothecene mycotoxicosis in animals and fish (Yuan et al. 2014).

9.3.2.2 Effect on Animal and Human Health

- Aflatoxins

Aflatoxins can result in huge health problems as they are carcinogenic, teratogenic, and mutagenic; also, they negatively affect growth and immune system (Sarma 2010). The International Agency for Research on Cancer (IARC) has classified total aflatoxins (AFB1 + AFB2 + AFG1 + AFG2) and AFM1 as Group 1 category carcinogenic compounds, but the highest carcinogenic potential is of AFB1 (Bhetariya et al. 2011). The primary target organ of aflatoxins is the liver, and the primary target organ of hepatitis virus infection (hepatitis B or C) is the liver too; therefore, the chances of hepatocellular carcinoma (HCC) increase too many times in hepatitis patients having aflatoxin exposure through diet, and it was reported that both viral infections and aflatoxins are more prevalent in the developing countries (Wang and Tang 2005; Wogan et al. 2012; Wu 2013). Moreover, aflatoxins might cause male and female infertility. Besides infertility, exposure to aflatoxins might reduce male fertility by suppression of sperm production (Kudayer et al. 2019; Murad et al. 2015). As for females, aflatoxins can cause problems in sexual maturation via their negative effects on follicles and hormones, besides their impact on fetus growth (Kourousekos et al. 2015).

In addition, as a result of the similar structures between AFB1 and steroid hormones, AFB1 affects the receptors of steroid hormones, and hence interferes with the production of estrogen and progesterone and reduces females' fertility. Also, aflatoxins in maternal blood can lead to problems in birth weight and height, stillbirth, and jaundice (Abdulrazzaq et al. 2004; Sadeghi et al. 2009; Moslemi and Tavanbakhsh 2011). Furthermore, immunosuppression induced by aflatoxins leads to more vulnerability to infectious diseases as the cell-mediated immunity is reduced (Yin et al. 2016). Additionally, aflatoxins showed deleterious effects on certain biochemical parameters related to the peripheral and central nervous system of humans and animals (Marchese et al. 2018). It was reported that AFB1 is a neurotoxic agent that interferes with the normal functioning of the brain and causes nervous system disorders such as encephalopathy, demyelinating disease, neuropathies, and neurocognitive deficits. AFB1 exposure also resulted in cerebral edema, coma, convulsion, and might cause death (Kawata et al. 2016). In the same vein, it was demonstrated that aflatoxins increased lipid peroxidation and generation of free radicals, and thus facilitate brain damage, and long-term AFB1 exposure might be the reason for cytotoxicity via its impacts on blood-brain barrier (Chen et al. 2008). Also, it was shown that exposure to aflatoxins and renal cell carcinoma are closely related (Bbosa et al. 2013; Sharma et al. 2011).

- Ochratoxin

Because ochratoxin A (OTA) is a chemically stable substance, it cannot be significantly reduced in foods and beverages by standard food processing methods. OTA, as a strong renal carcinogen in a number of animal species, primarily targets the kidney (IARC 1993; World Health Organization 2002; Duarte et al. 2011).

Immunotoxicity, suppression of macromolecular synthesis, elevated lipid peroxidation, and suppression of mitochondrial respiration are additional negative consequences of OTA (Bondy and Pestka 2000; Marquardt and Frohlich 1992). Based on evidence of carcinogenicity in animal studies, the International Agency for Research on Cancer (IARC) has classified OTA as a Group 2B possible human carcinogen (IARC 1993; Fazekas et al. 2005); nevertheless, the carcinogenicity of OTA has not been definitively established in humans. According to a risk assessment study on OTA, OTA-DNA adduct levels were low and not typical of genotoxic carcinogens, and it was negative in genotoxicity assays with high specificity (Haighton et al. 2012).

- Zearalenone (ZEN)

In the temperate climate zone, *Fusarium* species, which are ZEN main producers, are found on a wide variety of plants, primarily corn and many other grains. Thus, when exposed to high temperatures, ZEN does not deteriorate and shows noticeable stability during storage (Appell et al. 2017; Hueza et al. 2014; Zinedine et al. 2007). Zearalenone belongs to xenoestrogens, and its modified forms are frequently produced in fungi, animals, and humans by phase I reduction. Examples include α-zearalenol (α-ZEL), β-zearalenol (β-ZEL), α-zearalanol (α-ZAL), and β-zearalanol (β-ZAL). ZEN and its derivatives have chemical structures that resemble natural estrogens such as 17β-estradiol, and so, they can attach to estrogen receptors, disrupting the hormonal balance and causing a variety of reproductive system disorders (Kowalska et al. 2016). Furthermore, it has been demonstrated to be absorbed in mammals' bodies rather quickly, after which it undergoes metabolization to produce compounds ten times more toxic than α-zearalenol (Brodehl et al. 2014).

Deoxynivalenol (DON)
As a mycotoxin, DON can prevent the synthesis of proteins. High levels of this mycotoxin can cause diarrhea, vomiting, and gastrointestinal bleeding, and long-term exposure can impair immunity and growth in farm animals and poultry (Khoshal et al. 2019). Prior to systemic distribution, 3-ADON and 15-ADON in humans, pigs, and poultry undergo deacetylation to DON, which may cause similar acute and long-term impacts as DON (Ajandouz et al. 2016; Knutsen et al. 2017). Free or unmodified mycotoxins such as 3-acetyl deoxynivalenol (3-ADON) and 15-acetyl deoxynivalenol (15-ADON) are acetylated forms of DON; they are produced as toxic secondary metabolites by fungi like *Fusarium graminearum* and *F. culmorum* (Knutsen et al. 2017, Mayer et al. 2017, Rychlik et al. 2014). DOM-1 was shown to be nontoxic to a variety of cell types, including fibroblasts, lymphocytes, and intestinal epithelial cells (Payros et al. 2016). Experiments carried out on pig by using contaminated feed and adding bacteria capable of converting DON into DOM-1 have demonstrated the significantly lower toxicity in comparison with the parent toxin DON. Neither weight gain nor a reduction in food intake was observed in these experiments (Payros et al. 2016).

Fumonisins
These toxins are correlated with a number of health problems such as esophageal cancer as investigated in various countries (Zimmer et al. 2008; Yu et al. 2021). Fumonisins which were recorded in Europe (51%) and Asia (85%) are a very important issue worldwide (Corrêa et al. 2018). It was demonstrated that they are toxic to both liver and nephrons in experimental animals (Chu and Li 1994). In addition, fumonisin B1 (FB1) is involved in stimulating hepatocarcinoma and immunosuppression, in addition to nephrotoxicity. It was reported as a promoter of hepatocarcinoma (Gelderblom et al. 1988). In animal models (rainbow trout and rats), these toxins synergistically interact with aflatoxin B1 (AFB1, which have been

demonstrated through two stages, namely, the starting stage and advancement one of cancer (Qian et al. 2016; Xue et al. 2018; Carlson et al. 2001). According to the definition of the International Agency for Research on Cancer (IARC), FB1 is a Group 2B potential human carcinogen. In addition, it can be toxic to a number of animals, including mice, rats, and rabbits (IARC 2002). Furthermore, the joint Food and Agriculture Organization (FAO) and World Health Organization (WHO) have reported that a temporary maximum tolerable daily intake for fumonisins is 2 μg/kg bw/day, based on the absence of any observed adverse effects for nephrotoxicity in male rats (WHO 2019).

- Beauvericin (BEA) and Enniatins (ENNs)

In vitro, these two mycotoxins showed acute cytotoxicity (Gruber-Dorninger et al. 2016; Fraeyman et al. 2017), while by using rats and chickens (in vivo), they have no or low acute toxicity (Fraeyman et al. 2017). Immunotoxicity in mice was reported in several times exposure to BEA and ENNs, but there are no data regarding their long-term toxicity in agriculturally related species (Maranghi et al. 2018). Other emerging mycotoxins such as alternariol (AOH), alternariol monomethyl ether (AME), and tenuazonic acid (TeA) are produced by *Alternaria* species such as *A. alternata*. Exposure to *Alternaria* toxins has many adverse health impacts in mice, chickens, and humans, and resulted in liver and kidney damage in rats (Ostry 2008).

9.4 Detection Methods of Natural Toxins

9.4.1 Detection of Cyanotoxins

Various traditional analytical methods have been used for the detection of cyanotoxins including high-performance liquid chromatography (HPLC) (Cadel-Six et al. 2014; Campos and Vasconcelos 2010), and inhibition of protein phosphatase assay (Bieczynski et al. 2013; Covaci et al. 2012) and enzyme-linked immunosorbent assay (Lotierzo et al. 2012). Although these methods have preciosity, sensitivity, and professionalism, their detection procedures require more time and high cost (Suo et al. 2021). Therefore, new advanced techniques are needed to offer more efficiency in toxin detection and are more applicable to all the types and variants of cyanotoxins (Du et al. 2019).

9.4.1.1 Molecular Detection Methods

Molecular methods are based on the detection of genes found in cyanobacteria and their related toxins, using polymerase chain reaction (PCR) techniques such as the conventional PCR, multiplex PCR, terminal restriction fragment length polymorphism (T-RFLP), random amplified polymorphic DNA (RAPD), denaturing gradient gel electrophoresis (DGGE), real-time PCR, and non-PCR-based techniques such as fluorescence in situ hybridization (FISH) and DNA microarrays (Humbert et al. 2010; Picardo et al. 2019). The PCR process consists of amplifying a DNA sequence in vitro using particular primers to the target sequence of DNA (Jungblut et al. 2005). Furthermore, the real-time PCR technique has high sensitivity and can determine cyanotoxins (Moreira et al. 2011). Also, the amplification of the ribosomal RNA gene, phycocyanin operon, and the RNA polymerase B subunit gene was utilized to detect the cyanotoxins (Gaget et al. 2011). Biotoxins of MC congeners, NOD, STX (Kellmann et al. 2008), ANA, and CYN (Rantala-Ylinen et al. 2011) can be explored by the genes that encode these biosynthetic enzymes.

9.4.1.2 Biochemical Detection Methods

Biochemical approaches were based on enzyme inhibition, such as determining microcystins (MCs), nodularin (NOD), and anatoxin-a (ANA). The anatoxin-a group inhibits acetyl-cholinesterase, whereas MCs and NODs inhibit phosphatases. An and Carmichael (1994) have established the first technique to investigate microcystins and nodularin in water by a colorimetric protein phosphatase assay. Afterward, Lambert et al. (1994) detected the lowest level using an enzyme bioassay depending on the radiolabeled ^{32}P phosphate quantity. The HPLC was combined with colorimetric reaction to achieve high levels of sensitivity and specificity (Rapala et al. 2002). Campàs et al. (2007) have developed an electrochemical biosensor to detect MC depending on the protein phosphatase 2A (PP2A) inhibition through the enzyme immobilized using a polyvinyl alcohol pendant water-soluble photopolymer. Also, an electrochemical biosensor was developed to investigate microcystin LR relying on inhibition of the recombinant protein phosphate 1A (PP1A).

On the other hand, anatoxin-a (ANA) detection has been developed by a biosensor detector relying on acetylcholinesterase of electric eel (Villate et al. 2002). However, this system showed a limitation in detecting one mg/L ANA-a; thus, it used oxime reactivation to distinguish the toxin and possible insecticides in the sample. Devic et al. (2002) used an amended and sensitive sensor based on engineered acetylcholinesterase (AChE) with a detection limit below the nanomole/liter. However, it cannot discriminate among all AChE inhibitors in the same sample. Therefore, they used a 4-mutant group of AChE variants: 2 mutants susceptible to ANA-a and the other two responsive to the insecticides. Although enzyme inhibition has a quick response without sample preparation, the method has a loss of specificity and versatility.

9.4.1.3 Immunochemical Methods

Immunoassays use antibody and antigen affinity to improve the process of detecting natural poisons in water. The first polyclonal rabbit antibodies for MCs were recorded by Brooks and Codd (1987). After that, different antibodies were created for various immunoassays, including enzyme-linked immunosorbent test, which are widely used and accessible in commercial kits. Although certain antibodies can offer a broad specificity for cyanotoxin groups for quick screening, there may be limitations in specificity across congeners or structurally similar drugs. Therefore, Sheng et al. (2006) created a polyclonal antibody for direct competition to detect the MCs in water, which showed a limit of detection for MC-LR of 0.12 mg L^{-1}, with different types of microcystins and NOD. Also, Yang et al. (2016) developed monoclonal antibodies to investigate nodularin and eight microcystins with limit of detection (0.16 and 0.10 mg L^{-1}) and some cases were achieved in high selectivity.

Consequently, several immunosensors have been produced to monitor water toxins based on advanced nanomaterials to develop the sensitivity and robustness of tests. Biocompatible nanomaterials were used as a substrate to immobilize more antigens due to the three-dimensional villiform-like structure's enormous surface area, which provides significant electrochemical signals and has a linear detection to MC-LR from 0.005 to 50 mg/L with a limit of detection of 0.004 mg L^{-1} (Gan et al. 2016). Also, photoelectrochemical immunosensor (Tian et al. 2014; Qileng et al. 2018) and electrochemiluminescent immunosensor (Zhang et al. 2015a, b) were developed for the detection of MC-LR using nanomaterials. Generally, the main advantages of immunoassays are highly specific approaches, no sample preparation is required, and they respond quickly. However, the fundamental drawback of immunoassays is the interference reaction of structurally related molecules, which leads to low specificity. Matrix effects can significantly influence test sensitivity. Finally, difficulty in producing antibodies against highly hazardous chemicals limits their availability.

9.4.1.4 Application of DNA to Monitor Toxins

Currently, there are two strategies to monitor the MC-LR based on DNA, including a recognition element and signal processing (Suo et al. 2021). The first approach used nucleic acids, labeled as aptamers to define the MC-LR. Aptamers play a key role in analytical chemistry, and adequate sequences should be selected to fit modulations (Nakamura et al. 2001). Aptamers may also be produced to be linked with different nanomaterials, and this approach is diverse and robust enough to perceive accurate identification (Zhong et al. 2020).

The second technique demonstrates how DNA nanotechnology can be integrated with signal processing technologies to enhance analytical efficiency. Because of the enzyme-based signal processing kinetics, the immune-based MC-LR test is prone to low sensitivity and instability. The emergence of DNA nanotechnology has provided an alternate pathway for processing antibody-antigen binding signals. By functionalizing the antibody with DNA segments, the recognition signal can be converted into DNA signal, which can be then processed using rolling circle amplification (RCA) (Schweitzer et al. 2002; Yin et al. 2008), catalytic hairpin assembly (CHA) (Wei et al. 2018a, b; Xing et al. 2019), hybridization chain reaction (HCR) (Bui et al. 2017), and other signal amplification techniques.

9.4.1.5 Chemical Analysis

The chemical analysis utilized to detect toxins in water samples relies on separate methods, including liquid chromatography (LC) or gas chromatography (GC) linked to various detectors such as mass spectrometry (MS), fluorescence (FL), and ultraviolet (UV). These techniques are considered more accurate and sensitive to cyanobacterial toxin detection (Moreira et al. 2014). However, these methods require solvents, costly instruments, skilled personnel, complex operational processes, and much time (Vogiazi et al. 2019; He et al. 2019).

9.4.2 Detection of Mycotoxins

Monitoring the occurrence of mycotoxins in natural waters has received little attention. The quantification of DON and ZEN in natural aqueous samples at nanograms per liter level was first reported by Bucheli et al. (2008) using LC-MS/MS analysis, with the detection limit of 1.4 ng/L for DON, but 1.5 ng/L for ZEN. In another study, DON produced by the fungi belonging to the genus *Fusarium* was detected using the LC-MS/MS method (Awad et al. 2013). Escriva et al. (2017) used LC-MS/MS to report different extraction methods for the quantification of 11 mycotoxins (AFB1, AFB2, OTA, ZEA, AFG1, AFG2, beauvericine, enniatin-A, enniatin-B, enniatin-A1, and enniatin-B1) in waters. The optimized method was highly sensitive with a level of detection (LOD) in the range of 0.1–15 mg/L (Escriva et al. 2017). Similarly, Serrano et al. (2016) analyzed several *Fusarium* emerging mycotoxins in water using LC-MS/MS to achieve rapid analysis with LODs ranging from 0.06 to 0.17 mg/L, and levels of quantification (LOQs) ranging from 0.20 to 0.58 mg L^{-1}.

On the other hand, LC-ESI-MS/MS was used to analyze lower concentrations of mycotoxins with a detection limit of the nanogram/liter. Schenzel et al. (2010) developed a methodological approach for detecting multiple mycotoxins in both natural water sources and wastewater. This technique employs a combination of solid-phase extraction (SPE) with Oasis HLB cartridges, followed by liquid chromatography-electrospray ionization mass spectrometry (LC-ESI-MS/MS) to evaluate thirty-three different mycotoxins, which is tested by negative and positive ionization methods to broaden its applicability. However, it was noted that negative ionization led to increased challenges due to higher matrix effects. The method achieved high recovery rates for the majority of the substances, with LOD of 10 ng/L for 27 from the total 33 compounds. Notably, Beauvericin and nivalenol were measured for the first time in both river and drainage water samples, with different levels in ng/L, as reported by Schenzel et al. (2010). Mata et al. (2015) used a similar method to assess mycotoxins in bottled water for aflatoxins and OTA-A at 0.2 ng/L. The researchers found that the most popular fungal toxin was aflatoxin-B2, with a peak concentration of 0.48 ng/L, followed by aflatoxin-B1, aflatoxin-G1, and OTA. There was considerable research focused on identifying filamentous fungi, which are found in aquatic systems and can cause odor production (Gonçalves et al. 2006). They could also release toxins, highlighting the importance of considering fungal contamination in water quality (Oliveira et al. 2018).

9.5 Impact of Climate Change on Freshwater Toxins

9.5.1 Effect of Climate Change on Cyanotoxins

Experiments indicated that rising temperature degrees and carbon dioxide levels can accelerate microalga and cyanobacterium proliferation (Murray et al. 2015). However, many variables under environmental circumstances make it complicated to create a prediction model for how climate change would affect the features and level of aquatic poisons. Therefore, there are more harmful and diversified blooms and toxicity events than in the past. However, Hallegraef et al. (2015) postulated that mass spectrometry was common in most studies, due to it being relatively new and the inability to determine if climate change has a particular amount of effect on aquatic toxins. Botana (2016) claimed that the discovery of novel natural toxins and hazards will require more precise screening techniques, more specific regulations, and a thorough understanding of the toxicology of the substances, which is not always available for research.

Cyanobacterial toxins are commonly connected with freshwater (Botana 2016). The primary concerns about cyanobacterial toxins arise from drinking water, freshwater aquaculture, particularly shrimp and fish (tilapia, carp, and milkfish), and shellfish that consume cyanotoxin land waters (Smith 2014). After a bloom, cells die and release toxins into the water. However, the amount of biomass is not necessarily linked to the amount of toxins or to the toxin profile, as a very poisoning strain may produce a high amount of toxins from a small number of cells (Zamyadi et al. 2012; Zamyadi 2014). Higher temperatures have been associated with a high occurrence of hazardous strains (Joung et al. 2011). However, it is unknown how climate change would affect the development of analogs of these toxins in each population. For example, microcystin (*Microcystis aeruginosa*) increase in temperature was associated with the production of higher (Reichwaldt et al. 2015) and lower amounts (Rapala et al. 1997) of the toxins, while an increase in CO_2 was linked to the growth of nontoxic strains. Furthermore, Kardinaal et al. (2007) claimed that microcystin production is connected to the light available at the water surface.

9.5.2 Effect of Climate Change on Fungi and Their Mycotoxins in Freshwater

Climate (temperature, water availability, and the intensity and amount of light, besides severe drought, desertification, and changes in moisture levels between dry and humid periods) is the key environmental element affecting the growth phases of fungi, including their capability to infect plants, survive, and produce toxins (Paterson 2006).

Temperature, precipitation, and sea level are the most variables to be influenced by the future global change. These changes are anticipated to have a variety of effects on plants and their pathogens, including their metabolites such as mycotoxins (Paterson and Lima 2010).

According to Vaughan et al. (2016), as suitable weather conditions shift or expand geographically, the composition and distribution of organisms within a community will be altered. For example, there is an anticipation of more frequent insect outbreaks moving northward (van der Fels-Klerx et al. 2016).

Mycotoxigenic fungi grow in specific ecological circumstances, including temperature, humidity, and rainfall dispersion. These conditions are crucial for crop infection, host colonization, toxin production, and overall survival. Consequently, changes in climate will cause alteration in fungal communities and mycotoxin ratio. As a result, some fungal species may be more resistant and grow in extreme ecological conditions, leading to the prevalence of thermophilic fungi at latitudes and psychrophilic fungi in high-level latitudes (van der Fels-Klerx et al. 2016). Therefore, rising temperatures may lead to more heat-tolerant strains moving, introducing, and growing as consequences of climate change. However, at the same time, the increased occurrence of extreme weather events will reduce the number of species with low ability to adapt to new conditions (Vaughan et al. 2016). This implies that certain types of fungi may vanish from their current habitats and appear in new areas that were previously considered safe (Zingales et al. 2022).

Leggieri et al. (2015), Medina et al. (2015), and Piva et al. (2006) demonstrated that in northern Italy during the 2003/2004 and 2012 seasons, where maize was predominantly cultivated, the occurrence of very hot and dry weather led to a transition of *Fusarium verticillioides* and fumonisin pollution to *A. flavus* and aflatoxin (AF) contamination. *A. flavus*, a mycotoxigenic fungus that is tropical and tolerant to drought, can thrive within a broad temperature range (19–35 °C), with its optimal growth occurring at around 28 °C and AF production at 28–30 °C according to Medina et al. (2014). AFs pose a significant threat in regions like sub-Saharan Africa. However, as the Mediterranean region experiences rising temperatures and an increased likelihood of dry conditions such as in Spain and Italy countries, it is anticipated that alterations in fungal and mycotoxin distributions will occur more frequently (Fapohunda and Adewunmi 2019). A comprehensive study of mycotoxins in maize (from 2011 to 2021) was conducted by Locatelli et al. (2022), who revealed that the species of *Fusarium* was more frequent, with fumonisins in that the primary mycotoxin was detected. Nonetheless, the prevalence of aflatoxin contamination is on the rise and becoming more widespread.

In addition to Italy, *A. flavus* has been found to be more common on maize crops in the Balkan countries (Serbia, Croatia, and Hungary) (van der Fels-Klerx et al. 2016; Dovenyi-Nagy et al. 2020). *A. flavus* is expected to outcompete other fungi with lower ideal temperatures. For instance, in Brazilian pepper and nuts, it was isolated more often than *Aspergillus ochraceus* (Freire et al. 2000). Furthermore, in grapes grown in northern Portugal, *A. flavus* may outcompete *Aspergillus carbonarius*, making AF more dangerous than OTA (Serra et al. 2006a, b). We can therefore conclude that alternariol, DON, FUM, and OTA will not replace AF in staple crops like peanuts, maize, and wheat because it is unlikely that *A. flavus* will be less dominant than *Alternaria, Fusarium, Aspergillus conspecifics*, or *Penicillium* spp. once it has established itself (Paterson and Lima 2011). In the future, *Aspergillus* section Flavi may be an important fungus genus. In addition, *A. flavus. Aspergillus parasiticus, Aspergillus nomius, Aspergillus pseudotamarii, Aspergillus bombycis, Aspergillus toxicarius, Aspergillus parvisclerotigenus, Aspergillus minisclerotigenes, Aspergillus arachidicola, Aspergillus pseudonomius,* and *Aspergillus pseudocaelatus* are among more than 20 species that are known to produce AFs (Varga et al. 2011). Given that warmer environmental conditions can favor the fungi found in *A.* section Flavi, it is reasonable to assume that CC will also increase the detection of other aflatoxigenic species other than *A. flavus*. For instance, a study during

2012–2016 was conducted in Serbia by Nikolić et al. (2021), who found that *A. parasiticus* grows on corn kernels. Earlier studies in Serbia and in other European countries showed that it was rare to find *A. parasiticus* in corn field (Giorni et al. 2007).

Meanwhile, *F. culmorum* is predominant in the Nordic regions, and *F. graminearum* is thought to be the primary DON producer in central and southern Europe. However, in certain parts of northern and middle of Europe, there has been a reported decrease in *F. culmorum* with an increase in *F. graminearum* (Logrieco and Moretti 2008; Moretti et al. 2019). The ideal growth temperatures for these two *Fusarium* species are 20–25 °C for *F. culmorum* and 24–28 °C for *F. graminearum.* This disparity could help to explain why *F. culmorum* is more common in cooler climates like northwest Europe, while *F. graminearum* has historically been more common in areas with relatively hot summers, like the USA, Canada, Australia, and parts of continental Europe (West et al. 2012). According to the reports, the climate in northern Europe is expected to become more humid and milder by 2050, which will favor the growth of *F. graminearum* due to the CC effects that follow (Paterson and Lima 2011; Moretti et al. 2019).

Similarly, hot nights can promote the prevalence of *F. graminearum* over *F. poae* in grains like barley and wheat, according to a recent study conducted in Argentina by Martínez et al. (2022). Grain quality will decrease as a result, and mycotoxin contamination will rise sharply. A rise in *Fusarium* head blight (FHB) on wheat in central China (Zhang et al. 2014) and southern England (West et al. 2012) has also occurred in line with the current change in favor of *F. graminearum* distribution.

Other changes in the dynamics of the fungi include *Fusarium langsethiae*, with more incidence of T-2 and HT-2 toxins, which became more common in the United Kingdom (Paterson and Lima 2017), while *Aspergillus steynii* and *Aspergillus westerdijkiae* mainly produce OTA in coffee (Gil-Serna et al. 2014). *Aspergillus niger* is abundant in onions and grapes at Hungarian, while *A. niger* and *A. tubingensis* prevailed *A. carbonarius* in grapes at southern Spanish (Farkas et al. 2011; Paterson et al. 2018). In addition, the highly aggressive and toxic *F. asiaticum* (3-acetyldeoxynivalenol) was substituted by the nivalenol (NIV) in China and the most toxic *F. graminearum* 3-A DON was replaced by 15-acetyldeoxynivalenol in North America (Valverde-Bogantes et al. 2019).

Moreover, according to Magan and Aldred (2008) and Leong et al. (2011), *Wallemia sebi*, *Xeromyces bisporus*, and *Chrysosporium* spp., which belonged to xerophilic fungi, might become more significant fungal colonies in food resources, where they might thrive in arid environments, leading to the emergence of new harmful substances. Walleminol and walleminone are created by the *W. sebi* fungus and have the potential to be toxic to all organisms and people (Fapohunda and Adewunmi 2019). On the other hand, Paterson and Lima (2010, 2011) declared that certain fungi capable of producing mycotoxins might face extinction, which could be seen as a positive outcome from a climate change perspective. For instance, the temperature in Pakistan has risen to 53.7 °C (Iqbal et al. 2011); such extreme conditions might lead to the disappearance of certain species of fungi or hinder the growth of xerophilic fungi. Therefore, the possibility of disappearing of aflatoxins fungal species from hot areas could be highly beneficial for tropical nations, as their food could have lower aflatoxin levels and be more easily exported (Paterson and Lima 2017). However, high temperature levels over 40 °C also represent a threat to agriculture crops. If fungi species that produce aflatoxins were to disappear, it wouldn't stop other thermophilic fungi from taking their place. The "Parasites Lost Phenomenon" theory suggests that crops might experience a reduction in pest pressure as they shift toward new areas, which could be beneficial in terms of fungal enemies and mycotoxin contamination (Mitchell and Power 2003; Paterson and Lima 2011). Nonetheless, if temperatures don't rise too high and drought conditions become more frequent, then there could be an increase in aflatoxin contamination (Fapohunda and Adewunmi 2019). All these changes in fungal distribution and prevalence will be reflected in the waterbodies which receive agricultural discharge from the crop fields.

Recently in Egypt, Abdel-Wareth and Sayed (2023) carried out a study on the distribution of fungal genera in different watercourses of Giza and Gharbia governorates. They reported on the appearance of fungal genera such as *Aphanoascus* sp., *Phoma* sp., *Pythium* sp., *Drechslera* sp. and *Absidia* sp. that were not recorded in these watercourses back in 2012 in the same watercourses (Saad et al. 2014).

9.6 Mitigation and Adaptation Strategies

Mitigation and adaptation efforts should be global and equitable and recognize that the health of the planet is integrally connected to human health. In 2001, some climate change adaptation policies were established by the Conference of the Parties to the United Nations Framework Convention on Climate Change (UNFCCC). The National Adaptation Programmes of Action (NAPAs) were intended for least developed countries (LDCs) to coordinate and communicate priority actions that allow access to adaptation funding mechanisms (Dabbadie et al. 2018). It is important to realize the natural toxins in an engagement in climate policy plans and in planning and action.

Mitigation and adaptation efforts should be focused on surveillance schedules, more laws, and comprehension of the toxicity of the toxic substances.

Silvarrey Barruffa et al. (2021) described the adaptation practices focused on treating cyanobacterial harmful algal bloom (CyanoHAB) presence under climate change conditions. They indicated that it plays an important role in filling the gap between a major environmental problem and potential solutions. The treatment of the widespread algal blooms in watercourses depends on the application of efficient nutrient reduction programs, and good controlling systems for catchment and water health. Thus, adaptation process to keep water quality good should be depended on agreement about the interaction between climate change and CyanoHABs. They stated that the methodology application of adaptations consists of four stages: the first was around understanding the link between cyanobacteria and climate change; the second was the sensibility in developing countries to climate change adaptation;

the third one was targeting adaptation practices; and the fourth stage focused on identification of potential adaptation practices. Thus, regarding the CyanoHAB adaptation process, implementation should be designed according to reducing the risks and the negative effects. Therefore, adaptation requires participation from those of economic sectors and involves many levels of decision-makers.

Moreover, mitigation strategies for cyanoblooms and cyanotoxins were discussed by Rastogi et al. (2015) who stated in their review that the approaches proceeded for the elimination or reduction of these blooms in the environment should be safe and sustainable for the aquatic ecosystems. Therefore, environmental sustainability is required considering the chemical, physical, biological, and other achieved approaches for mitigating the harmful cyanobacterial bloom occurring. Thus, henceforth, there is a need for advanced approaches or development of a new technology to prevent the harmful cyanobacterial blooms for good environmental health.

Regarding the aquaculture, cyanobacteria may have the impact on water quality metrics. In addition, changes in climatic condition impacts on the fish farms are more harmful on the maintained fishes in these aquacultures due to high sensitivity of these fishes to various kinds of biotic and abiotic stress that directly affect the growth, reproduction, physiology, and behavior of fishes. Thus, climatic parameters with cyanobacterial toxins may affect the aquaculture directly by influencing the fish stocks or indirectly altering the primary and secondary productivity. More recently, Saleh et al. (2024) in their research considered that cyanobacteria may have a considerable effect on the fish *Oreochromis niloticus* in the fish ponds at the Al-Abbassa Fish Hatchery in Sharkia Governorate in Egypt. *Anabaena, Gloeotrichia, Lyngbya, Oscillatoria,* and *Phormidium* were the five genera of cyanobacteria discovered in their study. The major genus in the chosen ponds was *Oscillatoria* sp. Significant seasonal differences were found in the physicochemical properties of collected water samples. After 72 h, the survival rate of *O. niloticus* was lowest when exposed to an extract from *Anabaena* sp. The collected *O. niloticus* from fish ponds showed an increase in biochemical markers than the control group. They suggested that water quality and *O. niloticus* fish populations were negatively impacted by the cyanobacterial genera that were discovered in the fish ponds; they recommended some adaptation and mitigation process should be needed. Similarly in India, Adhikari et al. (2018) found that fish farmers carried out some strategies to adapt and mitigate the impacts of climate change in their fish farms. Some of the farmers (43%) pumped freshwater to decrease the temperature of fish culture ponds. Other farmers applied oxygen tablets during higher summer. Also, 60% of them pumped water into fish ponds to maintain the water level in case of low rainfall.

All hands must be on deck in order to develop models capable of predicting the effect of climatic factors in the presence of natural toxins on the transmission of the infectious diseases and its distributions, as well as mapping the areas or regions in which climate change will have a profound impact on the disease's transmission and spatial distribution of the pathogens.

9.7 Research Gaps and Future Directions

This chapter highlights the gaps in our understanding of how natural toxins and climate change as co-stressor that must be addressed in order to develop management plans that adequately protect fisheries, aquaculture, aquatic ecosystems, and human health. These may be summarized as follows:

- Future researches should be focused on describing and assessing the new natural toxins and their new risks on human health and ecosystem.
- Advanced detection techniques to detect cyanotoxins are needed, as the current traditional methods can't detect all cyanotoxin types.
- A deeper understanding of the interaction between the harmful algal blooms (HABs) and climate change stressors which affect aquatic life in freshwater ecosystems is needed. There is a need to formulate management plans that protect human health, fisheries, and aquaculture.
- It is important to perform further studies to identify the impact of climate changes and natural toxins on human health problems.
- Modeling using artificial intelligence could be used to develop new technologies and applications for prediction of natural toxins caused by infectious diseases.
- Projections of the burden of human diseases resulted from freshwater toxins under different climate change scenarios must be included in the research strategies.
- Collaboration/communication with meteorological authority working on prediction of the future climatic variables that may elevate ecological health risks for identification of the climate variables' ranges or weather conditions that will be suitable for certain disease pathogens, hosts/vectors, or transmission of the diseases should be encouraged.

References

Abdel-Wahaab MA, Hasan AM, Aly SE, Mahrous KF (2005) Adsorption of sterigmatocystin by montmorillonite and inhibition of its genotoxicity in the Nile tilapia fish (*Oreachromis nilaticus*). Mutat Res Genet Toxicol Environ Mutagen 582:20–27

Abdel-Wareth MTA, Sayed SSM (2023) Association pattern between freshwater snails and fungi in relation to water quality parameters in two Egyptian governorates. Egypt J Aquatic Biol Fisher 27:179–201

Abdulrazzaq YM, Osman N, Yousi ZM, Trad O (2004) Morbidity in neonates of annals of tropical. Paediatrics 24:145–151

Adhikari S, Keshav CA, Barlaya G et al (2018) Adaptation and mitigation strategies of climate change impact in freshwater aquaculture in some states of India. J Fisher Sci 12:16–21

Ajandouz EH, Berdah S, Moutardier V et al (2016) Hydrolytic fate of 3/15-acetyldeoxynivalenol in humans: specific deacetylation by the small intestine and liver revealed using in vitro and ex vivo approaches. Toxins 8:1–18. https://doi.org/10.3390/toxins8080232

Aklakur M, Bakli S, Deo AD et al (2023) Cyanobacteria toxicity in aquaculture system and its impact on fish physiology. J Aquac Mar Biol 12:28–33

An J, Carmichael WW (1994) Use of a colorimetric protein phosphatase inhibition assay and enzyme linked immunosorbent assay for the study of microcystins and nodularins. Toxicon 32(12):1495e1507

Aneja KR, Dhiman R, Aggarwal NK, Aneja A (2014. Article ID 758942) Emerging preservation techniques for controlling spoilage and pathogenic microorganisms in fruit juices. Int J Microbiol 2014:1. https://doi.org/10.1155/2014/758942

Anfossi L, Giovannoli C, Baggiani C (2016) Mycotoxin detection. Current Opinion in Biotechnology. 2016. 37:120–126

Appell M, Wang LC, Bosma WB (2017) Analysis of the photophysical properties of zearalenone using density functional theory. JOL 188:551–557

Alshannaq A, Yu JH (2017) Occurrence, toxicity, and analysis of major mycotoxins in food. Int J Environ Res Pub Heal 14(6):632

Astoreca A, Magnoli C, Barberis C et al (2007) A. Ochratoxin a production in relation to ecophysiological factors by *Aspergillus* section *Nigri* strains isolated from different substrates in Argentina. Sci Total Environ 388:16–23

Awad W, Ghareeb K, Böhm J, Zentek J (2013) The toxicological impacts of the *Fusarium* mycotoxin, deoxynivalenol, in poultry flocks with special reference to immunotoxicity. Toxins 5:912–925. https://doi.org/10.3390/toxins5050912

Banack SA, Murch SJ (2018) Methods for the chemical analysis of β-N-Methylamino-L-A lanine: What is known and what remains to be determined. Neurotox Res 33:184–191

Baumann H, Wallace RB, Tagliaferri T et al (2015) Large natural pH, CO2 and O2 fluctuations in a temperate tidal salt marsh on diel, seasonal, and interannual time scales. Estuaries and Coasts. 38:220–231. https://doi.org/10.1007/s12237-014-9800-y

Bazin E, Mourot A, Humpage AR et al (2010) Genotoxicity of a freshwater cyanotoxin, cylindrospermopsin, in two human cell lines. Caco-2 and HepaRG

Bbosa GS, Kitya D, Lubega A et al (2013) In: Razzaghi-Abyaneh M (ed) Review of the biological and health effects of aflatoxins on body organs and body systems. Aflatoxins-Recent Advances and Future Prospects; In Tech. https://doi.org/10.5772/2500

Bennett JW, Klich M (2003) Mycotoxins. Clin Microbiol Rev 16:497–516. https://doi.org/10.1128/CMR.16.3.497-516.2003

Berry J (2013) Cyanobacterial toxins in food-webs: implications for human and environmental health. Current topics in public health. In Tech

Bhetariya PJ, Madan T, Basir SF et al (2011) Allergens/antigens, toxins and polyketides of important *Aspergillus* species. Indian J Clin Biochem 26(2):104–119

Bieczynski F, Bianchi VA, Luquet CM (2013) Accumulation and biochemical effects of microcystin-LR on the Patagonian pejerrey (Odontesthes hatcheri) fed with the toxic cyanobacteria Microcystis aeruginosa. Fish Physiol Biochem 39(5):1309–1321

Biswas S, Kranz WL, Shapiro CA et al (2017) Effect of rainfall timing and tillage on the transport of steroid hormones in runoff from manure amended row crop fields. J Hazard Mater 324:436–447. https://doi.org/10.1016/j.jhazmat.2016.11.009

Bondy GS, Pestka JJ (2000) Immunomodulation by fungal toxins. J Toxicol Environ Health B 3:109–143

Boretti A, Rosa L (2019) Reassessing the projections of the world water development report. NPJ Clean Water 2:1–6

Botana LM (2016) Toxicological Perspective on Climate Change: Aquatic Toxins. Chem Res Toxicol 29:619–625. https://doi.org/10.1021/acs.chemrestox.6b00020

Bouaïcha N, Miles CO, Beach DG et al (2019) Structural diversity, characterization and toxicology of microcystins. Toxins 11:714

Brodehl A, Moller A, Kunte H-J et al (2014) Biotransformation of the mycotoxin zearalenone by fungi of the genera *Rhizopus* and *Aspergillus*. FEMS Microbiol Lett 359:124–130. https://doi.org/10.1111/1574-6968.12586

Brooks WP, Codd GA (1987) Immunological and toxicological studies on Microcystis aeruginosa peptide toxin. Br Phycol J 22:301

Breitburg D, Levin LA, Oschlies A et al (1994) Declining oxygen in the global ocean and coastal waters. Science. 2018. 6371:p.eaam7240
Bruno M, Ploux O, Metcalf JS et al (2017) Anatoxin-a, homoanatoxin-a, and natural analogues. Handbook of cyanobacterial monitoring and cyanotoxin analysis. John Wiley & Sons, Ltd., Chichester, pp 138–147
Bucheli TD, Erbs M, Hartmann N et al (2005) Estrogenic mycotoxins in the environment. Mitt Lebensmittelunters Hyg 96:86–403
Bucheli TD, Wettstein FE, Hartmann N et al (2008) *Fusarium* mycotoxins: overlooked aquatic micropollutants? J Agric Food Chem 56:1029–1034. https://doi.org/10.1021/jf073082k
Bui H, Miao V, Garg S et al (2017) Design and analysis of localized DNA hybridization chain reactions. Small 13:1602983
Buratti FM, Manganelli M, Vichi S et al (2017) Cyanotoxins: producing organisms, occurrence, toxicity, mechanism of action and human health toxicological risk evaluation. Arch Toxicol 91:1049–1130
Cadel-Six S, Moyenga D, Magny S et al (2014) Detection of free and covalently bound microcystins in different tissues (liver, intestines, gills, and muscles) of rainbow trout (*Oncorhynchus mykiss*) by liquid chromatography tandem mass spectrometry: method characterization. Environ Pollut 185:333–339
Campàs M, Szydłowska D, Trojanowicz M et al (2007) Enzyme inhibition-based biosensor for the electrochemical detection of microcystins in natural blooms of cyanobacteria. Talanta 72:179–186
Campos A, Vasconcelos V (2010) Molecular mechanisms of microcystin toxicity in animal cells. Int J Mol Sci 11:268–287
Carlson DB, Williams DE, Spitsbergen JM et al (2001) Fumonisin B1 promotes aflatoxin B1 and N-methyl-N0-nitro-nitrosoguanidine-initiated liver tumors in rainbow trout. Toxicol Appl Pharmacol 172:29–36
Carmichael WW, Boyer GL (2016) Health impacts from cyanobacteria harmful algae blooms: Implications for the North American Great Lakes. Harmful Algae 54:194–212
Chen L, Liu L, Huang S (2008) Cadmium activates the mitogen-activated protein kinase (MAPK) pathway via induction of reactive oxygen species and inhibition of protein phosphatases 2A and 5. Free Radic Biol Med 45:1035–1044
Chen Y, Shen D, Fang D (2013) Nodularins in poisoning. Clin Chim Acta 425:18–29
Chen L, Chen J, Zhang X et al (2016) A review of reproductive toxicity of microcystins. J Hazard Mater 301:381–399. https://doi.org/10.1016/j.jhazmat.2015.08.041
Chorus I, Bartram J (1999) Toxic cyanobacteria in water: a guide to their public health consequences, monitoring and management. World Health Organization Report, London
Chu FS, Li GY (1994) Simultaneous occurrence of fumonisin B1 and other mycotoxins in moldy corn collected from the People's Republic of China in regions with high incidences of esophageal cancer. Appl Environ Microbiol 60:847–852
Cimbalo A, Alonso-Garrido M, Font G, Manyes L (2020) Toxicity of mycotoxins in vivo on vertebrate organisms: A review. Food Chem Toxicol 137:111–161
Colas S, Duval C, Marie B (2020) Toxicity, transfer and depuration of anatoxin-a (cyanobacterial neurotoxin) in medaka fish exposed by single-dose gavage. Aquat Toxicol 222:105422
Corrêa JAF, Orso PB, Bordin K et al (2018) Toxicological effects of fumonisin B1 in combination with other *Fusarium* toxins. Food Chem Toxicol 121:483–494
Covaci OI, Sassolas A, Alonso GA et al (2012) Highly sensitive detection and discrimination of LR and YR microcystins based on protein phosphatases and an artificial neural network. Anal Bioanal Chem 404:711–720
Cullen A, Pearson LA, Mazmouz R et al (2018) Heterologous expression and biochemical characterisation of cyanotoxin biosynthesis pathways. Nat Prod Rep 36:1117–1136
Dabbadie L, Aguilar-Manjarrez J, Beveridge MCM et al (2018) Chapter 20: Effects of climate change on aquaculture: drivers, impacts and policies In: Impacts of climate change on fisheries and aquaculture Synthesis of current knowledge, adaptation and mitigation options. FAO, pp 449–463. ISBN 978-92-5-130607-9 http://www.fao.org/fishery/statistics/en

De Boevre M, Van PC, Ediage EN et al (2018) Ultra-high-performance supercritical fluid chromatography as a separation tool for *Fusarium* mycotoxins and their modified forms. J AOAC Int 101:627–632. https://doi.org/10.5740/jaoacint.17-0336
Devic E, Li D, Dauta A et al (2002) Detection of anatoxin-a (s) in environmental samples of cyanobacteria by using a biosensor with engineered acetylcholinesterases. Appl Environ Microbiol 68:4102–4106
Diab AM, Salem RM, El-Keredy MSA et al (2018) Experimental ochratoxicosis A in Nile tilapia and its amelioration by some feed additives. Int J Vet Sci Med 6:149–158. https://doi.org/10.1016/j.ijvsm.2018.09.004
Díez-Quijada L, Prieto AI, Guzmán-Guillén R et al (2019) Occurrence and toxicity of microcystin congeners other than MC-LR and MC-RR: a review. Food Chem Toxicol 125:106–132
Díez-Quijada L, Medrano-Padial C, Llana-Ruiz-Cabello M et al (2020) Cylindrospermopsin-microcystin-LR combinations may induce genotoxic and histopathological damage in rats. Toxins 12:348
Ding CYG, Pang LM, Liang ZX et al (2018) MS/MS-based molecular networking approach for the detection of aplysiatoxin-related compounds in environmental marine cyanobacteria. Mar Drugs 16:505
Dittmann E, Fewer DP, Neilan BA (2013) Cyanobacterial toxins: biosynthetic routes and evolutionary roots. Microbiol Rev 37:23–43
Dovenyi-Nagy T, Racz C, Molnar K et al (2020) Pre-harvest modelling and mitigation of aflatoxins in maize in a changing climatic environment—a review. Toxins 12(12):768. https://doi.org/10.3390/toxins12120768
Drobac Backović D, Tokodi N, Marinović Z et al (2021) Cyanobacteria, cyanotoxins, and their histopathological effects on fish tissues in Fehérvárcsurgó reservoir, Hungary. Environ Monit Assess 193:554. https://doi.org/10.1007/s10661-021-09324-3
Du L, Liu H, Yuan L et al (2019) The diversity of cyanobacterial toxins on structural characterization, distribution and identification: a systematic review. Toxins 11:530
Duarte SC, Pena A, Lino CM (2011) Human ochratoxin A biomarkers–From exposure to effect. Crit Rev Toxicol 41:187–212
Duran-Riveroll LM, Cembella AD (2017) Guanidinium toxins and their interactions with voltage-gated sodium ion channels. Mar Drugs 15:303
El-Sayed YS, Khalil RH (2009) Toxicity, biochemical effects and residue of aflatoxin B1 in marine water-reared sea bass (*Dicentrarchus labrax* L.). Food Chem Toxicol 47:1606–1609
Emídio ES, Da Silva CP, De Marchi MRR (2020) Estrogenic mycotoxins in surface waters of the Rico Stream micro-basin, São Paulo, Brazil: occurrence and potential estrogenic contribution. Ecletica Química 45:23–32. https://doi.org/10.26850/1678-4618eqj.v45.2.2020.p23-32
Escriva L, Manyes L, Font G, Berrada H (2017) Mycotoxin analysis of human urine by LC-MS/MS: a comparative extraction study. Toxins 9(10):330
European Commission (2006) Commission recommendation of 17 August 2006 on the presence of deoxynivalenol, zearalenone, ochratoxin A, T-2 and HT-2 and fumonisins in products intended for animal feeding (2006/576/EC). Official J Eu Union 49:7–9
Fapohunda SO, Adewunmi AA (2019) Climate change and mycotoxins—the African experience. Croat J Food Sci Technol 11(2):283–290. https://doi.org/10.17508/CJFST.2019.11.2.09
Farkas J, Beczner J, Mohacsi-Farkas C (2011) Potential impact of the climate change on the risk of mycotoxin contamination of agricultural products in Southeast Central Europe. Acta Universitatis Sapientiae-Alimentaria 4:89–96. http://www.acta.sapientia.ro/acta-alim/C4/alim4-8.pdf
Fazekas B, Tar A, Kovacs M (2005) Ochratoxin a content of urine samples of healthy humans in Hungary. Acta Vet Hung 53:35–44
Fraeyman S, Croubels S, Devreese M, Antonissen G (2017) Emerging *Fusarium* and *Alternaria* mycotoxins: occurrence, toxicity and toxicokinetics. Toxins 9:1–30. https://doi.org/10.3390/toxins9070228

Freire F, Kozakiewicz Z, Paterson R (2000) Mycoflora and mycotoxins in Brazilian black pepper, white pepper and Brazil nuts. Mycopathologia 149(1):13–19. https://doi.org/10.1023/A:100724182793

Gaget V, Gribaldo S, Tandeau de Marsac N (2011) An rpoB signature sequence provides unique resolution for the molecular typing of cyanobacteria. Int J Syst Evol Microbiol 61:170–183

Gan C, Ling L, He Z et al (2016) In-situ assembly of biocompatible core–shell hierarchical nanostructures sensitized immunosensor for microcystin-LR detection. Biosens Bioelectron 78:381–389

Gareis M, Wolff J (2000) Relevance of mycotoxin contaminated feed for farming animals and carryover of mycotoxins in food of animal origin. Mycoses 43:79–83

Gautam P, Dill-Macky R (2012) Free water can leach mycotoxins from *Fusarium* infected wheat heads. J Phytopathol 160:484–490. https://doi.org/10.1111/j.1439-0434.2012.01928.x

Gelderblom W, Jaskiewicz K, Marasas W et al (1988) Fumonisins-novel mycotoxins with cancer-promoting activity produced by *Fusarium moniliforme*. Appl Environ Microbiol 54:1806–1811

Gil-Serna J, Vazquez C, Sandino FG et al (2014) Evaluation of growth and ochratoxin A production by *Aspergillus steynii* and *Aspergillus westerdijkiae* in green-coffee based medium under different environmental conditions. Food Res Int 61:127–131. https://doi.org/10.1016/j.foodres.2014.01.058

Giorni P, Magan N, Pietri A et al (2007) Studies on *Aspergillus* section *Flavi* isolated from maize in northern Italy. Int J Food Microbiol 113:330–338. https://doi.org/10.1016/j.ijfoodmicro.2006.09.007

Gonçalves AB, Paterson RP, Lima N (2006) Survey and significance of filamentous fungi from tap water. Int J Hyg Environ Health 209(3):257e264

Gracia-Lor E, Zuccato E, Hernandez F, Castiglioni S (2020) Wastewater-based epidemiology for tracking human exposure to mycotoxins. J Hazard Mater 382:1–7. https://doi.org/10.1016/j.jhazmat.2019.121108

Griffith AW, Gobler CJ (2020) Harmful algal blooms: A climate change co-stressor in marine and freshwater ecosystems. Harmful Algae 91:101590

Gromadzka K, Waśkiewicz A, Goliński P, Świetlik J (2009) Occurrence of estrogenic mycotoxin–zearalenone in aqueous environmental samples with various NOM content. Water Res 43(4):1051–1059

Gruber-Dorninger C, Novak B, Nagl V, Berthiller F (2016) Emerging mycotoxins: beyond traditionally determined food contaminants. J Agric Food Chem 65:7052–7070. https://doi.org/10.1021/acs.jafc.6b03413

Haighton LA, Lynch BS, Magnuson BA, Nestmann ER (2012) A reassessment of risk associated with dietary intake of ochratoxin A based on a lifetime exposure model. Crit Rev Toxicol 42:147–168

Hallegraef G.M. Harmful marine algal blooms and climate change: progress on a formidable predictive challenge, in Climate Change and Marine and Freshwater Toxins Botana, L. M., Louzao, M. C., Vilariño, N 2015. pp 181–193, De Gruyter, Berlin

Hameed S (2013) Investigation of the production and isolation of bioactive compounds from cyanobacteria (Doctoral dissertation). Robert Gordon University, PhD thesis. https://rgu-repository.worktribe.com/OutputFile/295731

Harada K, Ogawa K, Matsuura K et al (1990) Structural determination of geometrical isomers of microcystins LR and RR from cyanobacteria by two-dimensional NMR spectroscopic techniques. Chem Res Toxicol 3:473–481

He D, Wu Z, Cui B et al (2019) A novel SERS-based aptasensor for ultrasensitive sensing of microcystin-LR. Food Chem 278:197–202

Heil CA, Muni-Morgan AL (2021) Florida's harmful algal bloom (HAB) problem: escalating risks to human, environmental and economic health with climate change. Front Ecol Evol 9:646080

Heisler J, Glibert PM, Burkholder JM (2008) Eutrophication and harmful algal blooms: a scientific consensus. Harmful Algae 8:3–13

Hinojosa MG, Guti'errez-Praena D, Prieto AI et al (2019a) Neurotoxicity induced by microcystins and cylindrospermopsin: a review. Sci Total Environ 668:547–565

Hinojosa MG, Prieto AI, Gutiérrez-Praena D et al (2019b) Neurotoxic assessment of Microcystin-LR, cylindrospermopsin and their combination on the human neuroblastoma SH-SY5Y cell line. Chemosphere 224:751–764

Hoerger CC, Wettstein FE, Hungerbühler K, Bucheli TD (2009) Occurrence and origin of estrogenic isoflavones in Swiss river waters. Environ Sci Technol 43(16):6151–6157

Hu Y, Chen J, Fan H et al (2016) A review of neurotoxicity of microcystins. Environ Sci Pollut Res 23:7211–7219

Hueza IM, Raspantini PCF, Raspantini LER et al (2014) Zearalenone, an estrogenic mycotoxin, is an immunotoxic compound. Toxins 6:1080–1095. https://doi.org/10.3390/toxins6031080

Hurtado-McCormick S, Sánchez L, Martínez J, Calderón C, Calvo D, Narváez D, Rodríguez Susa M (2016) Fungi in biofilms of a drinking water network: occurrence, diversity and mycotoxins approach. Water Sci Technol: Water Supply 16(4):905–914

Huisman J, Codd GA, Paerl HW et al (2018) Cyanobacterial blooms. Nat Rev Microbiol 16:471–483

Humbert JF, Quiblier C, Gugger M (2010) Molecular approaches for monitoring potentially toxic marine and freshwater phytoplankton species. Anal Bioanal Chem 397:1723–1732

IARC (1993) IARC monographs on the evaluation of carcinogenic risks to humans: some naturally. International Agency for Research on Cancer, Lyon, pp 489–521

Igwaran A, Kayode AJ, Moloantoa KM et al (2024) Cyanobacteria harmful algae blooms: causes, impacts, and risk management. Water Air Soil Pollut 235. https://doi.org/10.1007/s11270-023-06782-y

International Agency for Research on Cancer (IARC), World Health Organisation (WHO) (2002) *Fumonisin B1*, in IARC monographs on the evaluation of carcinogenic risks to humans, some traditional herbal medicines, some mycotoxins, naphthalene and styrene, vol 82. IARC Press, Lyon, pp 301–366

IPCC (2021) Climate Change The Physical Science Basis. Contribution of Working Group I to the Sixth Assessment Report of the Intergovernmental Panel on Climate Change. Available online: https://www.ipcc.ch/report/ar6/wg1/downloads/report/IPCC_AR6_WGI_SPM_final2021

Iqbal SZ, Paterson RRM, Bhatti IA, Asi MR (2011) Comparing aflatoxin contamination in chilies from Punjab, Pakistan produced in summer and winter. Mycotoxin Res 27(2):75–80. https://doi.org/10.1007/S12550-010-0078-Z/TABLES/2

Iqbal SZ, Nisar S, Asi MR, Jinaq S (2014) Natural incidence of aflatoxins, ochratoxin A and zearalenone in chicken meat and eggs. Food Control 43:98–103

Ito E, Satake M, Yasumoto T (2002) Pathological effects of lyngbyatoxin A upon mice. Toxicon 40:551–556

Jaja-Chimedza AD (2014) Contribution of Lipophilic Secondary Metabolites to the Toxicity of Strains of Freshwater Cyanobacterial Harmful Algal Blooms, Identified Using the Zebrafish (*Danio rerio*) Embryo as a Model for Vertebrate Development, 1535. FIU Electronic Theses and Dissertations

Janssen EML (2019) Cyanobacterial peptides beyond microcystins—a review on co-occurrence, toxicity, and challenges for risk assessment. Water Res 151:488–499

Jiang W, Zhou W, Uchida H et al (2014) A new lyngbyatoxin from the Hawaiian cyanobacterium *Moorea producens*. Mar Drugs 12:2748–2759

Joung SH, Oh HM, Ko SR et al (2011) Correlations between environmental factors and toxic and non-toxic Microcystis dynamics during bloom in Daechung Reservoir. Harmful Algae 10:188–193

Jungblut AD, Hawes I, Mountfort D et al (2005) Diversity within cyanobacterial mat communities in variable salinity meltwater ponds of McMurdo Ice Shelf, Antarctica. Environ Microbiol 7:519–529

Kagami M, Miki T, Takimoto G (2014) Mycoloop: chytrids in aquatic food webs. Front Microbiol 5:166

Kalaitzidou M, Petridou E, Economou V et al (2016) Toxic cyanobacteria. A biological hazard for animal and public health: a review. Air Water Borne Dis 5:1–4

Kardinaal WE, Tonk L, Janse I et al (2007) Competition for light between toxic and nontoxic strains of the harmful cyanobacterium Microcystis. Appl Environ Microbiol 73:2939–2946

Kawata K, Liu CY, Merkel SF, Ramirez SH, Tierney RT, Langford D (2016) Blood biomarkers for brain injury. What are we measuring?. Neurosci Biobehav Rev 68:460–473

Kellmann R, Michali TK, Neilan BA (2008) Identification of a saxitoxin biosynthesis gene with a history of frequent horizontal gene transfers. J Mol Evol 67:526–538

Khoshal AK, Novak B, Martin PGP et al (2019) Co-occurrence of DON and emerging mycotoxins in worldwide finished pig feed and their combined toxicity in intestinal cells. Toxins 11:1–17. https://doi.org/10.3390/toxins11120727

Knutsen HK, Alexander J, Barregård L, Bignami M, Brüschweiler B, Edler L (2017) Risks to human and animal health related to the presence of deoxynivalenol and its acetylated and modified forms in food and feed. EFSA Journal 15(9):e04718

Kolpin DW, Schenzel J, Meyer MT et al (2014) Mycotoxins: diffuse and point source contributions of natural contaminants of emerging concern to streams. Sci Total Environ 470:669–676. https://doi.org/10.1016/j.scitotenv.2013.09.062

Koop SH, van Leeuwen CJ (2017) The challenges of water, waste and climate change in cities. Environ Dev Sustain 19:385–418

Kourousekos GD, Theodosiadou EK, Lymberopoulos AG et al (2015) Effect of prolonged aflatoxin B1 administration on blood serum oestradiol-17β and progesterone concentrations of goats during the anoestrus period. Czeh J Anim Sci 60(4):164–170

Kowalska K, Habrowska-Górczynska DE, Piastowska-Ciesielska AW (2016) Zearalenone as an endocrine disruptor in humans. Environ Toxicol Pharmacol 48:141–149. https://doi.org/10.1016/j.etap.2016.10.015

Kudayer AM, Alsandaqchi AT, Saleh FM, Alwan NA (2019) Toxic effect of aflatoxin B1 on heart, lung, and testis of male albino rats: histopathology study. Mater Sci Eng A 571(1):12–19

Kumar P, Mahato DK, Kamle M et al (2017) Aflatoxins: A global concern for food safety, human health and their management. Front Microbiol 7:2170

Lambert TW, Boland MP, Holmes CF et al (1994) Quantitation of the microcystin hepatotoxins in water at environmentally relevant concentrations with the protein phosphatase bioassay. Environ Sci Technol 28:753–755

Lauwers M, De Baere S, Letor B et al (2019) Multi LC-MS/MS and LC-HRMS methods for determination of 24 mycotoxins including major phase I and II biomarker metabolites in biological matrices from pigs and broiler chickens. Toxins 11:1–30. https://doi.org/10.3390/toxins11030171

Lee J, Lee S, Jiang X (2017) Cyanobacterial toxins in freshwater and food: important sources of exposure to humans. Annu Rev Food Sci Technol 8:281–304

Leggieri MC, Bertuzzi T, Pietri A, Battilani P (2015) Mycotoxin occurrence in maize produced in Northern Italy over the years 2009–2011: Focus on the role of crop related factors. Phytopathol Mediterr 54(2):212–221. https://doi.org/10.14601/PHYTOPATHOL_MEDITERR-14632

Leong S, Lin L, Pettersson OV, Rice T, Hocking AD, Schnürer J (2011) The extreme xerophilic mould *Xeromyces bisporus*-Growth and competition at various water activities. Int J Food Microbiol 145(1):57–63. https://doi.org/10.1016/J.IJFOODMICRO.2010.11.025

Li J, Li R, Li J (2017) Current research scenario for microcystins biodegradation—a review on fundamental knowledge, application prospects and challenges. Sci Total Environ 595:615–632. https://doi.org/10.1016/j.scitotenv.2017.03.285

Liew WPP, Mohd-Redzwan S (2018) Mycotoxin: Its Impact on gut health and microbiota. Front Cell Infect Microbiol 8. https://doi.org/10.3389/fcimb.2018.00060

Locatelli S, Scarpino V, Lanzanova CA et al (2022) Multi-mycotoxin long-term monitoring survey on North-Italian maize over an 11-year period (2011–2021): The cooccurrence of regulated, masked and emerging mycotoxins and fungal metabolites. Toxins 14(8):520. https://doi.org/10.3390/TOXINS14080520

Logrieco AF, Moretti A (2008) Between emerging and historical problems: An overview of the main toxigenic fungi and mycotoxin concerns in Europe. In: Mycotoxins: detection methods, management, public health and agricultural trade 2008. CABI, pp 139–153. https://doi.org/10.1079/9781845930820.0139

Lotierzo M, Abuknesha R, Davis F et al (2012) A membrane-based ELISA assay and electrochemical immunosensor for microcystin-LR in water samples. Environ Sci Technol 46:5504–5510. https://doi.org/10.1021/es2041042

Lucci P, Moret S, Bettin S, Conte L (2017) Selective solid-phase extraction using a molecularly imprinted polymer for the analysis of patulin in apple-based foods. J Sep Sci 40(2):458–465

Magan N, Aldred D (2008) Environmental fluxes and fungal interactions: maintaining a competitive edge. British Mycological Society Symposia Series 27(C):19–35. https://doi.org/10.1016/S0275-0287(08)80044-6

Manolidi K, Triantis TM, Kaloudis T et al (2019) Neurotoxin BMAA and its isomeric amino acids in cyanobacteria and cyanobacteria-based food supplements. J Hazard Mater 365:346–365

Mantzouki E, Lürling M, Fastner J et al (2018) Temperature effects explain continental scale distribution of cyanobacterial toxins. Toxins 10:156

Maranghi F, Tassinari R, Narciso L et al (2018) In vivo toxicity and genotoxicity of beauvericin and enniatins. Combined approach to study in vivo toxicity and genotoxicity of mycotoxins beauvericin (BEA) and enniatin B (ENNB). EFSA Support 15:1–183. https://doi.org/10.2903/sp.efsa.2018.en-1406.Publ

Marchese S, Polo A, Ariano A et al (2018) Aflatoxin B1 and M1: biological properties and their involvement in cancer development. Toxins 10:214

Marie B, Blaha L, Maria CA et al (2016) Bioassay use in the field of toxic cyanobacteria. Handbook of cyanobacterial monitoring and cyanotoxin analysis 27:272–278

Marin S, Ramos AJ, Cano-Sancho G, Sanchis V (2013) Mycotoxins: Occurrence, toxicology, and exposure assessment. Food Chem Toxicol 60:218–237

Marquardt RR, Frohlich AA (1992) A review of recent advances in understanding ochratoxicosis. J Anim Sci 70:3968–3988

Martínez M, Biganzoli F, Arata A et al (2022) Warm nights increase *Fusarium* head blight negative impact on barley and wheat grains. Agric For Meteorol 318:108909. https://doi.org/10.1016/J.AGRFORMET.2022.108909

Martínez-Ruiz Cooper EBM, Fastner J, Szewzyk U (2020) Manganese-oxidizing bacteria isolated from natural and technical systems remove cylindrospermopsin. Chemosphere 238:124625

Massey IY, Yang F, Ding Z et al (2018) Exposure routes and health effects of microcystins on animals and humans: a mini-review. Toxicon 151:156–162

Mata AT, Ferreira JP, Oliveira BR et al (2015) Bottled water: analysis of mycotoxins by LC-MS/MS. Food Chem 176:455e464

Mayer E, Novak B, Springler A et al (2017) Effects of deoxynivalenol (DON) and its microbial biotransformation product deepoxy-deoxynivalenol (DOM-1) on a trout, pig, mouse, and human cell line. Mycotoxin Res 33:297–308. https://doi.org/10.1007/s12550-017-0289-7

McLaughlin JT (2013) Application of cytotoxicity assays for detection of potentially harmful bioactive compounds produced by freshwater bloom-forming algae, 1095. Electronic Thesis and Dissertation Repository

Medina A, Rodriguez A, Magan N (2014) Effect of climate change on Aspergillus flavus and aflatoxin B1 production. Front Microbiol 5:348

Medina A, Rodriguez A, Sultan Y, Magan N (2015) Climate change factors and *Aspergillus flavus*: effects on gene expression, growth and aflatoxin production. World Mycotoxin J 8(2):171–179. https://doi.org/10.3920/WMJ2014.1726

Meerpoel C, Vidal A, di Mavungu JD et al (2018) Development and validation of an LC–MS/MS method for the simultaneous determination of citrinin and ochratoxin a in a variety of feed and foodstuffs. J Chromatogr A 1580:100–109. https://doi.org/10.1016/j.chroma.2018.10.039

Meriluoto J, Spoof L, Codd GA (2017) Appendix 3: tables of microcystins and nodularins. Handbook of cyanobacterial monitoring and cyanotoxin analysis:526–537

Metcalf JS, Codd GA (2009) Cyanobacteria, neurotoxins and water resources: are there implications for human neurodegenerative disease? Amyotroph Lateral Scler 10:74–78

Mitchell CE, Power AO (2003) Release of invasive plants from fungal and viral pathogens. Nature 421(6923):625–627. https://doi.org/10.1038/nature01317

Mogensen JM, Frisvad JC, Thrane U, Nielsen KF (2009) Production of fumonisin B2 and B4 by *Aspergillus niger* on grapes and raisins. J Agric Food Chem 58:954–958

Morabito S, Silvestro S, Faggio C (2018) How the marine biotoxins affect human health. Nat Prod Res 32:621–631

Moreira C, Martins A, Azevedo J et al (2011) Application of real-time PCR in the assessment of the toxic cyanobacterium *Cylindrospermopsis raciborskii* abundance and toxicological potential. Appl Microbiol Biotechnol 92:189–197

Moreira C, Ramos V, Azevedo J et al (2014) Methods to detect cyanobacteria and their toxins in the environment. Appl Microbiol Biotechnol 98:8073–8082

Moretti A, Pascale M, Logrieco AF (2019) Mycotoxin risks under a climate change scenario in Europe. Trends Food Sci Technol 84:38–40. https://doi.org/10.1016/j.tifs.2018.03.008

Morseletto P, Mooren CE, Munaretto S (2022) Circular economy of water: definition, strategies and challenges. Circ Econ Sustain 2:1463–1477

Moslemi MK, Tavanbakhsh S (2011) Selenium–vitamin E supplementation in infertile men: effects on semen parameters and pregnancy rate. Int J Gen Med 4:99

Murad AF, Ahmed S, Abead S (2015) Toxicity effect of aflatoxin B1 on reproductive system of albino male rats. Pak J Biol Sci 18(3):107

Murray S., John U., Kremp A. Alexandrium spp.: genetic and ecological factors influencing saxitoxin production and proliferation, in Climate Change and Marine and Freshwater Toxins (Botana, L. M., Louzao, M. C., Vilariño, N) 2015. pp 125–154, De Gruyter, Berlin

Mutoti MI, Edokpayi JN, Mutileni N et al (2023) Cyanotoxins in groundwater; occurrence, potential sources, health impacts and knowledge gap for public health. Toxicon 226:107077

Nakamura C, Kobayashi T, Miyake M et al (2001) Usage of a DNA aptamer as a ligand targeting microcystin. Mol Cryst Liq Cryst 371:369–374

Nicholson P (2004) Mycotoxins in food. Woodhead Publishing Limited, Cornwall

Nikolić M, Savić I, Nikolić A et al (2021) Toxigenic species *Aspergillus parasiticus* originating from Maize Kernels grown in Serbia. Toxins 13(12):847. https://doi.org/10.3390/toxins13120847

Nomura H, Ogiso M, Yamashita M et al (2011) Uptake by dietary exposure and elimination of aflatoxins in muscle and liver of rainbow trout (*Oncorhynchus mykiss*). J Agric Food Chem 59:5150–5158

Nugumanova G, Ponomarev ED, Askarova S et al (2023) Freshwater cyanobacterial toxins, cyanopeptides and neurodegenerative diseases. Toxins 15:233

Nunes-Costa D, Magalhães JD, G-Fernandes M. et al (2020) Microbial BMAA and the Pathway for Parkinson's Disease Neurodegeneration. Front Aging Neurosci 7:12–26. https://doi.org/10.3389/fnagi.2020.00026

Ohtani I, Moore RE, Runnegar M (1992) Cylindrospermopsin: A potent hepatotoxin from the blue-green alga *Cylindrospermopsis raciborskii*. J Am Chem Soc 114:7941–7942

Oliveira BR, Mata AT, Ferreira JP et al (2018) Production of mycotoxins by filamentous fungi in untreated surface water. Environ Sci Pollut Res 25:17519–17528. https://doi.org/10.1007/s11356-018-1952-z

Olson NE, Cooke ME, Shi JH et al (2020) Harmful algal bloom toxins in aerosol generated from Inland Lake Water. Environ Sci Technol 54:4769–4780

Ostry V (2008) Alternaria mycotoxins: an overview of chemical characterization, producers, toxicity, analysis and occurrence in foodstuffs. World Mycotoxin J 1:175–188. https://doi.org/10.3920/wmj2008.x013

Ostry V, Malir F, Toman J, Grosse Y (2017) Mycotoxins as human carcinogens-The IARC Monographs classification. Mycotoxin Res 33:65–73

Paerl HW (2014) Mitigating harmful cyanobacterial blooms in a human-and climatically-impacted world. Life 4:988–1012

Pašková V, Adamovský O, Pikula J et al (2008) Detoxification and oxidative stress responses along with microcystins accumulation in Japanese quail exposed to cyanobacterial biomass. Sci Total Environ 398:34–47

Paterson RRM (2006) Identification and quantification of mycotoxigenic fungi by PCR. Process Biochem 41:1467–1474
Paterson RRM, Lima N (2010) How will climate change affect mycotoxins in food? Food Res Int 43(7):1902–1914. https://doi.org/10.1016/j.foodres.2009.07.010
Paterson RRM, Lima N (2011) Further mycotoxin effects from climate change. Food Res Int 44(9):2555–2566. https://doi.org/10.1016/j.foodres.2011.05.038
Paterson RRM, Lima N (2017) Thermophilic fungi to dominate aflatoxigenic/mycotoxigenic fungi on food under global warming. Int J Environ Res Public Health 14(2):199. https://doi.org/10.3390/ijerph14020199
Paterson RRM, Venancio A, Lima N et al (2018) Predominant mycotoxins, mycotoxigenic fungi and climate change related to wine. Food Res Int 103:478–491. https://doi.org/10.1016/j.foodres.2017.09.080
Payros D, Alassane-Kpembi I, Pierron A et al (2016) Toxicology of deoxynivalenol and its acetylated and modified forms. Arch Toxicol 90:2931–2957. https://doi.org/10.1007/s00204-016-1826-4
Pearson LA, Dittmann E, Mazmouz R et al (2016) The genetics, biosynthesis and regulation of toxic specialized metabolites of cyanobacteria. Harmful Algae 54:98–111
Persi NJ, Pleadin D, Kovacevi´ Scortichini G., Milone S. (2014) Ochratoxin A in raw materials and cooked meat products made from OTA-treated pigs. Meat Sci 96:203–210
Picardo M, Filatova D, Nunez O et al (2019) Recent advances in the detection of natural toxins in freshwater environments. Trends Anal Chem 112:75–86
Piva G, Battilani P, Pietri A (2006) Emerging issues in Southern Europe: Aflatoxins in Italy, in the mycotoxin factbook, food and feed topics. Wageningen Academic, pp 139–153
Ploux O, Combes A, Eriksson J et al (2017) β-N-methylamino-l-alanine and (S)-2,4-diaminobutyric acid. Handbook of cyanobacterial monitoring and cyanotoxin analysis 16:160–164
Puerto M, Prieto AI, Maisanaba S et al (2018) Mutagenic and genotoxic potential of pure Cylindrospermopsin by a battery of in vitro tests. Food Chem Toxicol 121:413–422
Qian G, Tang L, Lin S et al (2016) Sequential dietary exposure to aflatoxin B1 and fumonisin B1 in F344 rats increases liver preneoplastic changes indicative of a synergistic interaction. Food Chem Toxicol 95:188–195
Qileng A, Cai Y, Wei J et al (2018) Construction of CdS/B-TiO2 nanorods photoelectrochemical immunosensor for the detection of microcystin-LR using SiO2@ G-quadruplex as multi-amplifier. Sens Actuators B 254:727–735
Rantala-Ylinen A, Känä S, Wang H et al (2011) Anatoxin-a synthetase gene cluster of the cyanobacterium Anabaena sp. strain 37 and molecular methods to detect potential producers. Appl Environ Microbiol 77:7271–7278
Rapala J, Sivonen K, Lyra C et al (1997) Variation of microcystins, cyanobacterial hepatotoxins, in Anabaena spp. as a function of growth stimuli. Appl Environ Microbiol 63:2206–2212
Rapala J, Erkomaa K, Kukkonen J et al (2002) Detection of microcystins with protein phosphatase inhibition assay, high-performance liquid chromatography–UV detection and enzyme-linked immunosorbent assay: Comparison of methods. Anal Chim Acta 466:213–231
Rastogi RP, Madamwar D, Incharoensakdi A (2015) Bloom dynamics of cyanobacteria and their toxins: Environmental health impacts and mitigation strategies. Front Microbiol 6:1254
Reichwaldt E.S., Sinang S.C., Ghadouani A. Global warming, climate patterns and toxic cyanobacteria, in Climate Change and Marine and Freshwater Toxins (Botana, L. M., Louzao, M. C., Vilariño, N) 2015. pp 196–237, De Gruyter, Berlin
Rheeder JP, Marasas WF, Vismer HF (2002) Production of fumonisin analogs by *Fusarium* species. Appl Environ Microbiol 68:2101–2105
Righetti L, Paglia G, Galaverna G, Dall'Asta C (2016) Recent advances and future challenges in modified mycotoxin analysis: Why HRMS has become a key instrument in food contaminant research. Toxins 8(12):361
Rocha da MEB, Freire FDCO, Maia FEF, Guedes MIF, Rondina D (2014) Mycotoxins and their effects on human and animal health. Food Control 36(1):159–165

Rychlik M, Humpf HU, Marko D et al (2014) Proposal of a comprehensive definition of modified and other forms of mycotoxins including "masked" mycotoxins. Mycotoxin Res 30:197–205. https://doi.org/10.1007/s12550-014-0203-5

Saad AA, Ragab FMA, Mekawey AA, Abdel-Wareth MTA (2014) Seasonal variations of fungi associated with some freshwater snails in four Egyptian governorates. J Environ Sci 28(4):27–51

Sabart M, Crenn K, Perrière F et al (2015) Co-occurrence of microcystin and anatoxin-a in the freshwater lake Aydat (France): analytical and molecular approaches during a three-year survey. Harmful Algae 48:12–20. https://doi.org/10.1016/j.hal.2015.06.007

Sadeghi N, Oveisi MR, Jannat B et al (2009) Incidence of aflatoxin M1 in human breast milk in Tehran, Iran. Food Control 20(1):75–78

Saleh HA, Sayed SS, El-Gamal MA, Ismail NMM (2024) Effect of some freshwater cyanobacteria on Oreochromis niloticus (L.) fish and water quality in Egyptian aquaculture ponds. Int J Environ Stud 81(3):1132–1148. https://doi.org/10.1080/00207233.2022.2059257

Salmaso N, Bernard C, Humbert JF et al (2017) Basic guide to detection and monitoring of potentially toxic cyanobacteria. Handbook of cyanobacterial monitoring and cyanotoxin analysis 6:46–69

Sanseverino I, Antonio DC, Loos R et al (2017) Cyanotoxins: methods and approaches for their analysis and detection. Centre, JR, Ed. https://doi.org/10.2760/36186

Sarma PU (2010) Fascinating potential of aspergilli. Indian J Clin Biochem 25(4):331–334

Schenzel J, Schwarzenbach RP, Bucheli TD (2010) Multi-residue screening method to quantify mycotoxins in aqueous environmental samples. J Agric Food Chem 58:11207–11217. https://doi.org/10.1021/jf102737q

Schenzel J, Hungerbü K, Bucheli TD (2012) Mycotoxins in the environment: II. Occurrence and origin in Swiss river waters. Environ Sci Technol 46:13076–13084. https://doi.org/10.1021/es301558v

Schwartz P, Thorpe KL, Bucheli TD et al (2010) Short-term exposure to the environmentally relevant estrogenic mycotoxin zearalenone impairs reproduction in fish. Sci Total Environ 409:326–333. https://doi.org/10.1016/j.scitotenv.2010.10.017

Schweitzer B, Roberts S, Grimwade B et al (2002) Multiplexed protein profiling on microarrays by rolling-circle amplification. Nat Biotechnol 20:359–365

Serra R, Lourenço A, Alípio P, Venâncio A (2006a) Influence of the region of origin on the mycobiota of grapes with emphasis on *Aspergillus* and *Penicillium* species. Mycol Res 110(8):971–978. https://doi.org/10.1016/J.MYCRES.2006.05.010

Serra R, Mendonça C, Venâncio A (2006b) Ochratoxin A occurrence and formation in Portuguese wine grapes at various stages of maturation. Int J Food Microbiol 111:S35–S39. https://doi.org/10.1016/J.IJFOODMICRO.2006.03.007

Serrano AB, Font G, Manes J et al (2016) Dispersive liquid-liquid microextraction for the determination of emerging *Fusarium* mycotoxins in water. Food Anal Methods 9(4):856–862

Shah SH, Rahman T, Shah GM et al (2019) Toxic algae and their environmental consequences. Trends Telemed E-Health. https://doi.org/10.31031/TTEH.2019.01.000519

Sharma V, Sharma C, Paliwal R et al (2011) Ameliorative effects of *Curcuma longa* and *curcumin* on aflatoxin B1 induced serological and biochemical changes in kidney of male mice. Asian J Biochem Pharm Res 1(2):338–351

Sheng JW, He M, Shi HC et al (2006) A comprehensive immunoassay for the detection of microcystins in waters based on polyclonal antibodies. Anal Chim Acta 572:309–315

Silvarrey BA, Sposito V, Faggian R (2021) Climate change and cyanobacteria harmful algae blooms: adaptation practices for developing countries. Mar Freshw Res 72:1722. https://doi.org/10.1071/MF21081

Skafi M, Duy SV, Munoz G et al (2021) Occurrence of microcystins, anabaenopeptins and other cyanotoxins in fish from a freshwater wildlife reserve impacted by harmful cyanobacterial blooms. Toxicon 194:44–52. https://doi.org/10.1016/j.toxicon.2021.02.004

Smith PT (2014) Cyanobacterial toxins in aquaculture, 3rd edn. CRC Press, Boca Raton

Sonak S, Patil K, Devi P et al (2018) Causes, human health impacts and control of harmful algal blooms: A comprehensive review. Environ Pollut Prot 3(1):40–55

Soydan S, Erbaş O (2022) Human health impact of cyanobacterial toxins. J Exp Basic Med Sci 3:103–108

Stroka J, Maragos CM (2016) Challenges in the analysis of multiple mycotoxins. World Mycotoxin J 9(5):847–861

Stocker TF, Qin D, Plattner GK et al (2013) Climate change 2013: the physical science basis. Contribution of Working Group I to the Fifth Assessment Report of the Intergovernmental Panel on Climate Change. Cambridge University Press, Cambridge, New York, p 1535

Suo T, Sohail M, Xie S et al (2021) DNA nanotechnology: a recent advancement in the monitoring of microcystin-LR. J Hazard Mater 403:123418. https://doi.org/10.1016/j.jhazmat.2020.123418

Takadong JJT, Mouafo HT, Manet L et al (2020) Assessment of the presence of total aflatoxins and aflatoxin B1 in fish farmed in two Cameroonian localities. Int J Food Sci 2020. https://doi.org/10.1155/2020/2506812

Tamele IJ, Silva M, Vasconcelos V (2019) The incidence of marine toxins and the associated seafood poisoning episodes in the African countries of the Indian ocean and the red sea. Toxins 11:58

Taylor MS, Stahl TW, Redshaw CH, Osborne NJ (2014) Toxic alkaloids in Lyngbya majuscula and related tropical marine cyanobacteria. Harmful 31:1–8

Tian J, Zhao H, Quan X et al (2014) Fabrication of graphene quantum dots/silicon nanowires nanohybrids for photoelectrochemical detection of microcystin-LR. Sens Actuators B 196:532–538

Turner AD, Higgins C, Davidson K, Veszelovszki A, Payne D, Hungerford J, Higman W (2015) Potential threats posed by new or emerging marine biotoxins in UK waters and examination of detection methodology used in their control: brevetoxins. Marine Drugs 13(3):1224–1254

Tzima E, Serifi I, Tsikari I et al (2017) Transcriptional and behavioral responses of zebrafish larvae to microcystin-LR exposure. Int J Mol Sci 18. https://doi.org/10.3390/ijms18020365

UN Water (UN world water development report): water and climate change, UNESCO i-WSSM Water Reuse within a Circular Economy Context. UNESCO, Paris. 2020. Available at https://unesdoc.unesco.org/ark:/48223/pf0000374715.locale=en

Valverde-Bogantes E, Bianchini A, Herr JR et al (2019) Recent population changes of *Fusarium* head blight pathogens: drivers and implications. Can J Plant Pathol 42(3):315–329. https://doi.org/10.1080/07060661.2019.1680442

van der Fels-Klerx HJ, Liu C, Battilani P (2016) Modelling climate change impacts on mycotoxin contamination. World Mycotoxin J 9(5):717–726. https://doi.org/10.3920/WMJ2016.2066

Varga J, Frisvad JC, Samson R (2011) Two new aflatoxin producing species, and an overview of *Aspergillus* section *Flavi*. Stud Mycol 69(1):57–80. https://doi.org/10.3114/sim.2011.69.05

Vaughan M, Backhouse D, del Ponte EM (2016) Climate change impacts on the ecology of *Fusarium graminearum* species complex and susceptibility of wheat to *Fusarium* head blight: a review. World Mycotoxin J 9(5):685–700. https://doi.org/10.3920/WMJ2016.2053

Villatte F, Schulze H, Schmid R et al (2002) A disposable acetylcholinesterase-based electrode biosensor to detect anatoxin-a (s) in water. Anal Bioanal Chem 372:322–326

Vogiazi V, de la Cruz A, Mishra S et al (2019) A comprehensive review: development of electrochemical biosensors for detection of cyanotoxins in freshwater. ACS Sensors 4:1151–1173

Wang JS, Tang L (2005) Epidemiology of aflatoxin exposure and human liver cancer. Aflatoxin food safety:195–211

Waśkiewicz A, Bocianowski J, Perczak A, Goliński P (2015) Occurrence of fungal metabolites-Fumonisins at the ng/L level in aqueous environmental samples. Sci Total Environ 524:394–399

Wei J, Chang W, Qileng A, et al (2018a) Dualmodal split-type immunosensor for sensitive detection of Microcystin-LR: enzymeinduced photoelectrochemistry and colorimetry. Anal Chemis 90(15):9606–9613

Wei J, Xie X, Chang W, Yang Z, Liu Y, (2018b) Ultrasensitive photoelectrochemical detection of microcystin-LR based on hybridization chain reaction assisted excitonplasmon interaction and enzymatic biocatalytic precipitation. Sens Actuators B Chem 276:180–188

Weirich CA, Miller TR (2014) Freshwater harmful algal blooms: Toxins and children's health. Current Problems in Paediatric and Adolescent Health Care. 44(1):2–24
Wells ML, Trainer VL, Smayda TJ et al (2015) Harmful algal blooms and climate change: Learning from the past and present to forecast the future. Harmful Algae 49:68–93
West JS, Holdgate S, Townsend JA et al (2012) Impacts of changing climate and agronomic factors on *Fusarium* ear blight of wheat in the UK. Fungal Ecol 5(1):53–61. https://doi.org/10.1016/j.funeco.2011.03.003
Wogan GN, Kensler TW, Groopman JD (2012) Present and future directions of translational research on aflatoxin and hepatocellular carcinoma. A review. Food Addit Contam Part A 29(2):249–257
World Health Organization (WHO) (2002) Evaluation of certain mycotoxins in food. Fifty-sixth report of the Joint FAO/WHO expert committee on food additives, World Health Organization, Geneva, pp 27–35
World Health Organization (WHO) (2019) Safety evaluation of certain food additives and contaminants: Prepared by the seventy fourth meeting of the Joint FAO/WHO expert committee on food additives (JECFA). Available online: https://apps.who.int/iris/bitstream/handle/10665/171781/9789240693982_eng.pdf;jsessionid=79EFFBD803B75293027EDA351F998A18?sequence=3 (accessed on 7 June 2019)
Wu F (2013) Aflatoxin exposure and chronic human diseases: estimates of burden of disease. Aflatoxins: finding solutions for improved food safety. http://www.ifpri.org/sites/default/fles/publications/focus20.pdf
Wu F, Munkvold GP (2008) Mycotoxin in ethanol co-products: modelling economic impacts on the livestock industry and management strategies. J Agric Food Chem 56:3900–3911
Xing C, Dai J, Huang Y et al (2019) Active self-assembly of train-shaped DNA nanostructures via catalytic hairpin assembly reactions. Small 15:e1901795
Xue KS, Qian G, Lin S et al (2018) Modulation of pre-neoplastic biomarkers induced by sequential aflatoxin B1 and fumonisin B1 exposure in F344 rats treated with UPSN clay. Food Chem Toxicol 114:316–324
Yang H, Dai R, Zhang H et al (2016) Production of monoclonal antibodies with broad specificity and development of an immunoassay for microcystins and nodularin in water. Anal Bioanal Chem 408:6037–6044
Yin P, Choi HMT, Calvert CR et al (2008) Programming biomolecular self-assembly pathways. Nature 451:318–322
Yin H, Jiang M, Peng X et al (2016) The molecular mechanism of G2/M cell cycle arrest induced by AFB1 in the jejunum. Oncotarget 7(24):35592
Yu S, Jia B, Liu N et al (2021) Fumonisin B1 triggers carcinogenesis via HDAC/PI3K/Akt signalling pathway in human esophageal epithelial cells. Sci Total Environ 787:147405
Yuan G, Wang Y, Yuan X et al (2014) T-2 toxin induces developmental toxicity and apoptosis in zebrafish embryos. J Environ Sci 26:917–925
Zamyadi A (2014) Emerging toxic cyanobacterial issues in freshwater sources: influence of climate change. In: Botana LM (ed) Seafood and freshwater toxins: pharmacology, physiology and detection. CRC Press, Boca Raton, pp 105–120
Zamyadi A, McQuaid N, Prevost M et al (2012) Monitoring of potentially toxic cyanobacteria using an online multiprobe in drinking water sources. J Environ Monit 14:579–588
Zanchett G, Oliveira-Filho EC (2013) Cyanobacteria and cyanotoxins: from impacts on aquatic ecosystems and human health to anticarcinogenic effects. Toxins 5:1896–1917
Zhang X, Halder J, White RP et al (2014) Climate change increases risk of *Fusarium* ear blight on wheat in central China. Ann Appl Biol 164(3):384–395. https://doi.org/10.1111/AAB.12107
Zhang JJ, Kang TF, Hao YC et al (2015a) Electrochemiluminescent immunosensor based on CdS quantum dots for ultrasensitive detection of microcystin LR. Sens Actuators B 214:117–123
Zhang F, Lee J, Liang S et al (2015b) Cyanobacteria blooms and non-alcoholic liver disease: evidence from a county level ecological study in the United States. Environ Health 7:14–41

Zhang Y, Li H, Zhang J, Shao B (2019a) Determination of *Alternaria* toxins in drinking water by ultra-performance liquid chromatography tandem mass spectrometry. Environ Sci Pollut Res 26:22485–22493. https://doi.org/10.1007/s11356-019-05483-6

Zhang Y, Zhu Z, Teng X et al (2019b) Enzyme-free fluorescent detection of microcystin-LR using hairpin DNA-templated copper nanoclusters as signal indicator. Talanta 202:279–284

Zhao S, Li G, Chen J (2015) A proteomic analysis of prenatal transfer of microcystin-LR induced neurotoxicity in rat offspring. J Proteomics 114:197–213

Zhong Y, Shen L, Ye X et al (2020) Neurotoxic Anatoxin-a can also exert immunotoxicity by the induction of apoptosis on Carassius auratus lymphocytes in vitro when exposed to environmentally relevant concentrations. Front Physiol 11. https://doi.org/10.3389/fphys.2020.00316

Zimmer I, Usleber E, Klaffke H et al (2008) Fumonisin intake of the German consumer. Mycotoxin Res 24:40–52

Zinedine AJM, Soriano JC, Moltó J, Mañes. (2007) Review on the toxicity, occurrence, metabolism, detoxification, regulations and intake of zearalenone: an oestrogenic mycotoxin. Food Chem Toxicol 45:1–18

Zingales V, Taroncher M, Martino PA et al (2022) Climate change and effects on molds and mycotoxins. Toxins 14:445. https://doi.org/10.3390/toxins14070445

Role of Technology to Predict Impacts of Climate Change on Health Toxicity

10

Wagida Anwer, Amal Saad-Hussein, Mona M. Taha, Atef M. F. Mohammed, Mohamed Salama, Salwa M. Abdallah, and Moh Moh Thant Zin

Abstract

Climate change (CC) can lead, expectedly, to changes in the development, growth, and organism activity. In order to reduce possible dangers to human well-being, technology plays a critical role in forecasting how CC would affect health toxicity. In this chapter, the potential for real-time monitoring and response systems toward climate changes as well as biomarkers of exposures and biological changes on human health were discussed.

Traditional epidemiological methods have been the main approach to define the associations between hazardous substances' exposures and the development of chronic diseases such as cardiovascular and respiratory diseases and cancer.

W. Anwer (✉)
Department of Community, Environmental and Occupational Medicine, Faculty of Medicine, Ain Shams University, Cairo, Egypt

National Committee of Toxicology, Academy of Scientific Research and Technology (ASRT), Ministry of Higher Education and Scientific Research, Cairo, Egypt
e-mail: wagidaanwar@med.asu.edu.eg

A. Saad-Hussein
National Committee of Toxicology, Academy of Scientific Research and Technology (ASRT), Ministry of Higher Education and Scientific Research, Cairo, Egypt

Department of Environmental and Occupational Medicine, Institute of Environment and Climate Change Research, National Research Centre, Giza, Egypt
e-mail: al.hussein@nrc.sci.eg

M. M. Taha
Environmental & Occupational Medicine Department, Environment & Climate Change Research Institute, National Research Centre, Giza, Egypt
e-mail: mm.taha@nrc.sci.eg

A. M. F. Mohammed
Air Pollution Research Department, Environment & Climate Change Research Institute, National Research Centre, Giza, Egypt
e-mail: at.mohammed@nrc.sci.eg

A. Saad-Hussein et al. (eds.), *Climate Change Impacts on Toxins and Health Effects*,
https://doi.org/10.1007/978-981-96-3416-3_10

Molecular epidemiology is a promising branch in the epidemiological studies, as it is addressing the genetic susceptibility and the mechanisms of disease induction, through usage of reliable biodetectors in these investigations for prediction of diseases and implementation in disease prevention programs.

Thanks to technology, enormous volumes of information about the climate, the environment, and public health can be gathered. Environmental monitoring of air pollutants is an important sector in technological methods done to design the strategies for mitigation and adaptation to climate change. Monitoring suspended particulate matters (SPM) with different fraction sizes could be carried out at different indoor areas either in residential or work places. Monitoring gaseous pollutants in the indoor of residential or workplaces, such as carbon dioxide (CO_2), nitrous dioxide (NO_2), sulfur dioxide (SO_2), carbon monoxide (CO), and volatile organic compounds (VOCs) could be estimated. Each gaseous pollutant needs special sensor to detect the air concentrations. Environmental organic emissions of polycyclic aromatic hydrocarbons compounds (PAHs) could be collected and extracted for identification and evaluation of the concentration levels.

This data is processed and analyzed with the aid of sophisticated analytical tools in order to find patterns, trends, and possible connections between health toxicity and climate change.

Advanced climate models make use of technology to simulate different climate scenarios and forecast future patterns of the climate. These models aid in the comprehension of how variations in temperature, precipitation, and extreme weather can affect the environment and, in turn, affect the toxicity to human health.

Technologies that use remote sensing, such as drones and satellites, can gather useful data about the state of the environment over wide geographic areas. They can assist in determining the possible health concerns linked to climate change by monitoring changes in vegetation cover, air quality, water pollution, and land use.

Satellite images are pictures of Earth obtained via satellites from orbit. Numerous applications, including natural resource management, urban planning, and weather forecasting, are possible with these photos.

M. Salama
National Committee of Toxicology, Academy of Scientific Research and Technology (ASRT), Ministry of Higher Education and Scientific Research, Cairo, Egypt

Institute of Global Health and Human Ecology, The American University in Cairo, Cairo, Egypt
e-mail: Mohamed-Salama@aucegypt.edu

S. M. Abdallah
Center of Excellence for Toxicological Testing, Central Agricultural Pesticides Lab (CAPL), Agricultural Research Center (ARC), Giza, Egypt
e-mail: Salwa.abdallah@arc.sci.eg

M. M. T. Zin
Division of Plant Science and Technology, University of Missouri, Columbia, MO, USA
e-mail: mztg7@missouri.edu

Disease outbreaks, vector-borne infections, and other health-related indicators are tracked by health surveillance systems, which are developed and implemented with the use of technology. These technologies make it possible to identify new health risks associated with climate change early on, which enables prompt resource allocation and response.

To predict the spread of infectious illnesses, the growth of toxic algal blooms, and other health risks impacted by climate change, machine learning algorithms and other predictive modeling techniques are used. Policymakers and medical professionals can better anticipate and address future health issues by using these models.

With the use of technology, decision support systems can be developed to help policymakers decide on public health interventions and adaption plans. These technologies improve the efficacy of response initiatives to reduce health hazards linked to climate change by combining data from multiple sources and offering real-time updates.

Technology plays a vital role in increasing public awareness about climate change impacts on health and promoting behavior changes that reduce vulnerability. Digital platforms, social media, and interactive tools are utilized to disseminate information, engage communities, and empower populations to use proactive measures to safeguard the health impacts due to the changing climate.

In summary, technology serves as a cornerstone in understanding, predicting, and assessing the impacts of climatic changes on health toxicity by enabling data-driven analysis, modeling, surveillance, and decision-making processes. Risk Assessment (RA) tools that produce decisions are composed of four steps: hazard identification, hazard characterization, exposure assessment, and risk characterization. Its continued advancement is essential for enhancing our ability to define the complex challenges results from climate-related health risks.

Keywords

Technology · Prediction · Impacts of climate change · Heath toxicity

Abbreviations

ACGIH	American Conference of Governmental Industrial Hygienists
AI	artificial intelligence
ALI	Advanced Land Imager
ALOS	Advanced Land Observation Satellite
AVHRR	Advanced Very High-Resolution Radiometer
CA	chromosome aberrations
CF	characterization factor
CMIP	Coupled Model Inter-comparison Project
CO	carbon monoxide
CO_2	carbon dioxide

CS_2	Carbon disulfide
CZCS	Coastal Zone Color Scanner
DALYs	disability-adjusted life years
DMC	Disaster Monitoring Constellation
ECVs	essential climate variables
ENSO	El Nino-Southern Oscillation
EOS	Earth Observing System
EPA	US Environmental Protection Agency's
ERS	European Radar Satellite
ETM+	Enhanced Thematic mapper
FU	Functional unit
GC	Gas chromatography
GHGs	greenhouse gases
GOES	Geostationary Operational Environmental Satellite
GPP	gross primary production
GWP	Global Warming Potential report
HAPs	Hazardous Air Pollutants
HBM	human biomonitoring
HVS	high volume sampler
IoT	Internet of Things
IPCC	Intergovernmental Panel on Climate Change
IRS	Indian Remote Sensing Satellite
JERS	Japanese Earth Resources satellite
LAI	leaf area index
LCA	Life cycle assessment
LCIA	life cycle impact assessment
LULC	land use/land cover
MERIS	Medium spectral Resolution Imaging Spectrometer
ML	machine learning
MN	micronuclei induction
MODIS	Moderate Resolution Imaging Spectro-radiometer
MOPITT	Measurements of Pollution in the Troposphere
MRI	magnetic resonance imaging
MSS	multispectral scanner
NO_2	nitrous dioxide
NOAA	Polarorbiting National Oceanic and Atmospheric Administration
OLI	Operational Land Imager
PAHs	polycyclic aromatic hydrocarbons compounds
PEC	Predicted environmental concentration
PNEC	Predicted no effect concentration$PM_{0.1}$: ultrafine particulate matter 0.1μm
PM_{10}	particulate matter 10 μm, > 2.5 μm
$PM_{2.5}$	particulate matter 2.5 μm, > 0.1 μm
RA	risk assessment
RCR	Risk characterization ratio

SCE	sister chromatid exchange
SDGs	Sustainable Development Goals
SeaWiFS	Sea-Viewing Wide-Field-of-View Sensor
SIF	Sun Induced Chlorophyll Fluorescence
SO_2	sulfur dioxide
SPM	suspended particulate matters
SPOT	Systeme Pour l'Observation de Ia Terre
SRS	Satellite remote sensing
SRTM	Shuttle Radar Technology Mission
TIROS-1	Television and Infrared Observational Satellite-1
TLV	Threshold limit values
TM	Thematic Mapper
TSPM	Total suspended particulate matters
UN	United Nations
VIIRS	Visible and Infrared Imaging Radiometer Suite
VOCs	volatile organic compounds
WHO	World Health Organization

Scope Technology will play a part in anticipating how climate change may affect health toxicity, which entails arranging important ideas and data logically.

10.1 Introduction

A. *Brief Overview of Climate Change and Its Impact on Health*
B. *Importance of Predicting Health Toxicity in the Context of Climate Change*
C. *Statement Highlighting the Role of Technology in Enhancing Predictive Capabilities*

Climate change (CC) represents one of the most common challenges that humanity encounters. Its consequences, which are far-reaching, are observed evidently in the rising temperatures, events of extreme weather, the melting ice cap, and raising of sea levels. On humans, the effects of CC could be shown directly and/or indirectly. For instance, CC can lead, expectedly, to significant changes in the development, growth, and organism activity, like seasonal pests like insects, mice, mites, bacteria, fungi, and nematodes in addition to unwanted plants. That in turn will directly impact human health through nutrition alteration as well as variation in the pattern of diseases.

Effectively, much efforts must be done to combat the current global crisis; it is important to know the complex mechanisms that result from this CC. Biomarkers are defined as biological indicators of different environmental changes. These biomarkers had an important and efficient role in unraveling the complex web of climate dynamics. Furthermore, these biomarkers that were associated with medical diagnostics could provide insight about physiological, pathological, or biochemical state of an organism.

So, total exposure assessment is essential for the evaluation of the direct and indirect reliable effects of CC and for the requirement of assessment of its impacts on human health (Rocque et al. 2021). Health risk assessment represents a multidisciplinary methodology where molecular biology is one of its disciplines that can contribute to that evaluation. That is because xenobiotics could affect the genome, proteome, transcriptome, and metabolome through resulting in measurable changes, which may represent valid biomarkers. These changes can depend on individual genetic background to some extent; in turn, this polymorphic distribution in the exposed populations can act as an essential and basic feature to interpret adequate data.

Currently, the use of biomarkers has been extended beyond the traditional aim of prediction and diagnosis to be part of an exposome paradigm where the totality of external exposure throughout life is correlated to internal responses. For that several forms of biomarkers can be used and integrated in this holistic paradigm.

Furthermore, digital biomarkers are used due to their simultaneous ability in deriving biomarkers from the different digital health technologies. In various cases, the characteristics of medical condition in the form of the disease and host can be simultaneously consolidated from multiple digital health technologies to derive a biomarker. This ability to derive biomarkers (characterizing host and disease as well as medical condition) from differential digital health technologies could potentially help in providing additional context to enrich patient-specific baselines. Researchers had highlighted the role of digital health in CC, where digital health had shown a potential role in integrating the insights data to generate the health impacts of CC.

In our chapter, we will discuss the potential for real-time monitoring and response systems toward CC as well as biomarkers of exposures and biological changes on human health.

10.2 Traditional Approaches to Predicting Health Toxicity

10.2.1 Epidemiological Studies

Historically, the word "epidemiology" is a Greek word and can be divided into "epi," "demos," and "logos," which mean "upon," "people," and "the study of," respectively, and has been focused on communicable infectious diseases on population level, but today it has been changed a wider coverage to include also the non-communicable diseases, infant health, behavioral and environmental health, injuries, natural disasters, and any health-related issues (Hernandez and Kim 2024).

Therefore, epidemiology studies could be either descriptive or analytic studies. Epidemiology aims to develop foundational knowledge regarding the health event, mainly that which includes rates of populations affected, geographic timing, and the occurrences of specific health events, the natural history of the health problem, transmission modes, risk factors, and disparities of the health events, which can develop effective informative analytical studies (Hernandez and Kim 2024). The study of the prevalence and incidence rates of breast cancer in different age groups and countries, as an example of descriptive epidemiological study, is essential to define the etiologies, risk factors, and potential preventative measures (Kelsey and Horn-Ross 1993).

Analytic epidemiological studies are used for testing the validity through seeking to discover the contribution and the potential associations between factors and the outcome, using a control group to test the hypothesis of the possibility via observational or experimental studies (Hernandez and Kim 2024). The overarching goal of experimental studies is establishing a causal relationship between exposure and the outcome, which could be through controlled clinical trials with the implementation of placebos and counteracting any potential confounding variables or bias, e.g., the outcome of host survival or host extinction exposed to parasites (Ebert et al. 2000).

Epidemiological methods traditionally have always been the main approach to demonstrate the associations between exposure to hazardous substances and the development of chronic diseases, such as cardiovascular and respiratory diseases and cancers. The endpoints of traditional epidemiology are usually investigating mortality and disease incidence of chronic diseases in relation to chronic exposures (Hernandez and Kim 2024). But the approach of traditional epidemiology has several limitations that could be minimized through molecular epidemiology.

Molecular epidemiology is a promising branch in the epidemiological studies, as it is addressing genetic susceptibility, etiology, and induction of the disease through the usage of reliable biomarkers in these investigations for the prediction of diseases and implementation in disease prevention programs (Bonassi and Au 2002). Molecular epidemiology depends mainly on the reliability of biomarker databases through integration of information from the genome programs to expand the scientific frontiers on health risk prediction, etiology, and preventive methodologies of diseases, especially environmental related diseases (Bonassi and Au 2002).

Environmental epidemiology (natural experiments) is an observational science that studies the distribution and environmental determinants of disease, focusing on the disease risk factors such as physical, chemical, and biological environmental agents taken into consideration confounders, biases, exposure misclassification, random sampling error, or chance (Bloom 2019).

10.2.2 Environmental Monitoring

Environmental monitoring of air pollutants is an important sector in technological methods done to design the strategies for mitigation and adaptation to CC. Air monitoring studies are targeting the estimation of emissions from different sources to plan reduction strategies (Viana et al. 2020), in addition to the detection of the hazardous indoor or outdoor exposures, which are very essential for prediction of health problems related to environmental pollutants. Chojer et al. (2024) mentioned that the identification of air pollution sources is an essential step toward mitigating pollutants and associated health risks. It was recorded that each year about 3.2 million premature deaths are attributed to indoor air pollution (WHO 2022).

– *Monitoring of Suspended Particulate Matters (SPM)*

One of the most important common air pollutants is suspended particulate matters (SPM), which plays an important role as a risk factor for the development of chronic diseases, such as cardiovascular diseases (Beelen et al. 2014). The aerodynamic equivalent diameter of SPM affects human health: coarse PM_{10} (10μm, > 2.5μm), fine $PM_{2.5}$ (2.5μm, > 0.1μm), and ultrafine particles $PM_{0.1}$ (0.1μm) (Holme et al. 2020). The CEL-712 Casella Micro Dust Pro Particulate Monitor may be used to monitor suspended particulate matters (SPM) with various fraction sizes (PM_1, $PM_{2.5}$, and PM_{10}) in various indoor environments, whether in homes or workplaces (Photo 10.1).

It is possible to collect total suspended particulate matters (TSPM) from indoor air at workplaces. Samples might be collected by using high volume sampler (HVS) and glass fiber filters of the Whatmann GFA type. To evaluate the weight of TSPM, loaded filters should be weighted both before and after sampling. Once in the laboratory, the TSPM samples need to be kept in unopened standard plastic bags. To assess the suspended contaminants, all samples must be air-dried at room temperature in dark environment (Hassan et al. 2020). It is also crucial to discuss the mean concentrations of the various SPM sizes on a seasonal and yearly basis.

– *Monitoring of Gaseous Pollutants*

Monitoring gaseous pollutants in the indoors of residential or workplaces, such as sulfur dioxide (SO_2), nitrous dioxide (NO_2), carbon monoxide (CO), carbon dioxide (CO_2), and volatile organic compounds (VOCs), could be estimated through the use of Aeroqual's 200 portable air quality monitors (Photo 10.2).

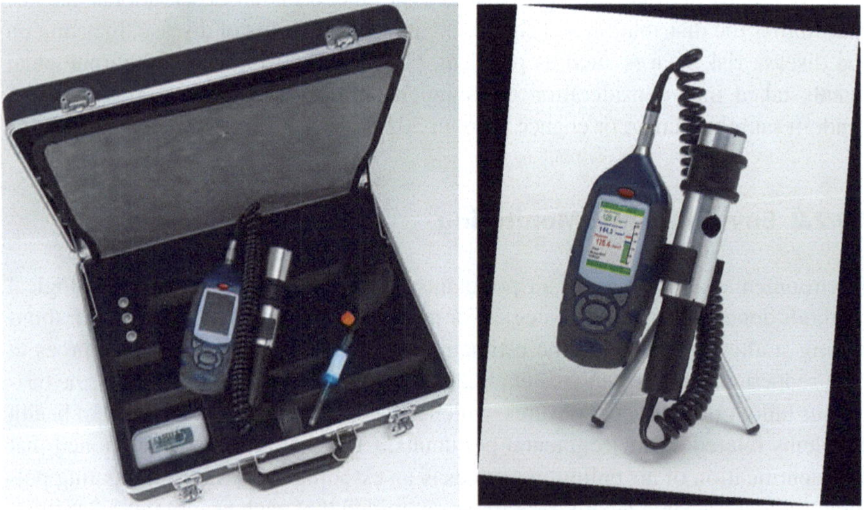

Photo 10.1 Micro-Dust Pro, CEL-712 Casella for monitoring particulate

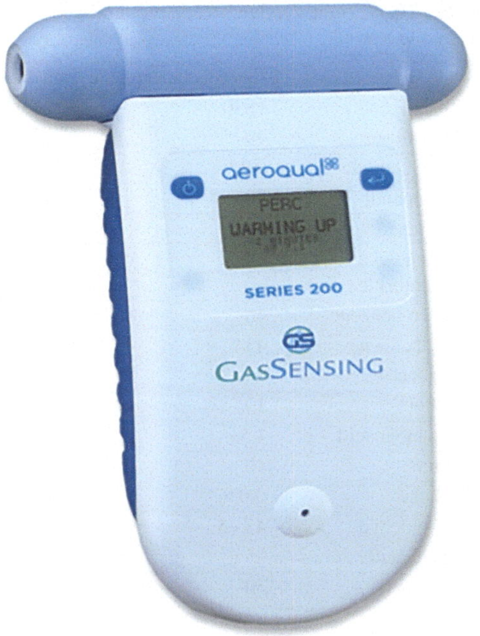

Photo 10.2 Aeroqual's 200 portable air quality monitors

The most important gaseous pollutants are SO_2, NO_2, CO, CO_2, and these are irritant gases that cause irritation of mucous membranes and the respiratory system, as well as their contribution to global warming. VOCs are often categorized as Hazardous Air Pollutants (HAPs) according to EPA (2017). They are formed by the volatilization of organic compounds in the treatment process. VOCs are associated with health effects and also cause safety problems due to common characteristics such as flammability. Each gaseous pollutant needs special sensor to detect the air concentrations. Monitoring of gaseous pollutants must be seasonal and annual to provide a trend that can be used in prediction of health hazards and to design strategies for reduction.

The most common method used for collecting VOCs samples is collection on activated charcoal tubes of type ORBOTM-32 activated coconut charcoal (20/40). Atmospheric air is drawn through tubes by using calibrated vacuum pump with flow rate of 0.2 l/min (low volume sampler method). Sample tubes are stored into special plastic bags and saved at a freezer till processed, not more than 15 days. Carbon disulfide (CS_2) is used for VOC extraction. A 2 ml of CS_2 are added to loaded charcoal, and then a mechanical shaker is used for shaking. A gas chromatography (GC) technique (limit of detection ranged from 0.001 to 0.02 mg/m^3) is used to identify and evaluate the concentration levels of the individual VOCs (Table 10.1).

– *Monitoring of Polycyclic Aromatic Hydrocarbons Compounds (PAHs)*

Another important air pollutant is polycyclic aromatic hydrocarbons compounds (PAHs), which are transporters to be inhaled with the ultrafine particles (Asweto

Table 10.1 The Egyptian acceptable limit and the TLV of ACGIH concentration levels of individual VOCs

VOCs	Egyptian acceptable limit (mg/m^3)	TLV-ACGIH (mg/m^3)
Ethanol	1880	1884
Chloroform	49	10
Formaldehyde	0.37	0.37
Isopropanol	983	492
p-Xylene	434	434
Carbon tetrachloride	31	31
Chloro benzene	46	46
Ethylbenzene	434	9
benzene	1.6	1.6
Phenol	19	19
Propanoic acid	1.5	30
Styrene	40	85
Toluene	0.036	75
Acetaldehyde	45.75	45
Acetic acid	25	25
Dimethyl formamide	30	30
Ethylene glycol	100	63
pyridine	16	16
Trichloro ethylene	269	134
Benzaldehyde	–	0.22
Pregnene	–	1180
Phenoxyethanol	–	0.17

TLV Threshold limit values, *ACGIH* American Conference of Governmental Industrial Hygienists

2018). There are hundreds of PAHs in the environment, but only 16 of them are considered to be on the list of priority pollutants according to the US Environmental Protection Agency's (EPA) (EPA 2002). The 16 priority PAHs identified by the EPA may be classified into three groups; low molecular weight (2 + 3-rings), intermediate (4-rings), and high molecular weight (5 + 6-rings) (Howard et al. 2021). Since the ambient load of PAHs in indoor and outdoor air comes from several sources, it is the most significant method of dispersing PAHs and is widely dispersed throughout the environment (Abdel-Shafy and Mansour 2016).

Environmental organic emissions of PAHs could be collected on Whatmann grade GF/A glass microfiber filters through low volume sampler method with an average flow rate equal to 10–12 l/min (Moubarz et al. 2023), but the filters must be impregnated in acetone for 24 h to remove all organic contamination, and then heated at 400 °C for four hours. After collecting SPM samples, sample filters must be covered with aluminum foil and stored for 24 h in desiccators in darkness. Each filter must be put in a glass flask. Ultrasonic bath and DCM/n-hexane (10 ml of 1:1, v:v) will be used for extraction of PAHs, three times for 10 min at room temperature. A column filled with silica gel (10 gm, 70–230 mesh, ASTM purchased from

Table 10.2 Chemical identity of some 16 EPA PAHs

Chemical identity PAHs	Abbreviation	Chemical formula	Molecular weight	Number of rings
Naphthalene	NAP	$C_{10}H_8$	128.00	Two
2-Bromonaphthalene	2-BNAP	Br-$C_{10}H_7$	208.00	Two
Acenaphthylene	ACY	$C_{12}H_{18}$	152.20	Three
Acenaphthene	ACE	$C_{12}H_{10}$	154.21	Three
Fluorene	FLU	$C_{13}H_{10}$	166.20	Three
Phenanthrene	PHE	$C_{14}H_{10}$	178.20	Three
Anthracene	ANT	$C_{14}H_{10}$	178.20	Three
Fluoranthene	FLT	$C_{16}H_{10}$	202.26	Four
Pyrene	PYR	$C_{16}H_{10}$	202.26	Four
Benz[a]anthracene	BAA	$C_{18}H_{12}$	228.30	Four
Chrysene	CRY	$C_{18}H_{12}$	228.30	Four
Benzo[b]fluoranthene	BBF	$C_{20}H_{12}$	252.30	Five
Benzo[a]pyrene	BAP	$C_{20}H_{12}$	252.30	Five
Dibenz[a,h] anthracene	DBA	$C_{22}H_{14}$	278.35	Five
Benzo[ghi]perylene	BGP	$C_{22}H_{12}$	276.34	Six
Indeno[1,2,3-cd] pyrene	IND	$C_{22}H_{12}$	276.3	Six

The standard PAHs mixture (2000µg/ml for each) containing: *NAP* naphthalene, *ACY* acenaphthylene, *ACE* acenaphthene, *FLU* fluorene, *PHE* phenanthrene, *ANT* anthracene, *FLT* fluoranthene, *PYR* pyrene, *BAA* benz[a]anthracene, *BkF* benzo[k]fluoranthene, *CRY* chrysene, *BBF* benzo[b]fluoranthene, *BAP* benzo[a]pyrene, *DBA* dibenz[a,h]anthracene, *BGP* benzo[ghi]perylene, *IND* indeno[1, 2, 3 – cd] pyrene. Standard solution mixture of 16 PAH compounds (Supelco, Inc., Bellefonte, PA, 2000 µg/ml for each PAH). TLV for PAHs must be less than 0.2 mg/m^3 = 200000 ng/m^3 (ATSDR 2012)

Merck) and 2 gm of anhydrous sodium sulfate will be used for extraction clean up. GC technique (limit of detection ranged from 0.50 to 3.91 ng/m^3) is used for identifying and evaluating concentration levels of the 16 EPA PAHs (Table 10.2). According to Drotikova (2020), diluted standard solution of EPA 16 PAH compounds (Supelco, Inc., Bellefonte, PA) should be calibrated with GC.

10.2.3 Challenges in Accurately Predicting Health Impacts Using Traditional Approaches

One major challenge for the use of traditional approaches to predict health outcomes is based on the complexity of human nature, which includes both exposure and outcome. Exposure is being exposed to variable agents on different time scales, while outcome is given the complex nature of many diseases that result from the interaction between environment and genes. Therefore, to embrace such complexity, adopting multidisciplinary approaches is needed to merge traditional tools and the new advances in technology (Nienaber-Rousseau 2025).

10.3 The Emergence of Technology in Climate Change Prediction

10.3.1 Introduction to Technological Advancements

One major challenge that faces environmental assessment is the lack of accurate retrospective assessment. Traditionally, this was based on surveys and questionnaires that hold the drawback of possible recall bias. The introduction of remote sensing data to the arena of environmental health monitoring will provide the opportunity to have cumulative exposure data for the studied persons based on their addresses. This can be revolutionary to the field of exposome and environmental monitoring (Hegazi et al. 2024).

10.3.2 Role of Sensors and Data Collection Devices

Satellite images are pictures of Earth obtained via satellites from orbit. Numerous applications, including natural resource management, urban planning, and weather forecasting, are possible with these photos. Satellite cameras in Earth's orbit are used to take pictures of the planet. Understanding the wide variety of technology systems and approaches used in this field is crucial to addressing the complexity of environmental monitoring in a comprehensive manner. These include:

1. *Fixed Stations*: Stations on the ground that communicate with satellites by sending and receiving signals.
2. *Satellite Images*: Using satellite cameras, images of Earth are captured from orbit.
3. *Mobile Stations*: Used in mobile satellite communication systems, such as satellite phones and GPS equipment, mobile stations are those that are mobile.
4. *Remote Sensing*: The technique of gathering information about the Earth from a distance, usually with the use of satellites or airplanes. Applications for remote sensing include mapping land usage and environmental monitoring.
5. *Earth Observation*: The monitoring and study of the Earth's surface, atmosphere, and seas via the use of satellites and other platforms. There are several uses for earth observation, including farm management, disaster response, and weather forecasting.

Climate change has an enormous impact on human society and creates major barriers to global sustainable development. Data on important climatic variables (ECVs) that accurately reflect the major changes that have occurred on Earth are needed to assess the impact of CC. Satellite remote sensing (SRS) technology, which provides multi-scale data on ECVs independent of in situ measurements and model simulations, has ushered in a new era of observations. This contributes to the development of CC policies by expanding our understanding of CC as observed from space. But because of the intricate nature of the climate system, remotely retrieving ECVs continues to be difficult.

As a result of human activity, there has been a worldwide warming trend since the middle of the 1800s, with temperatures rising to levels not seen in previous millennia. The term "climate change" refers to this trend and is defined as a change in the patterns of the climate that often lasts for several decades (IPCC 2021). Global sustainable development faces significant hurdles due to CC, which also has detrimental repercussions on human society. According to Salvia et al. (2019), a few of these effects include severe droughts and flooding, destructive fires, melting ice and rising sea levels, and dwindling biodiversity. Addressing CC and its effects immediately is one of the seventeen global Sustainable Development Goals (SDG 13) established by the United Nations (UN) Agenda 2030 (UN 2019). Reducing greenhouse gas emissions is one of the primary solutions for slowing down global warming (GHGs), which are the main cause of global warming, and adapting to the effects of CC with the least amount of resources possible are the main ways to counteract the effects of CC and better battle climate-related dangers. To support policymakers in addressing CC and to improve our understanding of it, both solutions include tracking and measuring significant changes that have occurred on Earth. When they are not available, modeling is a backup plan for gathering information. Measuring and observing are direct techniques. For instance, we can model future changes even though we cannot see them. The Coupled Model Intercomparison Project (CMIP) encourages research on CC and backs the IPCC assessment by offering simulated historical and current datasets in addition to projected climate data for the future report (Meehl et al. 2007; Eyring et al. 2016; Touzé-Peiffer et al. 2020). While site observations are limited in scope and cannot yield enough information on spatial variability, especially related to meteorological conditions, they present an obstacle to global observations obtained using conventional in situ measurements. However, observations continue to be essential.

– *Satellite Remote Sensing Application in the Atmosphere*

Greenhouse gases (GHGs) produce and absorb radiant energy in the thermal infrared spectrum, which is responsible for the greenhouse effect, which is essential to life as we know it on Earth. Therefore, it is essential that we identify the compositions of GHGs in the atmosphere and the patterns in their changes in order to better understand the feedback between changes in GHGs and climate. Despite the creation of various global networks to monitor GHGs since the 1990s, it remains difficult to completely understand the characteristics and change mechanisms between GHG sources and sinks due to a lack of ground station observations. SRS fills in some of these data gaps in ground network monitoring and may provide information on global atmospheric GHG (Rayner and O'Brien 2001; Yang et al. 2018; Xiao et al. 2019) (Fig. 10.1).

– *Satellite Remote Sensing (SRS) Application in the Ocean Ecosystem*

Seventy-one percent of the Earth's surface is made up of the oceans, which are crucial for biogeochemical cycling. Specifically, through the ocean carbon cycle,

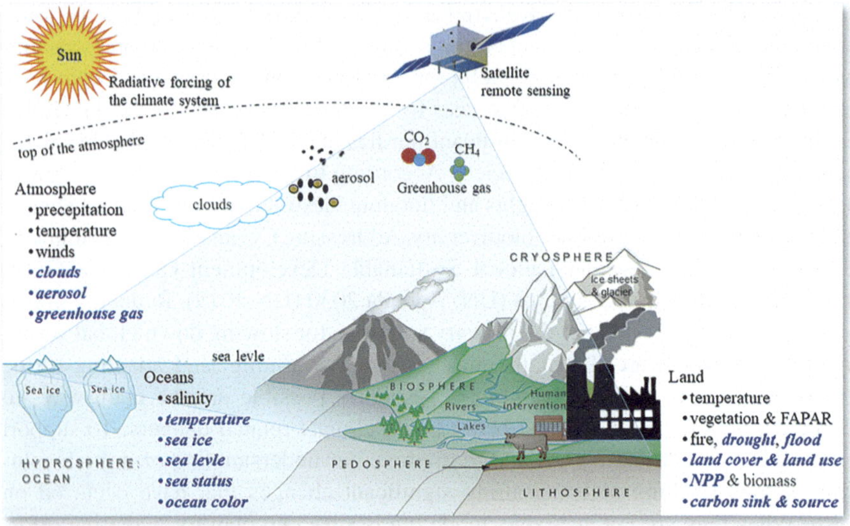

Fig. 10.1 The Earth's simplified climate system along with common variables used in climate research. (Source: Zhao et al. 2023)

the ocean contributes significantly to the regulation of atmospheric CO2 levels (Henson et al. 2022). This leads to complex feedback between the ocean and atmosphere (Eyring et al. 2021). The ocean is a massive carbon pool. El Nino-Southern Oscillation (ENSO) occurrences and SST anomalies are two common examples, both of which can result in significant climate-related risks, including drought or flood (Cai et al. 2021). Additionally, rising sea levels, melting sea ice, altered ocean currents, and changes in the physical and chemical characteristics of the ocean are all caused by elevated SST (Saba et al. 2016). Humans, coastal areas, and the marine ecology may all be at risk due to these changes (Ramírez et al. 2017). This is why research on CC requires the oceanic data that was previously mentioned. Traditional in situ measurements, such as argon floats and time-series monitoring programs using moored platforms and unmanned vehicles (Henson et al. 2022), have limited coverage because the ocean covers a large portion of the Earth's surface and some areas are inaccessible (or the measurements are expensive and take a short time). Despite this, these programs have the potential to provide important oceanic information. However, SRS may provide timely and consistent data on the world's oceans. A selection of the ocean color sensors that have been launched since the late 1970s include the Coastal Zone Color Scanner (CZCS) (NASA/USA, 1978–1986), the Medium Spectral Resolution Imaging Spectrometer (MERIS) (European Space Agency (ESA), 2002–2012), the Sea-viewing Wide-Field-of-View Sensor (SeaWiFS) (NASA/USA, 1997–2010), MODIS (NASA/USA, 1999–present), and the Visible and Infrared Imaging Radiometer Suite (VIIRS) (NOAA/USA, 2011–present) (Zhao et al. 2023) (Fig. 10.2).

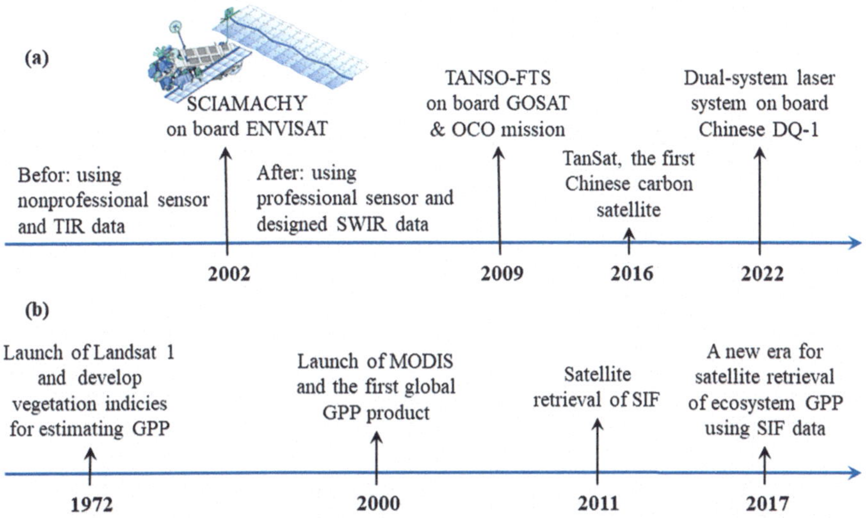

Fig. 10.2 Significant advancements in satellite remote sensing of common key climate variables: (**a**) atmospheric CO_2 column, and sun-induced chlorophyll fluorescence (SIF) and (**b**) gross primary production (GPP). (Source: Zhao et al. 2023)

The following table lists the main remote sensing methods that can be used to solve the concerns related to urban CC (Table 10.3).

10.3.3 Remote Sensing Technologies for Environmental Monitoring

We may learn more about many facets of our environment through environmental monitoring, including biodiversity, soil health, water and air quality, and climatic patterns. It offers the framework for the creation of successful environmental policies and evidence-based decision-making. We can spot patterns, spot new problems, and keep an eye on the effectiveness of conservation and cleanup initiatives by keeping an eye on environmental indicators. Environmental monitoring is acknowledged to be important; yet it has a number of drawbacks that reduce its efficacy. Manual sampling is a common component of traditional monitoring techniques, but it is expensive, time-consuming, and only covers a small area over a short period of time. Our capacity to gather data in real time and make precise assessments of the surrounding environment is hampered by this method.

These cutting-edge sensors are made to measure environmental characteristics with greater sensitivity, accuracy, and precision. They provide real-time data, automated analysis, and predictive capabilities by utilizing cutting-edge technologies such as nanotechnology, the Internet of Things (IoT), artificial intelligence (AI), and machine learning (ML) algorithms. Additionally, these sensors may integrate seamlessly with current monitoring systems and are frequently more affordable and

Table 10.3 Lists of the main remote sensing techniques that can be used to address urban climate change challenges (Milesi and Churkina 2020)

Urban climate change issues			Remote sensing capabilities			
Urban issue linked to climate change	Quantity needed	Quantity obtainable or available	Sensor	Spatial resolution [m]	Temporal resolution	Coverage
Urban population	Total population, density, and their change rates	Density, number, and type of dwelling units	Various Commercial VHR Sentinel-1 Sentinel-2 TerraSAR-X Global Urban Footprint Landsat 8 Gaofen 1 and 6	0.4–1 10 10–20 12 15–30 16	on-demand/ irregular 6–12 days 5–10 days 2011/2012 16-day 2–4 days	Irregular Global Global Global Global Global Global

Urban extent and structure	Urban area and its composition Building height	Urban area Urban land uses Fractional vegetation cover	Various Commercial VHR Sentinel-1 Sentinel-2 TerraSAR-X Global Urban Footprint Landsat 8 Gaofen 1 and 6 MODIS Vegetation Indices MODIS land cover VIIRS	0.4–1 10 10–20 12 15–30 16 250–1000 500 750	On-demand/ irregular 6–12 days 5–10 days 2011/2012 16-day 2–4 days 16-days Yearly 2001–2017 Monthly and yearly composites	On demand/ Irregular Global Global Global Global Global Global Global Global
		Digital Urban Canopy Height	Various Commercial VHR, KOMPSAT-3 and other stereo images COSMO SkyMed Sentinel-1 ALOS World 3D TanDEM-X DEM SRTM ASTER GDEM V2	0.5–1 1–100 10–100 5–30 12–30–90 30 30	3 days 4–16 days 6–12 days 2015 11 days 2000 Yearly 2000–2009	Global on demand Global Global Global Global Global Global
Energy Usage	Consumption of energy Emissions of waste energy (heat)	RGB photography	International Space Station Astronaut photography	5–200	Irregular	Irregular

(continued)

Table 10.3 (continued)

Urban climate change issues			Remote sensing capabilities			
Urban issue linked to climate change	Quantity needed	Quantity obtainable or available	Sensor	Spatial resolution [m]	Temporal resolution	Coverage
		Multispectral Nighttime Light emissions	EROS-B Jilin-1 (JL1-3B) Jilin-1 (JL1-07/08)	0.7 0.9 0.9	On demand On demand On demand	Tasked Tasked Tasked
		Panchromatic Nighttime Light emissions	LuoJia1-01 VIIRS DMSP/OLS	130 750 2700	15 days Daily Yearly composite	Global Global Global
		Thermal sensors	Landsat 8 ASTER VIIRS MODIS Sentinel-3	100 90 750 1000 1000	16 days 16 days and on demand Daily Daily Daily	Global Global Global Global Global
Greenhouse gasses (CO_2, CH_4, N_2O)	Emissions & concentrations	Concentrations of CO2 and CH4 in an air column	OCO-3 Sentinel 5P TanSAT TANSO-FTS TES	3500 7000 1000 × 2000 10500 5300 × 8500	Irregular Daily 16 days 3 days 2 days	80 × 80 km area hotspots Snapshot Area Map (SAM) Global Global Global Global

Other air pollutants	Emissions & concentrations of aerosols, ground level ozone, NOx, PM 2.5	Concentrations of ozone, NO2, SO2 in an air column	Sentinel-5P TES OMI	7000 5300 × 8500 13000 × 24000 13000 × 13,000 zoom	Daily 2 days Daily	Global Global Global
		Aerosol Optical Depth	Sentinel-2 Landsat 8 VIIRS CALIPSO MODIS	10 30 750 6000 5000 500, 1000, 3000, 10000	5, 10 days 16 day Daily Daily 1–2 days	Global Global Global Global Global
Surface urban heat island and heat waves	Air temperature Land surface temperature Water temperature	Land Surface Temperature Water Surface Temperature Sea Surface Temperature (SST)	ASTER Landsat 8 VIIRS MODIS Sentinel-3 AHI (SST only)	90 100 750 1000 1000 2000	16 days and on demand 16 days Daily 4 times/day ~14 days Sub-daily	Global Global Global Global Global Global

Where: *AHI* Advanced Himawari Imager, *ALOS* Advanced Land Observing Satellite, *ASTER* Advanced Spaceborne Thermal Emission and Reflection Radiometer, *CALIPSO* Cloud-Aerosol LiDAR and Infrared Pathfinder Satellite Observation, *COSMO* Constellation of Small Satellites for Mediterranean basin Observation, *DEM* digital elevation model, *DMSP/OLS* Defense Meteorological Satellite Program's Operational Linescan System, *EROS* Earth Resources Observation Satellite, *GDEM* Global Digital Elevation, *MODIS* Moderate Resolution Imaging Spectroradiometer, *OCO* Orbiting Carbon Observatory, *OMI* Ozone Monitoring Instrument, *SAR* Synthetic Aperture Radar, *SRTM* Shuttle Radar Topography Mission, *TANSO-FTS* Thermal and Near infrared Sensor for Carbon Observation—Fourier Transform Spectrometer, *TES* Technology Experiment Satellite, *VIIRS* Visible Infrared Imaging Radiometer Suite, *VHR* Very High Resolution scalable. Over the years, sensor technology has seen a remarkable transformation that has changed how we monitor and perceive our surroundings.

– *Development and Advancements in Environmental Research Using Remote Sensing Satellites and Sensors*

There are eight different remote sensing eras, some of which occur concurrently with other historical periods but which differ greatly in terms of technology, data use concepts, scientific applications, and data properties (e.g., Table 10.4) (Marcelo-Peña et al. 2007).

1. *Airborne Remote Sensing Era:* The First and Second World Wars saw the evolution of the airborne remote sensing period (Avery and Berlin 1992). In this period, military surveillance, mapping, surveying, and reconnaissance were the primary uses of remote sensing.
2. *The Early Days of Spaceborne Satellite Remote Sensing:* The space-borne remote sensing era began with the launch of "test of concept" rudimentary satellites, such as the American Explorer 1 and the Russian Sputnik 1, at the end of the 1950s (Devine 1993; House et al. 1986). Shortly afterward, in the late 1950s, the United States launched Television and Infrared Observational Satellite-1 (TIROS-1), the first meteorological satellite (House et al. 1986).
3. *Spy Satellite Remote Sensing Era*: Spy satellites like Corona (Dwayne 1988) were extensively used during the height of the Cold War. Nearly all of the data was gathered for military use. The information was created as physical copies rather than digitally. Nonetheless, the offshoot of the remote sensing technology established for military uses throughout the aforementioned three centuries gradually extended to mapping and then to applications involving the environment and natural resources.
4. *Meteorological Satellite Sensor Remote Sensing Era:* The Geostationary Operational Environmental Satellite (GOES) and the Polar Orbiting National Oceanic and Atmospheric Administration (NOAA) Advanced Very High-Resolution Radiometer (AVHRR) were the two early meteorological satellite sensors (Kramer 2002). During this time, data became digitally accessible and was analyzed with specialized computer gear and software. Additionally, environmental applications were feasible, and worldwide coverage became plausible during this time.
5. *Landsat Era*: Launched in 1972 with a multispectral scanner (MSS) sensor, Landsat-1 (formerly known as the Earth Resources Technology Satellite) marked the beginning of the Landsat era. Subsequently, Landsat satellites 2 through 3 carrying MSS and 4 and 5 carrying Thematic Mapper (TM) arrived, which additionally provided path finding. Landsat 7 is equipped with the Enhanced Thematic Mapper (ETM+) sensor. The Landsat-6 had a problem during launch. The Landsat-8 carrying Operational Land Imager (OLI) is slated for launch in 2011. The French Systeme Pour l'Observation de Ia Terre (SPOT) and the Indian Remote Sensing Satellite (IRS) are also good examples of sun-synchronous land

Table 10.4 Satellite sensor data characteristics (Melesse et al. 2007)

Sensor	Spatial (m)	Spectral (#)	Radiometric (bit)	band range (μm)	band widths (μm)	Irradiance (W $m^{-}2sr^{-}1\ \mu m^{-}1$)	Data Points (# per hectares)	Frequency of revisit (days)
A. Coarse Resolution Sensors								
1. AVHRR	1000	4	11	0.58-0.68	0.10	1390	0.01	Daily
2. MODIS	250; 500; 1000	36/7	12	0.62-0.67	0.05	1528.2	0.16; 0.04; 0.01	Daily
B. Multi Spectral Sensors								
3. Landsat-1; 2; 3 MSS	56X79	4	6	0.5-0.6	0.1	1970	2.26	16
4. Landsat-4; 5 TM	30	7	8	0.45-0.52	0.07	1970	11.1	16
5. Landsat-7 ETM+	30	8	8	0.45-0.52	0.65	1970	44.4, 11.1	16
6. ASTER	15; 30; 90	15	8	0.52-0.63	0.11	1846.9	44.4, 11.1, 1.23	16
7. ALI	30	10	12	0.048-0.69	0.64	1747.8	11.1	16
8. SPOT-1	2.5-20	15	16	0.50-0.59	0.09	1858	1600, 25	3-5
9. IRS-1C	23.5	15	8	0.52-0.59	0.07	1851.1	18.1	16
10. IRS-1	23.5	15	8	0.52-0.59	0.07	1852.1	18.1	16
11. IRS-P6-AWiFS	56	4	10	0.52-0.59	0.07	1857	3.2	16
12. CBERS -2	20		11	0.51-0.73	0.22	1934.03	25	25
C. Hyper-Spectral Sensor								
1. Hyperion	30	196	16	196	10		11.1	16
D. Hyper-Spatial Sensor								
1. IKONOS	1-4	4	11	0.445-0.516	0.71	1930.9	10000, 625	5
2. QUICKBIRD	0.61-2.44	4	11	0.45-0.52	0.07	1381.79	14872, 625	5

(continued)

Table 10.4 (continued)

Sensor	Spatial (m)	Spectral (#)	Radiometric (bit)	band range (μm)	band widths (μm)	Irradiance (W $m^{-}2sr^{-}1$ $\mu m^{-}1$)	Data Points (# per hectares)	Frequency of revisit (days)
3. RESOURSESAT	5.8	3	10	0.52 - 0.59	0.07	1853.6	33.64	24
4. RAPID EYE - A	6.5	5	12	0.44-0.51	0.07	1979.33	236.7	1-2
5. WORLDVIEW	0.55	1	11	0.45-0.51	0.06	1996.77	40000	1.7-5.9
6. FORMOSAT-2	2-8	5	11	0.45-0.52	0.07	1974.93	2500, 156.25	Daily
7. KOMPSAT-2	1-4	5	10	0.5-0.9	0.4	1379.46	10000, 625	3-28

satellites from the Landsat era (Jensen 2000). These satellites have a high resolution (nominal 2.5–80 meters) and can cover the whole planet. At this resolution, only Landsat is currently gathering data with complete world coverage. When it comes to the history of the really global and local application of remote sensing data for environmental objectives, this is by far the most significant time.

6. *Earth Observing System Era:* Since the Terra satellite was launched in 1999, the Earth Observing System (EOS) era (Stoney 2005; Bailey et al. 2001; Jensen 2000) has been characterized by high levels of processing (e.g., georectified, at-satellite reflectance), global coverage, frequent repeat coverage, and easy and mostly free access to data. The Moderate Resolution Imaging Spectro-radiometer (MODIS) and the Measurements of Pollution in the Troposphere (MOPITT) are two examples of the sensors carried by the Terra Aqua satellites, which visit and process data on a regular basis. The use of sensor data has expanded and has numerous uses. People and organizations that have never used remote sensing are starting to show interest in it. Furthermore, the processed data's accessibility in the form of products like the land use/land index and the leaf area index (LAI). It is now commonplace to be able to access the processed data in the form of products like the LAI and land use/land cover (LULC). At the moment, MODIS offers more than 40 items. Radar technology was also used in active spaceborne remote sensing instruments at this time, and during the Landsat era, with the launch of the European Radar Satellite (ERS), the Japanese Earth Resources satellite (JERS), Radarsat, and the Advanced Land Observation Satellite (ALOS). The data collection for digital elevation was done with the Shuttle Radar Technology Mission (SRTM).
7. *New Millennium Era*: The phrase "new millennium era" (Bailey et al. 2001) alludes to extremely advanced "test-of-concept" satellites placed into orbit around the same time as the EOS period, despite the concepts and ideas being unique. These are essentially the satellites and sensors of the future. Earth Observing-1 is one among them; it sent the first hyperspectral data into space ever. Examining Advanced Land Imager (ALI) as a more cost-effective and cutting-edge technological substitute for Landsat is also quite alluring.
8. *Private Industry Era*: The private industrial age began at the turn of the millennium and continued into this one (Stoney 2005). There are a lot of inventions in this creative era. Firstly, data collection at a very high resolution (<10 meters). IKONOS and Quickbird are two examples of these satellites. Second, a creative approach to data collection. A good example of this is the Rapideye satellite constellation, which comprises five satellites and covers almost the whole planet in five spectral bands, including a red-edge band, with a resolution of 6.5 meters. The third advancement is the launch of small satellites, some of which are a part of the Disaster Monitoring Constellation (DMC). These satellites were created and launched by Surrey Satellite Technology Ltd. for the USGS, the UK, Nigeria, China, Turkey, and other nations. Fourth, Google Earth (http://earth.google.com) makes it possible for non-specialists to zoom and pan remote sensing data with ease, and it can immediately obtain VHRI data for any location in the world using streaming technology.

- *Features and Advantages of Sensors in Environmental Monitoring*
  1. *Advanced Data Collection Capabilities*
     The sophisticated data-gathering features of next-generation sensors allow them to compile an abundance of environmental data. These sensors provide a wider range of data and have the capacity to measure numerous parameters at once, making it possible to fully comprehend the environmental conditions.
  2. *Increased Accuracy and Precision*
     Precision and accuracy are essential components of environmental monitoring. In order to produce extremely accurate and exact readings, next-generation sensors make use of cutting-edge technology, including upgraded sensors, calibration methods, and signal processing algorithms. This guarantees that the information gathered is trustworthy and safe to utilize for resource management, policymaking, and scientific study.
  3. *Real-Time Monitoring and Remote Access*
     One significant advancement in environmental sensors is real-time monitoring. The ability of next-generation sensors to offer continuous and instantaneous data updates enables prompt reactions to environmental condition changes. Furthermore, these sensors frequently provide remote access via mobile apps or web-based interfaces, allowing users to keep an eye on the surroundings at all times and from any location. Researchers, environmental authorities, and stakeholders are empowered to make timely judgments and take appropriate action through real-time and remote monitoring.
  4. *Integration with the Internet of Things (IoT)*
     The Internet of Things (IoT) ecosystem is seamlessly integrated with next-generation sensors. These sensors join a broader network of networked devices through Internet of Things connections, facilitating easy data sharing, analysis, and communication.
  5. *Cost-Effectiveness and Scalability*
     Scalability and cost-effectiveness are essential factors in the broad deployment of environmental monitoring systems. These issues are addressed by next-generation sensors, which provide reasonably priced solutions that may be expanded to cover wide areas or installed in several locations. These sensors frequently take advantage of developments in wireless communication, low power consumption, and downsizing to save installation and maintenance costs. A wider range of customers, including researchers, governmental organizations, and even private citizens interested in keeping an eye on their local environment, can utilize them thanks to their affordability and scalability.

10.3.4 Life Cycle Assessment (LCA)-Based Climate Change Impact on Human Health

Life cycle assessment (LCA) does not directly predict health toxicity; it can quantify emissions of climate change pollutants. Then, these emissions can indirectly influence human health through their contribution to climate change-related impacts.

However, environmental fate and transport of chemicals are not considered in LCA. Moreover, the human exposure route of chemicals is also lacking in LCA (Zin and Lim 2023). Meanwhile, risk assessment (RA) focuses on the human health impact by considering fate, transport, and exposure route. However, RA associated with climate change carries several inherent uncertainties due to the long-term nature of anthropogenic climate change, especially at the local or national level (Campbell-Lendrum and Woodruff 2006). Therefore, the combined use of LCA with RA can enhance the prediction of human health risks.

Life Cycle Assessment (LCA) Based on Effects of Climate Change on Human Health

The release of CO_2, greenhouse gases, aerosols, and ozone precursors has an impact on climate change. By combining GHGs produced by human activity into a single unit, such as CO_2-equivalent, LCA studies can quantify the contributions of GHGs to climate change. Climate change factors are computed using the Global Warming Potential (GWP) for well-mixed greenhouse gases from the Intergovernmental Panel on Climate Change (IPCC) assessment reports (Levasseur et al. 2016).

Risk Assessment (RA) Based on Climate Change Impact on Human Health

Air pollution, changes in climate, and human health are interconnected to each other. Changes in climate can directly impact human health by weather extremes, such as heatwaves and an increase in the production of some air pollutants and aero-allergens (Orru et al. 2017). Climate change can also indirectly impact human health by increasing the toxicity of chemicals (Hooper et al. 2013). Global warming in conjunction with ozone and PM exposures causes cardio-respiratory diseases and mortality (Noyes et al. 2009). Toxin-induced climate changes have the power to modify human response to stresses associated with global climate change by physiological, behavioral, or biochemical reactions such as growth (Hooper et al. 2013). The World Health Organization (WHO) established a comparative risk assessment approach to evaluate the disease burden. Estimates of the number of deaths and disability-adjusted life years (DALYs) linked to each risk factor were evaluated by the assessment (Campbell-Lendrum and Woodruff 2006; WHO 2011).

10.4 Applications of Technology in Health Toxicity Prediction

10.4.1 Satellite-Based Air Quality Monitoring Systems

Air pollution is considered one of the most important environmental stressors with serious implications for ecosystems and human health (EPA 2023). About 99% of the population worldwide are living in air pollution areas that are exceeding the WHO air guidelines levels (WHO 2023). Populations with anemia, cardiovascular, or respiratory diseases mostly experience health effects upon exposure to high CO levels (Dey and Dhal 2019). Moreover, 6.7 million deaths globally in 2019 were found to be among air pollution exposures (Fuller et al. 2022).

A short-lived climate forcing agent is considered to be CO_2, because CO_2 is indirectly contributing to CC through chemical reactions in the atmosphere that produce global warming gases. Hence, CO_2 emission reductions are prompting strategies for mitigation of global warming gases. Moreover, the suspended PM in air is resulting in smoke production and dust containing harmful compounds. The major component of air pollutants is PM10 and the respirable particles (PM2.5), which threaten environment and human health, causing mainly cardiovascular and respiratory diseases and early death (Guo et al. 2017).

Remote sensing imagery is defined as an alternative to regular monitoring techniques because it is less costly and time-effective method compared to the regular techniques that are used for provision of appropriate data about the air quality over large areas. Therefore, remote sensing can be employed in mapping and retrieving air quality parameters at a dynamic scale and regular intervals that will support decision-makers (Kaplan and Avdan 2020; Soleimany et al. 2020). However, there is an urgent need for continuous spatiotemporal monitoring using in situ measurements for monitoring air quality to provide information about small spatial scales at a few selected small locations. In situ measurements need a robust spatial and temporal modeling for detection of the most affected areas per day, per month, or per season (Gulia et al. 2015). Monitoring of air pollution using remote sensing data was launched with the development of the Total Ozone Monitoring Instrument (TOMS) in 1978, the Global Ozone Monitoring Experiment (GOME) in 1995, and the Ozone Monitoring Instrument (OMI) in 2004 (Matci et al. 2022).

The latest satellite-based sensor is Sentinel-5P TROPOMI, which provides new opportunities for remote-sensed data utilization for air quality monitoring (Ialongo et al. 2020), and covers the wavelength ranges between infrared and ultraviolet waves to enable air pollution observations, for example, CO, NO_2, O_3, SO_2, HCHO, CH_4, and particulate matter (PM2.5 and PM10) (Soleimany et al. 2020).

10.4.2 Predictive Modeling Using Artificial Intelligence

The high failure rate in clinical trials based on generated data due to the initial failure in prediction stimulated the emergence of artificial intelligence (AI). The advances in AI could be able to accelerate successful understanding of human exposure models and developing therapeutic and diagnostic tools accordingly. Although not new, the AI and its related deep learning are now enabled by the advances in technology, e.g., high-power computers, new frameworks for deep neural networks, and more. This has been reflected on unequaled accuracy in some applications, e.g., voice recognition and automatic cars (Zhavoronkov 2018).

Recently, one aim of the extensive data collection is to be able to develop predictive models that will enable us to act proactively and anticipate health outcomes. For developing predictive models, the use of machine learning (ML) is an excellent tool. Machine learning is a part of the artificial intelligence large scheme. By definition, ML is to give computers the ability to learn without being programmed; in other words, building predictive algorithms to identify patterns in data automatically. This revolutionary tool will help researchers in the field of environment-health interaction to accommodate the huge data produced and provide meaningful interpretation (Rashidi et al. 2021).

One detrimental point in using ML is to identify the ideal setting on when to switch from conventional statistical approaches to the advanced ML algorithm. The use of ML should be directed to large studies with big data sets that cannot be interpreted traditionally (Ghassemi and Mohamed 2022).

10.4.3 Integration of LCA

LCA standards are part of the ISO 14000 family of standards (ISO14000 2005). Using LCA, decision-makers can determine a product system's possible environmental effects throughout the course of its entire life cycle such as production, usage, and disposal. It is frequently used to direct environmental policies, to design more environmentally familiar products, etc. (European Union 2010).

Goal and scope, inventory analysis, life cycle impact assessment (LCIA), and interpretation are the four main steps of an LCA. The first step establishes the boundaries of the system and its functional unit (FU). In the second phase, every input and output throughout the life cycle stages is quantified. During the third phase, the LCIA multiplies these data by the characterization factor (CF) to convert them into possible environmental impacts such as global warming. The CFs are accessible using several ways (e.g., IMPACT 2002+, USEtox, or ReCiPe). These techniques work with both endpoint and midpoint. The midway displayed the changes resulting from the natural surroundings. While the endpoint indicator is for the damaging effects. The final stage is to interpret the results (Levasseur et al. 2016) (Fig. 10.3).

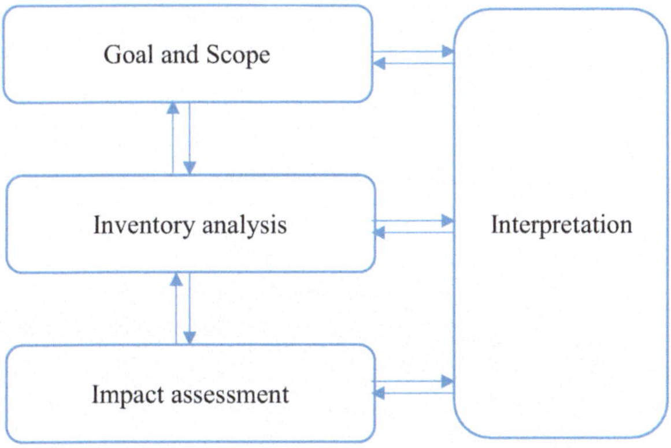

Fig. 10.3 The frameworks of life cycle assessment

$$CF_j = \sum_i CF_{ij} = \sum_i M_i . W_{ij}$$

Where M_i *represents the quantity of pollutants* $_i$
W_{ij} *denotes the emission i's characterization factor inside the category* $_j$
CF_{ij} *represents the pollutants i's contribution to category* $_j$
CF_j *denotes the characterized value of the category* $_j$

Risk Assessment (RA)
The risk assessment (RA) combines the fate and movement of chemicals in the environment, as well as the exposure route to human/ecological systems. Hazard identification, hazard characterization, exposure assessment, and risk characterization are the four phases that make up RA (FAO & WHO 2006). The main goals of hazard identification are what needs to be safeguarded, how the effects will be assessed, and how much of an impact is acceptable. The fate and movement of chemicals in the air, water, and land are then examined using the data. And this data is combined with data on exposed species. Both direct toxicity and indirect effects resulting from interactions between species are taken into account in this assessment of consequences. The projected no-effect concentration and the predicted ambient concentration are then compared. Risk managers use this data to determine if a risk is acceptable or mitigated. The threshold method, applied by RA, states that a specific pollutant's threshold values in an environment must not be exceeded (Muazu et al. 2021) (Fig. 10.4).

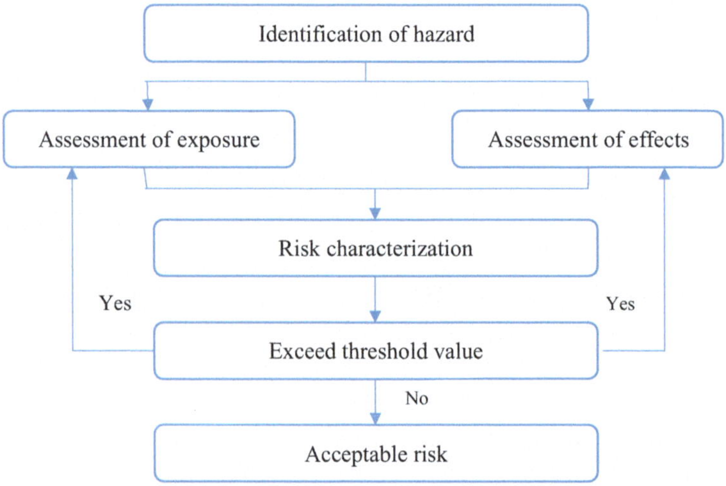

Fig. 10.4 The framework for risk assessment

Table 10.5 Comparison of LCA and ERA (Muazu et al. 2021)

Evaluation of the environment	LCA	RA
Fundamental	Less is preferable	Above the threshold
Objective	Evaluation of potential impacts	Environmental risks estimation
Scope	Life cycle viewpoint	Substances discharged into the environment
Measurements	Independent of time and space	Dependent of time and space
Focus	Mass	Concentration
Approach	Realistic for comparing products	Worst scenario for preliminary risk evaluation
Typical indicators	Potential for global warming	Likelihood, risk quotient (RQ)

The impact assessment of ERA is as follows:

RCR = PEC/PNEC
RCR = risk characterization ratio
PEC = predicted environmental concentration
PNEC = predicted no-effect concentration

Integration of LCA and RA

LCA considers the input and output of both harmful and non-toxic chemicals in a system. LCA compares the environmental performance of several items under normal operating and steady-state conditions. This suggests that the time and space spreading of environmental stressors, the significance of interspecific variation to a particular stressor, and the geographical and temporal variation. In the meanwhile, RA contrasts how ecological systems are exposed to particular stressors (such as single chemicals, chemical combinations, or numerous stressors) in a given environmental situation—typically the worst-case scenario that is plausible. Since LCA typically does not take background concentrations into account, LCA is unable to address threshold problems or accurately quantify risk. LCA is helpful in environmental management systems for evaluating impacts at the global (macro) scale, whereas RA helps evaluating impacts at the local (micro) scale (Simpson et al. 2021). Predicting human health risks can be improved due to the complementary of LCA and RA (Table 10.5).

10.5 Case Studies and LCA Examples

10.5.1 Case Studies and LCA Examples

The weakness of LCA is the failure to consider the fates and transportation of pollutants. To reduce these uncertainties, the fates in environmental, transports, and the

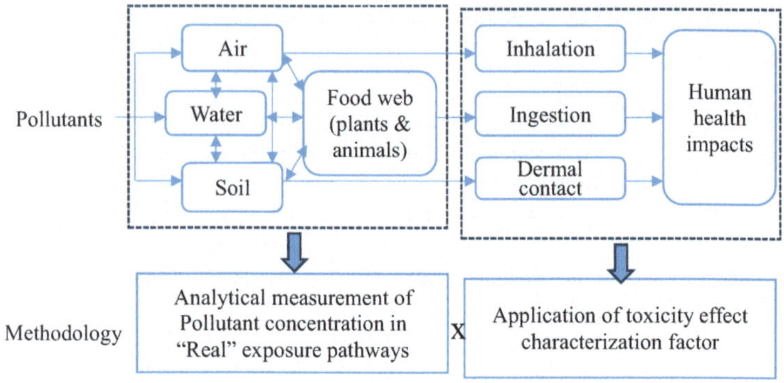

Fig. 10.5 A methodology to reduce the uncertainty of LCA. (Source: Zin and Lim 2023)

concentrations of chemicals in human exposure were considered in the study of Zin and Lim (2023). They first measure the amount of contaminants in seafood by assuming oral exposure. Secondly, the carcinogenic and non-carcinogenic toxicity potential of pollutants was calculated by using the USEtox. Then, the important pollutants were prioritized according to the carcinogenic potential. After that, pollutant release inventory data were used to understand the industries releasing those pollutants and the amount of those contaminants in the various environmental channels. Their modified methodology yielded different results from quantity- and LCIA-based methods. It proved that the modified method helps account for exposure concentrations that accurately reflect actual circumstances according to the fates and transportation of contaminants (Fig. 10.5).

Integration of LCA and RA could be used in evaluating food safety. First, the goal and scope of the cereal production were set according to LCA. Pollutants (such as CH_4, CO_2, heavy metals) can be emitted from synthetic fertilizer, organic fertilizer, energy, land application, water consumption, and processing in industries. LCA was used to understand the midpoint impact (acidification, climate change, human toxicity) by using The ReCiPe. The quantity of greenhouse gas emissions that a product emits over its life can be measured using its carbon footprint, which is expressed as CO_2 equivalent (ISO14067 2018).

A large number of compounds in the intermediate impact categories have some connection, either direct or indirect, to the field of protecting human health.

The application of sewage sludge on croplands during the primary production stage is calculated under the human toxicity midway category, but because of the heavy metal presence, it also has an impact on the human health endpoint category. (Przydatek and Wota 2019). Both the process of producing electricity and applying fertilizer releases greenhouse gases. The terminal damage to human health can be caused by NH_4 emitted from cereal cultivation, particulate matter, volatile organic compounds, NO_x, and SO_2 from grain processing (Gu et al. 2014; Huijbregts et al. 2017).

Heimersson et al. (2014) did a study that similarly integrated LCA and RA. Initially, the pathogen risk was evaluated using a microbiological risk assessment approach that was expressed in DALY annually. They calculated a system's total pathogen risk in a way similar to the findings in LCA. Second, they conducted an LCA for the same

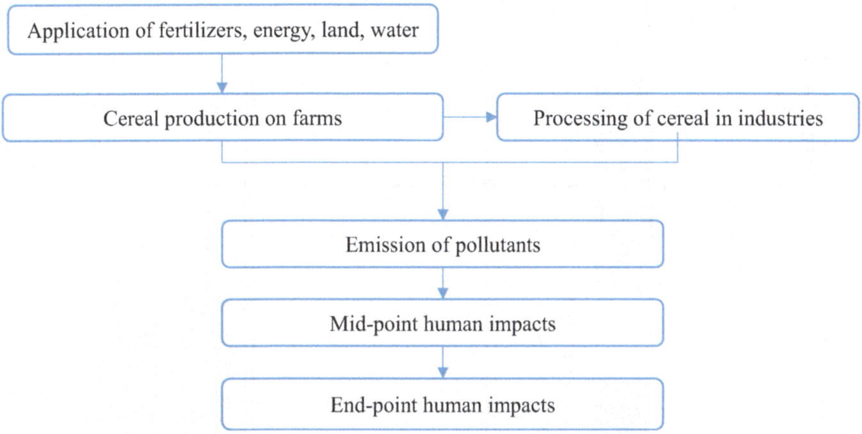

Fig. 10.6 Environmental effects of cereal production on climate change human health impact. (Source Feliciano et al. 2022)

sludge system that included effect categories that may be harmful to human health include particulate pollution, photochemical ozone generation, ionizing radiation, global warming, human toxicity, and stratospheric ozone depletion, by using USEtox and ReCiPe. The results were presented as DALYs for each functional unit. And then, the outcomes of the LCA and ERA were compared (Fig. 10.6).

Ribera et al. (2014) investigated both conventional and nanofiltration drinking water treatment methods, using a similar methodology. Initially, they conducted an LCA with a function unit of 1 m^3 of the finished drinking water. Using the ReCiPe, categories were determined as eutrophication, ozone depletion, and climate change. The risk of cancer related to the water treatment process was then assessed using a human health RA. A number of exposure routes, including ingestion, inhalation, and skin contact were carefully studied. In the end, the cost and advantages of the two water treatment techniques for the environment and human health could be reached.

10.5.2 Vulnerability of Densely Populated Areas to Air Pollution in Egypt Using Sentinel-5P Imageries

Sentinel-5P imageries was used for assessment of air pollution in Northern Nile Delta as a case study (Hassan et al. 2023). Northern Nile Delta region in Egypt was chosen for monitoring air pollution particularly in two coastal selected governorates: Kafr El Sheikh and El Behira. These two governorates cover a total area about 15,000 km^2, with total population of 10.2 million in 2021 (CAPMAS 2021).

Four main steps were applied in order to assess vulnerability of densely populated areas in the Nile Delta region to air pollution through Sentinel-5P remotely sensed data, including acquiring and manipulating data, validating retrieved air quality data, and mapping vulnerability to air pollution (Fig. 10.7).

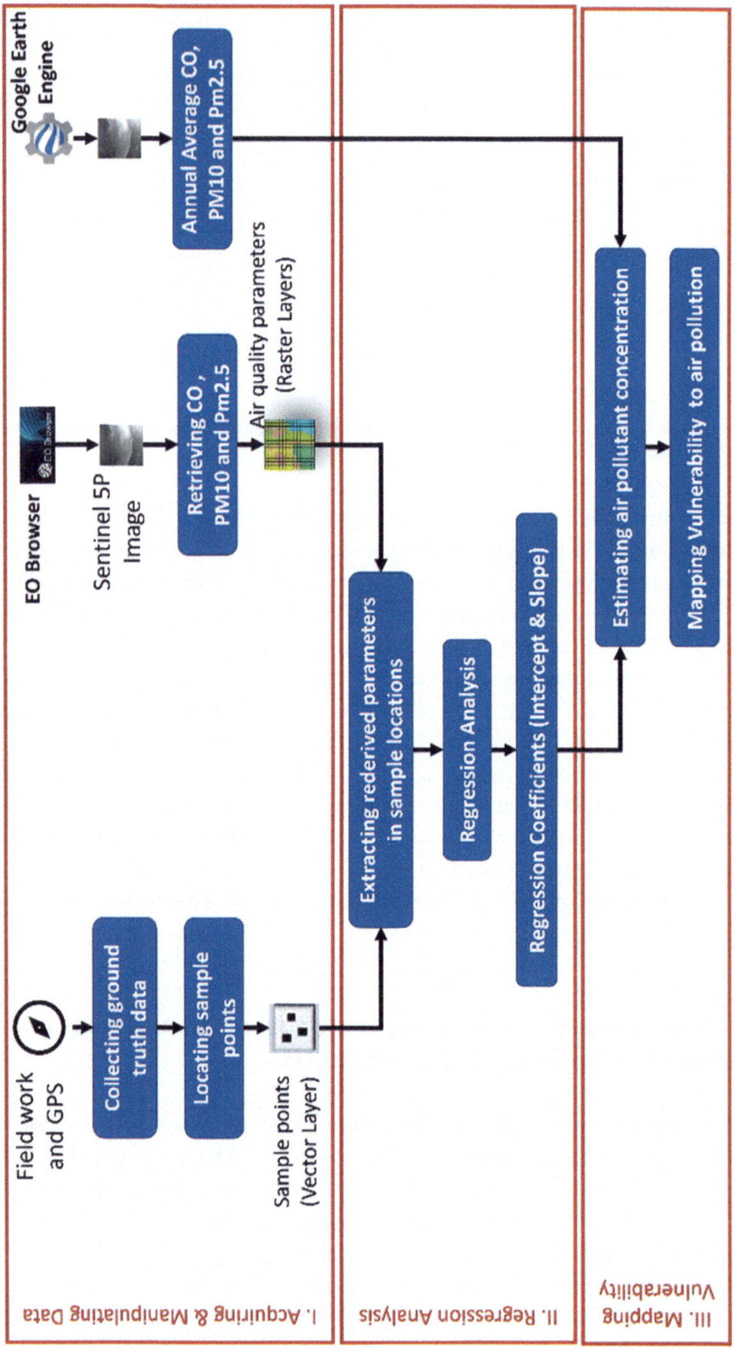

Fig. 10.7 Mythology of mapping vulnerability to air pollution through remote-sensed data. (Source: Hassan et al. 2023)

– *Manipulating Data and Regression Analysis*

For acquiring satellite-based observations for CO concentration and UVAI, two datasets for air quality parameters in this study were collected via Sentinel-5P imageries. The first dataset data were collected from Sentinel Hub EO Browser (SLGIS 2023). EO This dataset was used to calibrate the ground-truth data and the satellite-based observations relationships via regression analysis. The second dataset was Google Earth Engine, which was further employed to retrieve average CO and PM concentrations in the studied area through an Internet-based platform (Hassan et al. 2023). Google Earth Engine provides satellite imagery and cloud-based computing algorithms (Gorelick et al. 2017).

Regression Analysis Regression analysis was applied to study CO and UVAI concentrations relationship, that was retrieved from Sentinel-5P images and in situ measurement. Hassan et al. (2023) used various georeferenced sampling sites allocated and were used for extracting the retrieved CO, PM_{10}, and $PM_{2.5}$ concentrations at the selected areas through Spatial Analyst in ArcGIS (Extract Values to Points Tool). CO concentrations, as a dependent variable, and ground-truth data, retrieved from Sentinel-5P images were estimated through linear regression analysis. Also, the aerosol index and the ground truth data of PM_{10} and $PM_{2.5}$ relationships were studied (Hassan et al. 2023).

– *Mapping Vulnerability: Proposed Approach*

For the estimation of the average concentrations of air quality parameters in the different included areas in the study of Hassan et al. (2023), it was through the slope and intercept values that resulted from regression analysis applied to air quality datasets downloaded from Google Earth Engine. This is followed by mapping of the various air quality parameters and assessing vulnerability of air pollutions. $PM_{2.5}$ and CO are considered to be two primary indicators for vulnerable indexes. Therefore, the average concentrations of $PM_{2.5}$ and CO were estimated to detect the vulnerable areas of pollution in the studied areas through the Zonal Statistics Tool in ArcGIS. Through this methodology a direct-relationship can be established between pollutant concentrations and the population size have been exposed to such pollutant compared to those have low exposures and small population size (Hassan et al. 2023). The following formula was used to normalize the primary indicators:

$$\mathbf{N}\mathrm{x} = \mathbf{X} - \mathbf{X}\,\mathrm{min} / \mathbf{X}\,\mathrm{max} - \mathbf{X}\,\mathrm{min}$$

where N_x: is normalized value of primary indicator; X: is absolute value of primary indicator; X_{min}: is minimum absolute value of primary indicator; X_{max}: is maximum absolute value of primary indicator.

The composite index is a relative index ranging between 0 for least vulnerable localities and 1 for most vulnerable localities. The following formula is the vulnerability index:

$$V = \sum_{i=1}^{n} (Ni^*Wi)$$

where V: is vulnerability index; N_i: is normalized value of indicator i; W_i: is weight of indicator N_i

Therefore, it was proved that remote-sensing data can serve as a cost-effectiveness tool for monitoring air quality, and the Sentinel-5P data of the two major air pollutants ($PM_{25.}$ and CO) can be correlated strongly with the ground-truth data (Hassan et al. 2023). They also proved that Sentinel-5P data can be employed for retrieving and mapping $PM_{2.5}$ and CO concentrations, as was detected in the two included governorates of the Nile Delta region. Moreover, they found that the distribution pattern of CO concentrations was found to be associated with the distance to CO emission main sources, and the usage of the retrieved parameters accompanied with the population size determines the composite vulnerability index for the pollutants of concern and identifies the main hotspots (Hassan et al. 2023).

10.6 Future Trends and Innovations

10.6.1 Potential for Real-Time Monitoring and Response Systems

Worldwide, people were shown to be exposed to different kinds of chemicals. These chemicals were through diets, environment, consumer product use at the workplace. Chemical exposures were shown to be through several routes and pathways, i.e., through inhalation route as well as dermal and oral uptake. Moreover, aggregate exposure could take place through all routes by combined chemical exposure (Rodríguez-Carrillo et al. 2023).

So, human biomonitoring (HBM) represents an important approach to achieve a non-toxic and sustainable environment for people. Therefore, identification of high-risk environmental areas is the present of special chemical in high concentrations in this area, and that will initiate targeted measures, as the contribution of this chemical could be the cause of the development of several adverse health outcomes or metabolic disorders that might develop different chronic impairments (Rodríguez-Carrillo et al. 2023). This biomonitoring was achieved through measuring various biomarkers in different biological matrices in human (Dereumeaux et al. 2022).

Exposure biomarkers must be relevant and easy to measure as well as reliable to estimate particular environmental exposure. Many challenges in measuring these biomarkers were shown for detecting maximum levels and/or frequency in certain populations for the generation of reliable and interpretable data. So, this needs to determine the most appropriate and/or specific biomarkers (metabolite or parent compound), the suitable biological matrices for its determination, and the sufficient sensitive and specific analytical methods for that measurement (Vorkamp et al. 2021, Dereumeaux et al. 2022).

Included effect biomarkers will complete data of exposure with mechanistically-based information regarding early and late adverse effects. Using effect biomarkers can lead to the detection of early biological effects due to exposure to chemicals and identifying high-risk subgroups. In HBM, the development of such roadmap could confirm the usefulness of effect biomarkers and share in supporting one of the major HBM research objectives, linking exposure biomarkers to mechanistically valid effects as well as susceptibility biomarkers in order to get evident information for understanding public health implications in humans due to environmental exposure to chemicals (Rodríguez-Carrillo et al. 2023).

Evaluation of biomarkers incorporates tools that could aid in understanding the disease and/or health. The reasons, diagnosis, disease progress (advancement), relapse or outcome of treatment, and evaluation of hazard. This evaluation includes estimations on natural media straightforwardly (amniotic liquid, cerebrospinal liquid, blood, plasma, pleural and peritoneal fluid, saliva, seminal fluid, serum, sweat, and urine) (Khan 2022).

Moreover, serum or blood represents an invasive matrix that could not be preferred in the project study or survey. Hair (emerged from follicles and surrounded by capillaries) can potentially participate for chemical monitoring in blood (Alves 2014). Also, human hair has many significant advantages, as it represents a noninvasive sample, is low cost, is easy to be collected and transported, is stable when stored and ethically accepted. So, human hair represents an important tool and is essentially used in human biomonitoring development and application (Zhang et al. 2023).

Biomarker assessment should be achieved by a simple laboratory test, which must be rapid and specific. Second, it must be useful and specific to identify various causes and classes of disease. Third, it should be sensitive for early detection. Lastly, biomarker sensitivity and specificity should be high to reduce false results (Khan 2022).

In the human body, detection of minimally or noninvasive biomarkers provides additional abundant information to human health and performance monitoring (HHPM). Detection of this biomarker by existing or novel sensor technologies will share in emerging critical human monitoring research (Sim et al. 2022).

Important applications of real-time monitoring of biomarkers are supposed to be for immediate determination of disease progression or therapeutic effects to reach proper clinical decision in diagnostic field, prognosis, and therapy. Additionally, the development of these technologies must consider the requirements of patient safety due to frequent sample drawing, in situ analysis, and proper monitoring of online security (Kim et al. 2020).

Specific information and highly sensitive methods on a wide range of exogenous and external exposures can be provided through developments in analytical chemistry rapidly as well as other technology. As an example, using newer technology in mass spectrometry can enable (untargeted) screening of hundreds of chemicals at once in biological samples (Sobus et al. 2018). Moreover, the abundant increase of geospatial data in environments like air pollution, noise, and the built, natural, as well as social environments, may lead to much more characterization of external exposome factors (Baccarelli et al. 2023).

In vitro diagnostic technologies, which are shown traditionally, suffered from complex processes of samples with longer intervals of detection, as it can only detect single analytes. That is not practically useful in the management of home-based health. So, the advantage of wearable devices was shown in rapid detection process, compact size, easy integration, and consumption of a small sample, which can be more useful in replacing the endpoint detection. In this way, these devices will provide more information concerning human health comprehensively. Consequently, the development of wearable device that is based on biosensor of dual-channel for (real-time) monitoring of multiple metabolites continuously could address existent limitations (Yang et al. 2023).

Recently, digital biomarkers were shown to gain large interest, where they represent markers of health and disease of non-invasive properties. Typically, they could be used for monitoring continuously. In addition, they can be used for development of resilience biomarkers. In contrast to static biomarkers, these biomarkers can quantify precise imbalances in biological network that was in association with early progression toward diseases. The development of digital biomarkers with the concept of resilience combination may provide a novel kind of digital biomarkers (van den Brink et al. 2021).

10.6.2 Biomarkers (Exposures, Biological Changes)

Biomarker was defined as substance that could actually be an indicator of a typical biological, pathogenic process. Its measurement represents a potent index of abnormal and/or normal pathophysiology or pharmacological response to several regimens of treatment (Bodaghi et al. 2023) or responses to different exposures or interventions (Califf 2018).

Biomarkers could be useful and help in mitigating climate change, besides aiding to understand past and present climate conditions. For example, biomarkers could be used by scientists in assessing reforestation efforts effectiveness and monitoring ecosystems health. Scientists could make informed decisions for combating climate change effectively by harnessing biomarkers. In climate science field, biomarkers represent invaluable tools that may have a role in reconstructing past conditions of climate, predicting climate feedbacks in future, and developing strategies in order to mitigate climate change (Chen 2023).

An ideal biomarker was that indicator, which helps in explaining the occurrence of moderate proportion of diseases in the community. So, this biomarker must possess some qualities to be used in clinical diagnosis and prognosis. First, in testing a biomarker, it should be safe and easy to be conducted; if possible, it may be non-invasive and preferred to be measured in external body fluids (Khan 2022). Moreover, a biomarker may be biological, chemical, or physical. These could be cells, molecules, genes or gene products, hormones, or enzymes. Moreover, biomarkers may be produced by the body in response to disease or diseased organs, such as tumor or infection (Khan 2022). In human body, every tissue possesses distinct biomolecular makeup with specific features, namely levels or activities (which means genes or proteins that can perform specific functions in human body) of proteins or genes or other biomolecules (Ahmad et al. 2023).

Bernhardt et al. (2023) had classified seven categories of biomarkers regarding their clinical use:

1. Susceptibility or risk
2. Diagnostic
3. Prognostic
4. Monitoring
5. Predictive
6. Pharmacodynamic and treatment response
7. Safety

- Susceptibility or risk biomarker
  This suggests probable risk for disease developing in an individual who shows no clinically apparent disease.
- Diagnostic biomarker
  This is supposed to be used for detecting and/or confirming either disease presence or characteristics of certain disease aspects.
- Prognostic biomarker
  It is useful in the identification of a clinical event probability as well as disease recurrence or its rate of progression in diseased patients.
- Monitoring biomarker
  This category of biomarker refers to a repeated measured item for assessing disease status or quantification in case of exposure to a medical product or environmental agent.
- Predictive biomarker
  For identifying individuals who were more potent than comparable individuals without the biomarker to test ability of improper or favorable effect from exposure to a medical product or an environmental agent.
- Pharmacodynamics/response biomarker
  To illustrate biological response that occurs in individuals who were exposed to an environmental agent or a medical product.
- Safety biomarker
  For measuring before or after exposure to environmental agents to show even the limit or presence of toxicity as a result of such adverse effects (Bernhardt et al. 2023).

  Shah et al. (2020) proposed biomarker classification in two additional ways regarding bodily source and measurement type.

  Bodaghi et al. (2023) had suggested that biomarkers can be classified according to their characteristics into three main types: cellular, molecular, and imaging biomarkers. Molecular biomarkers can be discovered using genomics, transcriptomics, and proteomics, metabolomics, lipidomics, and epigenomics.

– Genomics
  Genomics can be defined as the study of alterations in deoxyribonucleic acid (DNA) (including the even structure, function, and expression of nucleotide sequences of cell or organism) and its association with disease onset and exacerbation as well as patient prognosis and response to pharmacotherapies (Bodaghi et al. 2023).
– Transcriptomics and Proteomics
  Transcriptomics and proteomics are analyzing two important parts of the gene expression, which are RNA and proteins. Many factors can impact gene expression such as age, developmental stage, sex, type of tissue/cell, health status and others. Transcriptomic can identify and quantify a set of all RNA species completely, e.g., messenger RNAs, non-coding RNAs, and small RNAs (small interfering RNA, piwi-interacting RNA, and microRNA) that respond to different environmental and/or biological stimulants. Proteomics are describing protein complement of genome. It represents a set of proteins that is encoded by genome and can reflect the complex and dynamic interaction between gene and environment.
– Metabolomics
  Metabolomics is referred to as all metabolite measurements in biological system (organism, tissue, or cell) that are produced as a response to certain environmental exposures or interactions. These metabolites represent low molecular weight byproducts of gene transcription and enzymatic reactions (like fatty acids, amino acids, sugars, lipids, nucleic acids, peptides, and organic acids) present in cells, tissues, organs, and biological fluids.
– Lipidomics
  Lipidomics could be defined as metabolomics subfield. Lipids represent the primary component of cellular membranes (with both functional and structural roles) and also mediators in some cellular mechanism like signal transduction, inflammation, and immunity. Lipidomics could help in understanding cell structure and function and/or organism's lipidome in addition to how diseases and treatment can affect lipoproteins (Donovan et al. 2019).
– Epigenomics
  Epigenomics describes all epigenetic changes in a cell in terms of epigenetic modifications affecting the genome with no change in DNA sequence. It represents potential reversible modifications of the chromatin as a result of environmental modifiers or normal cellular repair. These changes can be measured as levels of DNA methylation, noncoding ribonucleic acid (RNA), and histone modifications, which result in alterations in gene expression in the long term. These alterations can persist during cell division and could be inherited from generation to generation. Epigenomics could be used to study the environmental influence of different stressors that include, as example, simple or single exposure to chemicals and infections and its impact on disease outcome, which can generate epigenetic markers enduring for a long time. Also, epigenetic biomarkers can help in assessing environmentally mediated disease risk in order to be applied in using polygenic risk scores for evaluation of heritable disease risk and developing preventive strategies for individuals who are mostly susceptible (Wu et al. 2023).

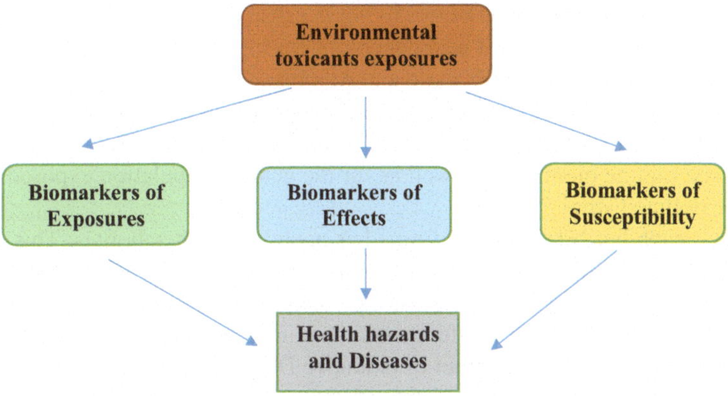

Fig. 10.8 Environmental toxicant exposures and biomarkers of exposure, effects, and susceptibility can integrate into health hazards and/or diseases

Turesky and Lu (2020) had classified biomarkers of environmental toxicants as follows:

10.6.2.1 Biomarkers of Exposure

This class of biomarkers involves quantification of particular chemical residues, specific products formed as a result of an interaction between defined chemicals and different target cells, molecules, or fractions of them and/or metabolites of xenobiotic compounds detected in various biological matrixes (Abdelzaher et al. 2022) as shown in Fig. 10.8. Example of biomarkers of exposure:

1. DNA adducts formed from DNA oxidation (e.g., for different tobacco smoke, PAHs from traffic exhaust, and air pollutants). This oxidation could share in modifying multiple DNA repair enzymes, which could result in substitutions that could finally lead to the synthesis of altered protein products and DNA mutations (Poirier 2004).
2. Protein adducts represent covalent interaction products between certain proteins and a chemical compound (reactive form) such as:
   - Albumin represents the only serum protein that can bind with the aflatoxin, forming many adducts, e.g., AFB1, AFG1, AFB2, and AFM1 (Naeem et al. 2021).
   - Nonapeptide (NP) adduct formed through BChE-adduct proteolysis, and it represents the most important and sensitive biomarkers in pesticide exposure (John et al. 2021).
   - The protein adducts are formed with serum albumin and hemoglobin. Its assessment was shown in the case of environmental, smoking-related, or occupational exposure to PAHs. The relevant protein adducts formed with benzo[a]pyrene (BaP) (the carcinogenic metabolite of BaP) (r-7,t-8 dihydroxyt-9,t-10-epoxy-7,8,9,10-tetrahydrobenzo[a]pyrene), BPDE. This BaP adducts reported higher levels in individuals who lived in rural or suburban

areas in comparison with others living in urban residences. These higher levels could be as a result of heat generation during organic combustion in these former regions (Poblete-Naredo and Albores 2016).

3. Other adduct: 1-hydroxypyrene (1-OHP) (main pyrene metabolite). Urinary 1-OHP is used for tobacco smokers and workers exposed to PAHs occupationally as a routine biomarker as well as for the general population exposed to airborne polluted areas.
4. 1 and 2 naphthol NT: naphthol formed as naphthalene metabolites. This adduct represents biomarker PAH exposure in case of its air or inhalation exposure.
5. Oxidized nucleobases. As well-known previously, ROS can lead to modification in DNA nucleobases that produce mutations in case of lacking repair by repair mechanisms. So, a large number of oxidative DNA modifications may be generated, e.g., 8-oxo-7,8-dihydroguanine (8-oxoGua), was generated as a result of guanine oxidation, in addition to other deoxynucleosides like 8-oxo-7,8-dihydro-2'- deoxyguanosine (8-oxodG). These two products are considered biomarkers of oxidative DNA damage and the most commonly measured end products.
6. DNA damage: The single-cell gel electrophoresis assay, or the comet assay, represents an easy, fast, and sensitive method for DNA transient damage evaluation at single- and double-stranded break levels, and DNA replication and repair intermediates (Poblete-Naredo and Albores 2016).
7. Exposure to heavy metals was reported to be associated with serious human diseases anywhere. In children, biomarkers of heavy metal exposure exhibited particular importance in monitoring exposure and health risk assessment (Turesky and Lu 2020).

10.6.2.2 Biomarkers of Effects

Usually, this class of biomarkers involves quantifiable changes in physiologic and/or biochemical parameters. Biomarkers of effects are subclassified into (a) biomarkers of adverse effects, (b) biomarkers of combined effects or biological activity, as shown in Fig. 10.9.

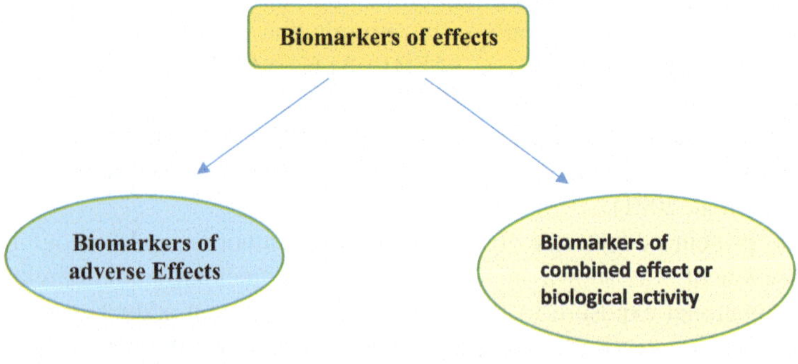

Fig. 10.9 Classification of biomarkers of effects

– *Biomarkers of Adverse Effects*

Effect biomarkers are known as the physiological, biochemical, behavioral, or other alterations (that can be quantified) in an organism that could be associated with an established disease or potential health impairment that is dependent on its magnitude (Zare-Jeddi et al. 2021). In consequence, this can aid in predicting adverse effects at various levels of biological organization, i.e., molecules, cells, tissues, and organs or systems (such as brain, testes, and kidney function). In association with a range of environmental chemicals such as phthalates, bisphenols, perfluoroalkyl substances [PFAS], the pesticide chlorpyrifos, and heavy metals such as arsenic, among others, these biomarkers can provide relation between this internal exposure and the early as well as late health effects due to its exposure (Mustieles et al. 2020). Therefore, it may be considered as "early biomarkers of effect" owing to its early occurrence when interacting with a xenobiotic with some biological target (Rodríguez-Carrillo et al. 2023).

Examples include:

(a) *Epigenetic Markers*

- DNA adducts
- enzyme induction/inhibition,
- Measurement of oxidative stress markers, DNA strand breaks, and cytogenetic endpoints, like micronuclei induction (MN), sister chromatid exchange (SCE), and chromosome aberrations (CA) (Abdelzaher et al. 2022).

These effect biomarkers that measure changes at structural or functional levels ("altered function/structure") in affected tissues or systems or actual clinical disease are defined as "late biomarkers of effect" (for example, a rise in both serum glucose and lipid levels associated with metabolic disease). In consequence, selection of some different biomarkers may take place, ranging from biochemical biomarkers, e.g., hormone levels and metabolic parameters. Moreover, different markers could be able to provide other information (quantitative) about the human body, such as behavioral tests, magnetic resonance imaging (MRI), or anogenital distance (Ventura et al. 2021). So, we can conclude that these biomarkers can help us to identify early effects resulting from chemical stressors and therefore can improve risk assessment of a known chemical and/or mixture of complex chemicals in humans (Rodríguez-Carrillo et al. 2021).

(b) *Biomarkers of Combined Effect or Biological Activity*

This is defined in the case of exposure to a mixture of multiple chemicals *via* single or multiple sources of exposure and routes (Bopp et al. 2018). Also, an effect biomarker could be considered as an integrated biological activity, which measurements can take place by *in vitro* assays, with potent assessment of the combined action of its mixture on endocrine receptors or different physiological targets. Also, this type of biomarker can be called "*ex vivo* hormone activity biomarkers," "combined biological activity biomarkers," or "combined internal exposure biomarkers" (Rodríguez-Carrillo et al. 2021). The main advantage is that evaluation of real-world chemical mixtures can be done in different human biological matrices. So, new analytical methods must be selected, in this case, to isolate specific chemical families (like PFAS) for assessment of their combined effect by *in vivo* and/or *in vitro* assays that in turn will lead to an increase in their regulatory uptake (Bjerregaard-Olesen et al. 2019). Moreover, this approach could not be applied in the case of comparing differences in the potency of individual components of the mixture. Each individual component must be characterized separately.

For example, these biomarkers had been used in several epidemiological studies to assess the effects of "complex mixtures of persistent organic pollutants" on different human health outcomes, like the development of breast cancer in adults. These studies reported relation between the antiandrogenic or estrogenic effect of mixture of chemicals that was isolated from human biological matrices and the increased risk of male urogenital malformations or breast cancer (Vinggaard et al. 2021).

10.6.2.3 Biomarkers of Susceptibility

Examples of these biomarkers are metabolic phenotype and genetic polymorphism that can be helpful in providing key information and can show reflection on intrinsic characteristics toward an exposure adverse effect. However, we can suggest strategies that can aid to understand the effect of an exposure on biological pathways toward various forms of exposure, health changes that can lead to the development of disease, and provide more meaningful comprehensive risk assessment (Abdelzaher et al. 2022).

This class of biomarkers include phase I and II enzymes and GST and DNA repair gene polymorphisms as follows:

- *Phase I Enzymes.*

This class of biomarkers includes phase I and II enzymes, like conjugating enzymes and members of cytochrome P450 (CYP450) superfamily. Phase I metabolism includes a series of oxidative reactions that have a role in the biotransformation of a parent compound into detoxified as well as bioactive metabolites.

Examples of such phase I polymorphic enzymes are polymorphisms of CYP1A1, the most studied polymorphisms in human gene susceptibility (Yi et al. 2007). The association of this polymorphism with increased levels of urinary 1-OHP was studied and showed a distinct relation with air pollution (Petchpoung et al. 2011), PAH exposure (Chuang and Chang 2007) and light tobacco consumption.

Other examples of phase I polymorphic enzymes were epoxide hydrolase 1 (EPHX1) and aldo-keto reductases (AKR). They are related to metabolism of environmental toxicant. EPHX1 is encoding a protein that catalyzes water addition to epoxides in order to detoxify. There are two polymorphic variants of EPHX1, which are A15543G (H139R) and T8668C (Y113H).

In (AKR) family, a previous study revealed an increased lung cancer susceptibility that was associated with variant AKR1C3*2 (Gln5His) in the Chinese population following PAH exposure in smoky coals (Lan et al. 2004). Another study recorded an inverse relation between bladder cancer risk and this polymorphism in Caucasians (Figueroa et al. 2008), while another study found no relation in smokers with several levels of mRNA of some AKRs (AKR1A1, AKR1C1-AKR1C3) (Barrón-Vivanco et al. 2013).

– *Phase II Enzymes*

The phase II metabolizing enzymes include conjugation reactions in the form of increasing hydrophilicity of chemicals in order to excrete and remove or eliminate possible reactive metabolites more easily. Mainly, phase II enzymes include transferases, such as glutathione S-transferases (GST), UDP-glucuronosyltransferases (UGT), N-acetyltransferases (NAT) and sulfotransferases (SULT). N-acetyltransferase NATs (NAT1 and NAT2) represent the key enzymes for the conjugation of arylamine compounds. It displays high variability for inter-individual acetylation of certain drugs. This variability can result in classifying individuals as fast, intermediate, and slow "metabolizers" (Walker et al. 2009). Inducing this phase II metabolizing enzyme is through environmental pollutants at the transcriptional level (Srivastava et al. 2014). Other phase II enzymes at gene levels such as Quinone oxidoreductase-I (NQO1), NAD(P)H, and heme oxygenase-1 (HO-1) genes, were found to be upregulated by various airborne contaminants (Srivastava et al. 2014). Moreover, nuclear factor erythroid 2-related factor 2 (Nrf2) represents transcriptional regulator for the expression of antioxidant and phase II detoxifying genes. Nrf2, as transcription factor, can recognize AREs (antioxidant response elements) as specific DNA elements that are found in the enhancer region of target genes, like NQO1, HO-1, UGTs, GST subunits A1 and A2, and others.

– *Different Classes of GST Polymorphisms*

This class of glutathione-S-transferase (GST) polymorphisms was shown to be associated with an increase in health risk levels according to the modification in various biomarkers of exposure and effects of environmental pollutants. GSTM1 polymorphism is one of the studied GSTM polymorphisms (showing deleted alleles of high prevalence between populations). GSTM1 was relevant or in association with high concentrations of urinary 1-OHP (Sánchez-Guerra et al. 2012), 8-OHdG (Prasad et al. 2013), 1- and 2-NT (Nan et al. 2001) and levels of DNA adduct. Furthermore, it represents risk factor in preterm delivery that is associated with PM10 exposure (Suh et al. 2008).

Conversely, the GSTT null mutant had shown no association with an increase in levels of urinary 1-OHP (Sánchez-Guerra et al. 2012) and elevated levels of urinary 8-OHdG (Prasad et al. 2013) or high concentrations of DNA adduct, while in Chinese children who lived near to industrial plants, it is associated with higher predisposition to acute leukemia (Yang et al. 2009) and in smokers, it reported an association for coronary artery disease. Moreover, GSTP polymorphisms, the A2627G genotype (I105V), were associated with an increase or fall in urinary 1-OHP levels (Petchpoung et al. 2011) in occupationally exposed workers. Other genotype GSTP1 C5317T (A114V) showed relation with high urinary 1-OHP and OHPhe levels in German workers exposed to high PAH levels (Poblete-Naredo and Albores 2016).

– *DNA Repair Gene Polymorphisms*

In these repair genes, the polymorphisms present were considered variable factors that can predispose people to health hazards and/or outcomes in high-risk situations of toxicological exposure.

Individuals who are carrying variants of DNA repair gene polymorphisms such as X-Ray cross-complementing XRCC1 (399Q of XRCC1) and excision repair cross-complementing ERCC1 (N118N of ERCC1) and X-Ray cross-complementing XRCC3 variants (241T of XRCC3) are at higher risk for malignant mesothelioma development in highly asbestos-polluted areas (Roszak et al. 2014).

Other DNA repair gene polymorphic variants are 8-oxo-guanine-DNA glycosylase/AP lyase (hOGG1) 326C, which is considered a susceptibility factor. It revealed an association with high levels of markers of oxidative DNA damage in bus drivers exposed to PAHs and volatile compounds (Bagryantseva et al. 2010). Moreover, it is considered a higher risk factor for lung cancer in indoor PAH-exposed individuals (Lan et al. 2004). Another DNA repair gene is Xeroderma pigmentosum-D (XPD). In exon 23, the variant of (XPD) A35931C Lys751Gln was found to induce high DNA strand break levels in the aforementioned population (Bagryantseva et al. 2010). Also, in intron 9 (PAT+) of Xeroderma pigmentosum-C (XPC) polyAT insertion of 83 bp, its association with high levels of the adduct anti-BPDE-DNA was demonstrated in either high or low PAH exposure levels.

In workers exposed to very low doses of ionizing radiation (IR), Ledda and Rapisarda (2020) had studied the impact of the detoxification enzymes GSTM1 and GSTT1 (null genotype) and variants of the DNA repair genes XRCC1 (Arg194Trp, Arg280His, Arg399Gln), XRCC3 (Thr241Met), and XPD (Lys751Gln). They found a higher frequency of spontaneous micronucleus (MN) in workers with at least one allelic variant in XRCC1 than in carriers of the wild type.

– *Gene-Gene Interactions (G × E)*

In the vast majority of diseases, it is clear that combinations of antagonistic or synergistic factors are important to the risk of disease. In the context of human health, G × E is focusing on how environmental and particular genetic factors can increase disease risk synergistically. So, we can describe instances where environmental and genetic risk factors can potentiate disease independently. Then, examination of how these two factors can behave together to increase the risk of disease (Virolainen et al. 2023).

In humans, the characterization and identification of G × E interactions is crucial to combat human disease. Individuals are born with genetic variants inherited from parents, and these variants are not manipulated easily (Virolainen et al. 2023).

For example, the interaction of multiple genes with environmental pollutants was demonstrated in people who carried more than one polymorphism in order to suppose the genetic susceptibility complex scenario in the extreme. For example, the interaction of CYP1A1*2A (MspI) polymorphism and different phase II enzymes where CYP1A1*2 polymorphism interacts with GSTP1 (Ile/Val), GSTM1 null, and EPHX (the slow variants) resulted in higher DNA adduct levels (Demetriou et al. 2012). In addition, the combination of the slow acetylator genotype of NAT2 with the GSTM1 null correlated with higher urinary PAH metabolites besides elevated CAs in people exposed to environments of high PAH. Moreover, lung and breast cancer risk is increasing in individuals who carry the genotype of GSTT1 null, GSTM1 null, and NAT2 rapid acetylator (Poblete-Naredo and Albores 2016).

- *Exposome, Climate Change and Human Health*
  Exposome

The "exposome" could be defined as the study of all environmental hazards (exposure) and how these exposures can affect health and disease (Baccarelli et al. 2023). The exposome is considered a new vision in exposure science. It is supposed to expand in traditional practice for the assessment of single exposure to a few or one chemical at one time point to an approach addressing exposure measures to multiple stressors along lifespan (Bocato et al. 2019). Otherwise, the absolute aim of exposomic science is to study the individual exposures to chemical and non-chemical environmental materials accurately (e.g., physical activity, diet, psychosocial stressors, and toxicants) over the lifecourse with prenatal inclusion (Baccarelli et al. 2023). Furthermore, the necessity of exposomics will be augmented due to climate change consequences, with numerous ramifications for the environment and the following (ensuing) effects on human health (Zhang et al. 2021). So, it will be useful to present stronger relationship for an individual (phenotype) with good health or disease conditions (Bocato et al. 2019).

The exposome is composed of two main domains: internal and external environmental exposures (Abdelzaher et al. 2022). In order to understand exposome concept, recent study described three non-genomic categories to be evaluated:

(a) Internal (endogenous)
(b) Specific external
(c) General external

All categories were fairly explained by themselves with regard to internal exposomal factors that encompass inflammatory factors, oxidative challenges, hormones, allostatic load, microbiota, microbiota, metabolism, and others. Specific external factors showed additional variation, including essential and non-essential dietary factors such as vitamins [indeed all food], electromagnetic radiation, pathogens, noise, environmental pollutants, and chemicals, and a wide variety of lifestyle factors like pharmacologic agents, tobacco, alcohol, caffeine, betel nut, other xenobiotic substances, and phytochemicals (Lucock 2021).

General external factors seem to be vast and include economic, social, and psychological influences. Other simple factors like your location (in terms of geography, where you live) can be too crucial if your location is in a polluted area or in a city at high latitude with low seasonal ultraviolet light (UV) levels that limit synthesis of vitamin D. Therefore, all the three non-genomic categories could show apparent overlap—allostatic load due to stress, but it demonstrates itself through metabolism, so a) and c) categories overlap. In CC that is anthropogenic with its sequelae, we can record that it fits into all three categories. Also, the field of exposomics can reflect great complexity, given recent exponential advances in omics technologies (Lucock 2021).

Examination of the polar bear blood exposome to identify specific chemicals that could result in thyroid disruption was an important example that can help in considering exposomics as an important component of CC research (Zhang et al. 2021). These animal models are used for the simulation of human exposome conditions like combinations of occupational and dietary exposures as well as to recognize aquatic exposomes downstream of municipal wastewater (Neveu et al. 2020), in addition to real-world applications due to exposure to tire rubber-derived. Other studies in nonhuman exposome seem to be of great significance in protecting ecosystems and wildlife, as recently demonstrated for coho salmon, and will help in protecting humans under one health model (Husby 2020).

The human exposome would describe the match of the human genome, which is represented by chemicals (endogenous and exogenous) that are interacting with the biology of the human systems (Pleil and Sheldon 2011). This may explain the interaction between gene–environment that may result in driving health outcomes from chronic conditions (which include asthma, neurological disorders, and cardiopulmonary diseases), autoimmune diseases, and cancer (Wallace et al. 2016). Another important purpose reflecting the role of exposome is that it could link the presence of a substance at low levels unnaturally as much as it could be present at elevated levels. Particularly, this was supposed to be correct, especially in the case of food exposome. In other words, we can suppose that foods are seasonal, where this seasonality could result in variation in phytochemical contents as well as important vitamins like folate. The historical significance of this variation can be explained in terms of the seasonal cycle of abundance of that food rich in folate that might have regulated embryo viability through 677T-MTHFR selection factor. In conclusion, research in exposome needs harmony in techniques and sampling essentially in order to flourish (Lucock 2021).

Recent study has revealed that external exposure can induce functional changes that were reflected by metabolites as the end points of some proteins in vivo. For example, different metabolites of mercapturic acids like N-acetyl-L-cysteine S-conjugates represent urinary metabolites of occupational and environmental toxicants (Frigerio et al. 2019). These metabolites could be derived from some GSH S-conjugates catabolism and are supposed to include five MAs (Song et al. 2023).

These metabolites represent important indicator or index factors after exposure to carcinogens and volatile toxicants (Chen et al. 2020). They are known as MAs (HEMA, CEMA, HPMA, SPMA, and HMPMA). Mostly, N-acetyl-S-(2-cyanoethyl)-L-cysteine (2-cyanoethylmercapturic acid, CEMA) represents one of the main urinary metabolites identified after exposure to acrylonitrile (Bowe et al. 2018). Other metabolites are N-acetyl-S-(3-hydroxypropyl)-L-cysteine (3-hydroxypropylmercapturic acid, HPMA) (Lindon et al. 2010) molecules, while 2-carboxyethyl-mercapturic acid represents the major urinary acrolein metabolite (Chen et al. 2019). Furthermore, N-acetyl-S-(phenyl)-L-cysteine (S-phenylmercapturic acid, SPMA) represents a specific urinary metabolite for benzene exposure and it was commonly recommended as urinary biomarker for its exposure, (Schwedler et al. 2021). 3-Hydroxy-1-methylpropylmercapturic acid (HMPMA) represents urinary crotonaldehyde metabolite (Song et al. 2023).

Taken into consideration that for humans, the levels of the previous urinary biomarkers' levels are carcinogenic. One of the latter urinary biomarkers, which is N-acetyl-S-(2-hydroxyethyl)-L-cysteine (2-hydroxyethylmethacrylate, HEMA), was considered common metabolite of several xenobiotic chemicals that are different in structures (Song et al. 2023).

Other studies concluded significant relationship between the count of serum exosomes CD-81+ and air pollutants. In turn, this exosome is considered a risk factor in airway inflammation progression; moreover, in patients suffering from COPD, serum CD81+ exosome count was found to exhibit significant rise levels and was related to monthly exposure to $PM_{2.5}$ and daily exposure to PM_{10} (Soleimanifar et al. 2023).

- *Essential Trace Elements as Key Components of Exposome*
    (i) Living organisms require essential trace elements for normal development as well as to maintain the biological functions, while excess levels represent toxic. Maitre et al. (2022) had evaluated, in whole blood, nine essential elements (Co, Cu, Mo, Mn, Mg, Na, K, Se, and Zn), and recorded a marked and prominent number of exposure-omics associations. The most remarkable elements were child copper (Cu) and maternal molybdenum (Mo). In humans, Mo action was a co-factor of four enzymes that participated in several main reactions, for example, regulation of sulfur-containing amino acids like methionine (Schwarz 2016). They revealed that in childhood there is an association of maternal Mo with high methionine levels. Moreover, in children, Cu was found to be associated with high C-reactive protein (CRP) levels (inflammation marker).
    (ii) Previously, relations between Cu-associated CpGs and different diseases were observed, e.g., type 2 diabetes, obesity, and rheumatoid arthritis. During childhood, the association of another essential trace element was recorded with multiple molecular features, and no overlapping was found among them. This may be expected as a result of its substantial essential roles. As an example, zinc (Zn) showed a relation with a high transcription level of Carbonic anhydrase 1 (CA1), where its gene expression is Zn^{2+}

dependent and enzymatic activity is using Zn^{2+} as a cofactor (Lionetto et al. 2016).

(iii) In pregnant women, gut microbiome could result in alteration in metabolism of arsenic and in neurodevelopment susceptibility to its exposure. That illustrated the role of diet in delivering several anthropogenic chemicals to the body (in case of fish and fruit intake), where gut microbiome may alter biological effect (Coryell et al. 2018).

– *Environmental Exposures Effects on Several Aging Hallmarks (Linking aging Hallmarks and Exposome to Age-Related Disease Risks)*

(i) Widespread heavy metals exposure has been associated with different health hazards and diseases in humans, especially in children. So, biomarkers of exposure to heavy metals were important to monitor exposure as well as health risk assessment. In hair of children from the Mid-Ohio Valley, levels of Pb, Mn, As, and Cd were measured for determination of metal levels in hairs between and within subject predictors. Results demonstrated that levels of Mn and Pb in hair were ~10-fold greater than Cd and As levels in hair. Moreover, levels of metal recorded variations between female and male subjects with various profiles along hair segments (Turesky and Lu 2020).

(ii) In Alzheimer's disease (AD), the association of organic pollutant exposure and heavy metals was suggested with DNA methylation alteration and mitochondrial dysfunction (Martin and Fry 2018) while early embryonic exposure to 2,3,7,8-tetrachlorodibenzo-p-dioxin can change patterns of global DNA methylation with an increase in T2DM risk (Ding et al. 2022).

(iii) Association of high concentrations of bisphenol A with an increase senescence, inflammation In T2DM patients was shown as well as decrease in telomere lengths (Ding et al. 2022).

(iv) A positive association was detected between indoor particulate matter absorbance (PMabs) and parental smoking with acceleration in age for children (de Prado-Bert et al. 2021).

(v) In age-related diseases, an alteration in gene expression shown in histone modification, DNA methylation, and noncoding RNA-associated gene silencing represents primary targets of environmental toxicants that may be supposed to sustain and induce resulted epigenetic alterations (Ding et al. 2022).

– *Analytical Methods for Different Class of Biomarkers*

So, we need novel analytical and interpretation methods in monitoring the real-life co-exposures of a large number of chemicals simultaneously that are not physiologically found or present in human body (Luijten et al. 2023).

– The proteomics approach and technology were performed through mass spectrometry and liquid chromatography for the identification of greater than 1,000,000 gene products in the human body.

- Besides infrared spectroscopy, nuclear magnetic resonance, liquid chromatography-mass spectrometry, high-performance liquid chromatography, and gas chromatography-mass spectrometry were used in the metabolomics biomarkers analysis (Bodaghi et al. 2023).
- Accelerator mass spectrometry (AMS) represents a sensitive technique for measuring long-lived radionuclides that are present naturally in the environment. Reports were recorded for applications of radiocarbon tracer technology for assessment of low-dose of naphthalene-DNA adduct, benzo[a]pyrene in humans, and the exposure to triclocarban (antibacterial) and its endocrine effect (Turesky and Lu 2020).
- In almost all human tissues, mass spectrometry (MS) represents the most specific methodology for identification and quantification of DNA adducts regarding its molecular weight. (It requires complex and expensive instrumentation). Furthermore, immunoassays and immunohistochemistry techniques are using antisera for the identification of specific DNA adducts directly in studied tissues or in DNA samples.
- Research of protein adducts is based on sample number, time, and available methodologies. In large-scale studies concerning protein adducts research, recommendations of immunoassays are preferred, while chromatographic-based techniques are also suitable, like HPLC-FP or GC-MS.
- The level of guanine-oxidized products (e.g., the 8-oxoGua product) is mainly determined by commercial ELISA kits and by chromatographic methods (Poblete-Naredo and Albores 2016).
- In biofluids, the small molecule chemical profiles were assessed through untargeted high-resolution mass spectrometry (HRMS) approaches. These technologies can be central in linking internal dose (like biomarkers of exposure), biological response (in terms of variations in endogenous metabolic pathways), and disease (via shifting in metabolism related to the pathogenesis of disease) to external exposure (as measuring environmental chemicals that concern food) (Zhang et al. 2021)
- Within the exposome, additional omics layers must be incorporated to allow a systems biology framework to identify the exposure's effects and provide more comprehensive characterization to illustrate more toxicological mechanisms at population level potential (Zhang et al. 2021).
- For identifying the content of global DNA methylation, some methods like high-performance capillary electrophoresis (HPCE) and HPLC-MS are used after digestion of DNA to single oligonucleotides. Furthermore, microarrays represent the basis of other methods of genome-wide methylation and include varying technologies like immunoprecipitation of 5-methylcytosine, methylation-sensitive restriction enzymes and PCR amplification, bisulfite treatment coupled with hybridization to microarrays, and sequencing, among others (Poblete-Naredo and Albores 2016).
- The exposome overhauls provide novel methodological framework that can extensively address the complex, highly dynamic interactions of both endogenous and exogenous exposures and modifiable factors in humans. However,

measuring exposures of airborne pollutant and investigating the associated biomarkers in such susceptible elder people can result in the suboptimal assessment and management of health risks potentially. Therefore, a combination of extensive human biomonitoring, personal monitoring for airborne pollutants, and omics analysis (advanced), confounding information, and statistical methods was used (Tang et al. 2021).

– Mass spectrometry (LC-MS/MS) is widely used for detection of exosome-associated proteins (Ludwig et al. 2019).
– The flow cytometry technique represents a preferred method for counting some exosomes (Soleimanifar et al. (2023).

10.6.2.4 Biomarkers in Occupational Environment

Concerning occupational settings, different exposure biomarkers were applied, for example:

- Benzene represents a potent carcinogen that can cause hematotoxicity to humans. Its exposure usually occurs in industrial environments and/or occupations, as well as exposure to active and passive smoking (Wallace et al. 1987). Muconic acid and its metabolite derivatives are biomarkers for benzene exposure, which is linked to indoor airborne benzene levels, in addition to the urinary excretion of 8-OhdG, which is used as a biomarker for the oxidative damage of DNA (Lagorio et al. 1994).
- Formaldehyde (FA), also, represents another human carcinogen. It is a living and occupational pollutant. Formaldehyde had an epigenetic effect that is shown in the form of alterations in DNA global methylation in human bronchial cells among exposed workers in different occupations. Additionally, histone alterations where FA induced histone changes in the form of reduced histone lysine residue acetylation in cells of human lung. Regarding microRNA expression, microRNA expression dysregulation was recorded in adenocarcinoma cells of the nose as well as olfactory bulbs, human lungs, rodents white blood cells, and nonhuman primates. These overall findings can confirm epigenetic modifications role as supposed possible mechanisms of action in FA exposure, and more research should be explored to investigate its mechanisms (Ledda and Rapisarda 2020). Other proteins, adducts known as N6-formyllysine, are detected in nasal epithelium, in tissues of rats, and to some extent in trachea due to inhalation exposure, but not in distant lung tissues, white blood cells, or bone marrow (Edrissi et al. 2017).
- Phthalate exposure was reported to occur to a less extent through inhalation, while the main exposure through diet and dust ingestion. Metabolites of parabens and phthalates showed relation to sphingomyelins endogenous pathways. The sphingomyelins represent an important lipid constituent or element of cell membranes, which was implicated in different disorders and involved in signaling. Upregulation of sphingosine biosynthesis intermediates and valine was recorded among pregnant women who were exposed to parabens and phthalates (Maitre et al. 2022). Phthalate metabolites in urine (DEHP) are known as mono-

ethyl phthalate (MEP), monobenzylphthalate (MBzP), mono-n-butyl phthalate (MnBP), mono-noctyl phthalate (MnOP), mono-(3-carboxypropyl) phthalate (MCPP), and also biomarkers of methylation, protein nitroxidation, and nucleic acids oxidation (Pigini et al. 2022).

10.6.2.5 Digital Biomarkers and Digital Health Technologies (DHT)

Continually, clinical researchers are exploring new biomarkers and were focusing on non-traditional, digital markers recently. We have seen that digitization is increasing across the healthcare continuum, which in turn has resulted in rapid progress in the adoption and development of Digital Health Technologies (DHT) through healthcare ecosystem. Proliferation of DHT, in turn, had led to the use of the expression 'digital biomarker' for describing a broad array of measurements (Vasudevan et al. 2022).

These digital biomarkers often include neurological, biological, and socioeconomic as well as environmental data for creating an intermediary biomarker (Kyriazakos et al. 2021). The term digital biomarkers referred to objective, physiological, and behavioral data (that can be quantified) that can be collected for measurement through digital devices like wearables, portables, digestibles, or implantables. The role of collected data is for explaining, influencing, and/or predicting health-related outcomes (Mulinari 2023). In other words, it could measure behavior and/or physiology, which indicates or reflects biological or pathological process or a response to an exposure or derived intervention from digital measures. Clinical meaning of these digital biomarkers is confirmed by an actual relationship that is reliable to an existing, validated endpoint (Mulinari 2023).

Now, digital tools in the form of smartphones, smartwatches, and other wearable devices can measure main indicators relevant to health status that include activity patterns, vital signs, sleep, and skin temperature (Coravos et al. 2019). At the same time, non-invasive novel types of continuous as well as wearable molecular sensing technologies should be developed ongoing, such as quantification of biomolecules, which is non-invasive, in saliva, sweat, and other body fluids (Bandodkar et al. 2019). Eventually, these kinds of measurements should lead to reinforce current abilities of digital measurements (van den Brink et al. 2021). Inclusion of digital biomarkers has been shown in the neurology field, and its impact was seen in the great unmet need for non-invasive biomarkers and objectives. Therefore, testing some digital biomarkers must be implemented for its feasibility and reliability in Parkinson's and Alzheimer disease and clinical outcome assessments. Still, they are considered new and, in turn, are not implemented commonly in clinical practice and development. Due to its simplicity in use with low cost, it comes to the detriment of measuring distally from pathological events as well as producing complex large data, which will challenge data analytics. Several additional drawbacks are including data integrity, reliability and assessment, malfunctions of hardware, identifying relevant data, various interpretations, and determination of baseline (Babrak et al. 2019).

Therefore, biomarkers could not only help us in understanding present and past climate conditions but also play a critical role in climate change mitigation. For example, scientists use biomarkers in assessment effectiveness of reforestation efforts and in monitoring health ecosystems. By using biomarkers, scientists will be able to make informed decisions to reduce or prevent climate change more professionally. Biomarkers represent invaluable tools in the climate science field. These biomarkers may permit scientists in reconstructing past climate conditions, help in predicting climate feedbacks in the future, and help in developing different strategies in order to mitigate climate change. Digital biomarkers in contribution to digital health may aid in management of diseases related to climate change. This can be achieved through appropriate detection of outbreaks, reducing environmental risk factors exposure, and identifying environmental risks in addition to facilitating care delivery to under-resourced areas.

10.7 Ethical Considerations and Challenges

There are several ethical considerations and challenges for using technology to predict the impacts of CC on health toxicity:

1. Data Privacy and Security:
   Ethical concerns arise regarding the collection, storage, and use of sensitive health and environmental data, necessitating robust privacy protection measures (Culnane; 2020).
2. Equity and Access:
   There is a risk of exacerbating existing disparities in access to technology and healthcare services, highlighting the importance of ensuring equitable access to predictive models and interventions (Maani & Galea; 2021).
3. Fairness and Bias:
   Predictive models may perpetuate biases present in historical data, resulting in discriminatory or unfair outcomes, emphasizing the need for algorithmic fairness and bias mitigation strategies (Obermeyer et al. 2019).
4. Accountability and Transparency:
   Ensuring transparency in predictive modeling methodologies and decision-making processes is essential for accountability and trust-building among stakeholders (Veale and Binns; 2017).
5. Informed Consent and Participation:
   Ethical considerations regarding informed consent and voluntary participation of individuals in data collection and predictive modeling efforts underscore the importance of ethical data governance frameworks (Hinchliffe et al. 2018).

10.8 Conclusions

This chapter discusses the crucial role of technology in guiding and mitigating the health impacts of CC. It highlights the use of real-time monitoring systems, biomarkers, and advanced analytical tools to understand the connections between CC and health toxicity. Technologies such as environmental monitoring, climate modeling, remote sensing, and machine learning aid in predicting and tracking health risks associated with CC. These technologies also support decision-making processes for public health interventions and adaptation strategies. Overall, technology plays a pivotal role in raising awareness, enabling data-driven analysis, and informing decisions to address the complex challenges posed by climate-related health risks.

A. Recap of the role of technology in predicting CC impacts on health toxicity
B. Call to action for continued research, development, and implementation of technology
C. Emphasize the importance of a multi-disciplinary approach

References

Abdel-Shafy HI, Mansour MSM (2016) A review on polycyclic aromatic hydrocarbons: source; environmental impact; effect on human health and remediation. Egypt J Pet 25(1):107–123. https://doi.org/10.1016/j.ejpe.2015.03.011

Abdelzaher H, Tawfik SM, Nour A, Abdelkader S, Elbalkiny ST, Abdelkader M, Abbas WA, Abdelnaser A (2022) Climate change; human health; and the exposome: utilizing OMIC technologies to navigate an era of uncertainty. Front Public Health 10:973000. https://doi.org/10.3389/fpubh.2022.973000

Ahmad A, Imran M, Ahsan H (2023) Biomarkers as biomedical bioindicators: approaches and techniques for the detection; analysis; and validation of novel biomarkers of diseases. Pharmaceutics 15:1630. https://doi.org/10.3390/pharmaceutics15061630

Alves JED (2014) Demographic transition, age structure transition and aging. Portal de Divulgação Magazine 40

Asweto CO (2018) Cardiovascular health risk posed by polycyclic aromatic hydrocarbon and ultrafine particles. J Clin Exp Tox 1(1):46–50. http://www.alliedacademies.org/clinical-experimental-toxicology/

ATSDR (Agency for Toxic Substances and Disease Registry) (2012) Case studies in environmental medicine toxicity of polycyclic aromatic hydrocarbons (PAHs): report. https://www.atsdr.cdc.gov/csem/pah/docs/pah.pdf

Avery TE, Berlin GL (1992) Fundamentals of remote sensing and airphoto interpretation, 5th edn. Macmillan Publishing Company, New York, p 472

Babrak LM, Menetski J, Rebhan M, Nisato G, Zinggeler M, Brasier N, Baerenfaller K et al (2019) Traditional and digital biomarkers: two worlds apart? Digit Biomark 16;3(2):92–102. https://doi.org/10.1159/000502000

Baccarelli A, Dolinoy DC, Walker CL (2023) A precision environmental health approach to prevention of human disease. Nat Commun 14:2449. https://doi.org/10.1038/s41467-023-37626-2

Bagryantseva Y, Novotna B, Rossner P Jr, Chvatalova I, Milcova A, Svecova V et al (2010) Oxidative damage to biological macromolecules in Prague bus drivers and garagemen: impact of air pollution and genetic polymorphisms. Toxicol Lett 199:60–68. https://doi.org/10.1016/j.toxlet.2010.08.007

Bailey GB, Lauer DT, Carneggie DM (2001) International collaboration: the cornerstone of satellite land remote sensing in the 21st century. Space Policy 17(3):161–169

Bandodkar AJ, Jeang WJ, Ghaffari R, Rogers JA (2019) Wearable sensors for biochemical sweat analysis. Annu Rev Anal Chem 12:1–22. https://doi.org/10.1146/annurev-anchem-061318-114910

Barrón-Vivanco BS, Rothenberg SJ, Medina-Díaz IM, Robledo-Marenco L, Rojas-García AE et al (2013) AKRs expression in peripheral blood lymphocytes from smokers: The role of body mass index. Hum Exp Toxicol 32:418–426. https://doi.org/10.1177/0960327112455071

Beelen R, Stafoggia M, Raaschou-Nielsen O, Andersen ZJ, Xun WW, Katsouyanni K et al (2014) Long-term exposure to air pollution and cardiovascular mortality: an analysis of 22 European cohorts. Epidemiology 25(3):368–378

Bernhardt AM, Tiedt S, Teupser D, Dichgans M, Meyer B et al (2023) A unified classification approach rating clinical utility of protein biomarkers across neurologic diseases. EBioMedicine 89:104456

Bjerregaard-Olesen C, Bach CC, Long M, Wielsøe M, Bech BH, Henriksen TB, Olsen J et al (2019) Associations of fetal growth outcomes with measures of the combined xenoestrogenic activity of maternal serum perfluorinated alkyl acids in Danish pregnant women. Environ Health Perspect 127:17006. https://doi.org/10.1289/EHP1884

Bloom MS (2019) Environmental epidemiology. In: Nriagu J (ed) Encyclopedia of environmental health, 2nd edn. Elsevier, Oxford, pp 419–427

Bocato MZ, Bianchi Ximenez JP, Hoffmann C, Barbosa F (2019) An overview of the current progress; challenges; and prospects of human biomonitoring and exposome studies. J Toxicol Environ Health B Crit Rev 22(5–6):131–156. https://doi.org/10.1080/10937404.2019.1661588

Bodaghi A, Fattahi N, Ramazani A (2023) Biomarkers: promising and valuable tools towards diagnosis; prognosis and treatment of Covid-19 and other diseases. Heliyon 9(2):e13323. https://doi.org/10.1016/j.heliyon.2023.e13323

Bonassi S, Au WW (2002) Biomarkers in molecular epidemiology studies for health risk prediction. Mutat Res Rev Mutat Res 511:73–86

Bopp SK, Barouki R, Brack W, Dalla Costa S, Dorne JCM, Drakvik PE, Faust M et al (2018) Current EU research activities on combined exposure to multiple chemicals. Environ Int 120:544–562. https://doi.org/10.1016/j.envint.2018.07.037

Bowe B, Xie Y, Li T, Yan Y, Xian H, Al-Aly Z (2018) Particulate matter air pollution and the risk of incident CKD and progression to ESRD. J Am Soc Nephrol 29(1):218–230. https://doi.org/10.1681/ASN.2017030253

Cai W, Santoso A, Collins M, Dewitte B, Karamperidou C, Kug JS, Zhong W (2021) Changing El Niño–Southern oscillation in a warming climate. Nat Rev Earth Environ 2(9):628–644

Califf RM (2018) Biomarker definitions and their applications. Exp Biol Med (Maywood) 243(3):213–221. https://doi.org/10.1177/1535370217750088

Campbell-Lendrum D, Woodruff R (2006) Comparative risk assessment of the burden of disease from climate change. Environ Health Perspect 114(12):1935–1941

CAPMAS (Central Agency for Public Mobilization and Statistics) (2021) Statistical Yearbook-2021; Issue No. 112; Cairo; Egypt. http://www.capmass.gov.eg

Chen C (2023) The role of biomarkers in understanding climate change. J Biomark Drug Dev 4:22

Chen M, Carmella SG, Sipe C, Jensen J, Luo X, Le CT, Murphy SE, Benowitz NL, McClernon FJ, Vandrey R et al (2019) Longitudinal stability in cigarette smokers of urinary biomarkers of exposure to the toxicants acrylonitrile and acrolein. PLoS One 14(1):e0210104. https://doi.org/10.1371/journal.pone.0210104

Chen M, Carmella SG, Li Y, Zhao Y, Hecht SS (2020) Resolution and quantitation of mercapturic acids derived from crotonaldehyde; methacrolein; and methyl vinyl ketone in the urine of smokers and nonsmokers. Chem Res Toxicol 33(2):669–677. https://doi.org/10.1021/acs.chemrestox.9b00491

Chojer H, Branco PTBS, Martins FG, Alvim-Ferraz MCM, Sousa SIV (2024) Source identification and mitigation of indoor air pollution using monitoring data – current trends. Environ Technol Innov 33:103534

Chuang CY, Chang CC (2007) Urinary 1-hydroxypyrene level relative to vehicle exhaust exposure mediated by metabolic enzyme polymorphisms. J Occup Health 49:140–151. https://doi.org/10.1539/joh.49.140

Coravos A, Khozin S, Mandl KD (2019) Developing and adopting safe and effective digital biomarkers to improve patient outcomes. NPJ Digit Med 2:14. https://doi.org/10.1038/s41746-019-0119-8

Coryell M, Mcalpine M, Pinkham NV, Mcdermott TR, Walk ST (2018) The gut microbiome is required for full protection against acute arsenic toxicity in mouse models. Nat Commun 9:5424

Culnane C (2020) The ethics of data privacy in predictive analytics. Springer

de Prado-Bert P, Ruiz-Arenas C, Vives-Usano M, Andrusaityte S, Cadiou S, Carracedo A et al (2021) The early-life exposome and epigenetic age acceleration in children. Environ Int 155:106683

Demetriou CA, Raaschou-Nielsen O, Loft S, Moller P, Vermeulen R, Palli D et al (2012) Biomarkers of ambient air pollution and lung cancer: a systematic review. Occup Environ Med 2012(69):619–627. https://doi.org/10.1136/oemed-2011-100566

Dereumeaux C, Mercier F, Soulard P, Hulin M, Oleko A, Pecheux M et al (2022) Identification of pesticides exposure biomarkers for residents living close to vineyards in France. Environ Int 159:107013. https://doi.org/10.1016/j.envint.2021.107013

Devine R (1993) The Sputnik challenge. Oxford University Press, New York

Dey S, Dhal GC (2019) Deactivation and regeneration of hopcalite catalyst for carbon monoxide oxidation: a review. Mater Today Chem 14:100180

Ding E, Wang Y, Liu J et al (2022) A review on the application of the exposome paradigm to unveil the environmental determinants of age-related diseases. Hum Genomics 16:54. https://doi.org/10.1186/s40246-022-00428-6

Donovan BM, Bastarache L, Turi KN, Zutter MM, Hartert TV (2019) The current state of omics technologies in the clinical management of asthma and allergic diseases. Ann Allergy Asthma Immunol 123(6):550–557. https://doi.org/10.1016/j.anai.2019.08.460

Drotikova T, Ali AM, Halse AK, Reinardy HC, Kallenborn R (2020) Polycyclic aromatic hydrocarbons (PAHs) and oxy- and nitro-PAHs in ambient air of the Arctic town Longyearbyen; Svalbard. Atmos Chem Phys 20:9997–10014. https://doi.org/10.5194/acp-20-9997-2020

Dwayne AD (1988) Eye in the sky: the story of the Corona spy satellites; Logsdon JM, Latell B (eds) Smithsonian Books, Washington, DC pp 29–85; 143–156. ISBN: 1560988304 (ISBN1-56098-773-1 (paperback) or ISBN 1-56098-830-4 (hardcover))

Eyring V, Righi M, Lauer A, Evaldsson M, Wenzel S, Jones C, Williams KD (2016) ESMValTool (v1. 0)–a community diagnostic and performance metrics tool for routine evaluation of Earth system models in CMIP. Geosci Model Dev 9(5):1747–1802

Ebert D, Lipsitch M, Mangin KL (2000) The effect of parasites on host population density and extinction: experimental epidemiology with Daphnia and six microparasites. Am Nat 156(5):459–477

Edrissi B, Taghizadeh K, Moeller BC, Yu R, Kracko D, Doyle-Eisele M, Swenberg JA, Dedon PC (2017) N6-Formyllysine as a biomarker of Formaldehyde exposure: formation and loss of N6-Formyllysine in Nasal Epithelium in Long-Term; low-dose inhalation studies in Rats. Chem Res Toxicol 30(8):1572–1576. https://doi.org/10.1021/acs.chemrestox.7b00075

EPA (2023) E. P. A. Air Quality-National Summary. https://www.epa.gov/air-trends/air-quality-national-summary

EPA (Environment Protection Authority Victoria) (2017) Air monitoring report 2016 – Compliance with the National Environment Protection (Ambient Air Quality) Measure. Published September 2017. https://www.epa.vic.gov.au/about-epa/publications/1663

EPA (US. Environmental Protection Agency) (2002) Toxicological Review of Benzene (NONCANCER EFFECTS); CAS No. 71-43-2. In: Support of summary information on the integrated risk information system (IRIS). US.EPA, Washington, DC. October 2002. https://iris.epa.gov/static/pdfs/0276tr.pdf

European Union (2010) International reference life cycle data system (ILCD) handbook: framework and requirements for life cycle impact assessment models and indicators

Eyring V, Mishra V, Griffith GP, Chen L, Keenan T, Turetsky MR, Van der Linden S (2021) Reflections and projections on a decade of climate science. Nat Clim Chang 11(4):279–285

FAO & WHO (2006) Food safety risk analysis: a guide for national food safety authorities

Feliciano RJ et al (2022) Strategies to mitigate food safety risk while minimizing environmental impacts in the era of climate change. Trends Food Sci Technol 126:180–191

Figueroa JD, Malats N, García-Closas M, Real FX, Silverman D, Kogevinas M et al (2008) Bladder cancer risk and genetic variation in AKR1C3 and other metabolizing genes. Carcinogenesis 29:1955–1962. https://doi.org/10.1093/carcin/bgn163

Frigerio G, Mercadante R, Polledri E, Missineo P, Campo L, Fustinoni S (2019) An LC-MS/MS method to profile urinary mercapturic acids; metabolites of electrophilic intermediates of occupational and environmental toxicants. J Chromatogr B Anal Technol Biomed Life Sci 1117:66–76. https://doi.org/10.1016/j.jchromb.2019.04.015

Fuller R, Landrigan PJ, Balakrishnan K, Bathan G, Bose-O'Reilly S, Brauer M et al (2022) Pollution and health: a progress update. Lancet Planet Health 6:e535–e547

Ghassemi M, Mohamed S (2022) Machine learning and health need better values. NPJ Digit Med 5(1):51

Gorelick N, Hancher M, Dixon M, Ilyushchenko S, Thau D, Moore R (2017) Google earth engine: planetary-scale geospatial analysis for everyone. Remote Sens Environ 202:18–27

Gu B, Sutton MA, Chang SX, Ge CY, Chang J (2014) Agricultural ammonia emissions contribute to China's urban air pollution. Front Ecol Environ 12(5):265–266

Gulia S, Shiva Nagendra SM, Khare M, Khanna I (2015) Urban air quality management-A review. Atmos Pollut Res 6:286–304

Guo J, Xia F, Zhang Y, Liu H, Li J, Lou M et al (2017) Impact of diurnal variability and meteorological factors on the PM(2.5)—AOD relationship: Implications for PM(2.5) remote sensing. Environ Pollut 221:94–104

Hassan SK, Mohammed AMF, Khoder MI (2020) Characterization and health risk assessment of human exposure to PAHs in dust deposited on leaves of street trees in Egypt. Polycycl Aromat Compd 40(4):1013–1027. https://doi.org/10.1080/10406638.2018.1517810

Hassan MA, Abdallah SM, Shalaby EA, Ibrahim AA (2023) Assessing vulnerability of densely populated area to air pollution using Sentinel-5 P imageries: a case study of the Nile Delta; Egypt. Sci Rep 13:17406

Hegazi M, El-Jaafary S, Shebl N, El-Fawal H, Rizig M, Salama M (2024) Remote Sensing Data as a tool for studying environmental aspects of Parkinson's diseases. J Parkinsons Dis.; Pre-press:1–4. https://doi.org/10.3233/JPD-230382

Heimersson S, Harder R, Peters GM, Svanström M (2014) Including pathogen risk in life cycle assessment of wastewater management. 2. Quantitative comparison of pathogen risk to other impacts on human health. Environ Sci Technol 48(16):9446–9453. https://doi.org/10.1021/es501481m

Henson SA, Laufkötter C, Leung S, Giering SL, Palevsky HI, Cavan EL (2022) Uncertain response of ocean biological carbon export in a changing world. Nat Geosci 15(4):248–254

Hernandez JBR, Kim PY (2024) Epidemiology Morbidity and Mortality. [Updated 2022 Oct 3]. In: StatPearls [Internet]. StatPearls Publishing, Treasure Island. Available from: https://www.ncbi.nlm.nih.gov/books/NBK547668/

Hinchliffe S, Butcher S, Rahman MM (2018) Health data ethics: privacy; fairness; and precision public health. Annu Rev Public Health 39:401–413

Holme SAN, Sigsgaard T, Holme JA, Holst GJ (2020) Effects of particulate matter on atherosclerosis: a link via high-density lipoprotein (HDL) functionality? Part Fibre Toxicol 17:36. https://doi.org/10.1186/s12989-020-00367-x

Hooper MJ, Ankley GT, Cristol DA, Maryoung LA, Noyes PD, Pinkerton KE (2013) Interactions between chemical and climate stressors: a role for mechanistic toxicology in assessing climate change risks. Environ Toxicol Chem 32(1):32–48. https://doi.org/10.1002/etc.2043

House FB, Gruber A, Hunt GE, Mecherikunnel AT (1986) History of satellites missions and measurements of the Earth Radiation Budget (1957–1984). Rev Geophys 24:357–377. (ISSN 8755-1209)

Howard IC, Okpara KE, Techato K (2021) Toxicity and risks assessment of polycyclic aromatic hydrocarbons in river bed sediments of an Artisanal crude oil. Refining area in the Niger Delta; Nigeria. Water 13:3295. https://doi.org/10.3390/w13223295

Huijbregts MAJ, Steinmann ZJN, Elshout PMF et al (2017) ReCiPe2016: a harmonised life cycle impact assessment method at midpoint and endpoint level. Int J Life Cycle Assess 22:138–147. https://doi.org/10.1007/s11367-016-1246-y

Husby A (2020) On the use of blood samples for measuring DNA methylation in ecological epigenetic studies. Integr Comp Biol 60:1558–1566

Ialongo I, Virta H, Eskes H, Hovila J, Douros J (2020) Comparison of TROPOMI/Sentinel-5 Precursor NO_2 observations with ground-based measurements in Helsinki. Atmos Meas Tech 13:205–218

IPCC (2021) Climate Change 2021: the physical science basis. In: Contribution of Working Group I to the Sixth Assessment Report of the Intergovernmental Panel on Climate Change. https://www.ipcc.ch/report/ar6/wg1/. Last accessed 8 May 2023

ISO14000 (2005) Environmental management: life cycle assessment; requirements and guidelines. International Organization for Standardization

ISO14067 (2018) Greenhouse Gases: Carbon Footprint of Products: Requirements and Guidelines for Quantification. International Organization for Standardization

Jensen JR (2000) Remote sensing of the environment: an earth resource perspective, 2nd edn. Prentice-Hall, Upper Saddle River, p 544

John H, Richter A, Siegert M, Eyer F, Thiermann H (2021) Evidence of exposure to organophosphorus toxicants by detection of the propionylated butyrylcholinesterase-derived nonapeptide-adduct as a novel biomarker. Forensic Sci Int 323:110818

Kaplan G, Avdan YZ (2020) Space-borne air pollution observation from Sentinel-5p tropomi: relationship between pollutants; geographical and demographic data. Int J Eng Geosci 6:168

Kelsey JL, Horn-Ross PL (1993) Breast cancer: magnitude of the problem and descriptive epidemiology. Epidemiol Rev 15(1):7–16

Khan IA (2022) Biomarkers of human health. Biomed Res 33(4)

Kim DH, Paek SH, Choi DY, Lee MK, Park JN, Cho HM, Paek SH (2020) Real-time monitoring of biomarkers in serum for early diagnosis of target disease. BioChip J 14(1):2–17. https://doi.org/10.1007/s13206-020-4102-x

Kramer HJ (2002) Observation of the earth and its environment: survey of missions and sensors, 4th edn. Springer Verlag, Berlin. ISBN 3-540-42388-5.322

Kyriazakos S, Pnevmatikakis A, Cesario A, Kostopoulou K, Boldrini L, Valentini V et al (2021) Discovering composite lifestyle biomarkers with artificial intelligence from clinical studies to enable smart eHealth and digital therapeutic services. Fron Digit Health 3

Lagorio S, Tagesson C, Forastiere F, Iavarone I, Axelson O, Carere A (1994) Exposure to benzene and urinary concentrations of 8-hydroxydeoxyguanosine, a biological marker of oxidative damage to DNA. Occup Environ Med 51(11):739–743. https://doi.org/10.1136/oem.51.11.739

Lan Q, Mumford JL, Shen M, Demarini DM, Bonner MR, He X et al (2004) Oxidative damage-related genes AKR1C3 and OGG1 modulate risks for lung cancer due to exposure to PAH-rich coal combustion emissions. Carcinogenesis 25:2177–2181. https://doi.org/10.1093/carcin/bgh240

Ledda C, Rapisarda V (2020) Biomarkers and human biomonitoring in occupational medicine. Appl Sci 10(18):6181. https://doi.org/10.3390/app10186181

Levasseur A, Cavalett O, Fuglestvedt JS, Gasser T, Johansson DJA, Jørgensen SV et al (2016) Enhancing life cycle impact assessment from climate science: review of recent findings and recommendations for application to LCA. Ecol Indic 71:163–174

Lindon JC, Holmes E, Bollard ME, Stanley EG, Nicholson JK (2010) Metabonomics technologies and their applications in physiological monitoring; drug safety assessment and disease diagnosis. Biomarkers 9(1):1–31. https://doi.org/10.1080/13547500410001668379

Lionetto MG, Caricato R, Giordano ME, Schettino T (2016) The complex relationship between metals and carbonic anhydrase: new insights and perspectives. Int J Mol Sci. https://doi.org/10.3390/ijms17010127.39

Lucock MD (2021) A brief introduction to the exposome and human health. Explor Res Hypothesis Med 6(1):18–23. https://doi.org/10.14218/ERHM.2020.00070

Ludwig N, Whiteside TL, Reichert TE (2019) Challenges in exosome isolation and analysis in health and disease. Int J Mol Sci 20(19):4684. https://doi.org/10.3390/ijms20194684
Luijten M, Vlaanderen J, Kortenkamp A, Antignac JP, Barouki R, Bil W, van den Brand A et al (2023) Mixture risk assessment and human biomonitoring: lessons learnt from HBM4EU. Int J Hyg Environ Health 249:114135. https://doi.org/10.1016/j.ijheh.2023.114135
Maani N, Galea S (2021) Public health and precision medicine: addressing equity; ethics; and social determinants. Annu Rev Public Health 42:233–250
Maitre L, Bustamante M, Hernández-Ferrer C et al (2022) Multi-omics signatures of the human early life exposome. Nat Commun 13:7024. https://doi.org/10.1038/s41467-022-34422-2
Marcelo-Peña JL, Reynel-Rodríguez C, Zevallos-Pollito P, Bulnes-Soriano F, Pérez-Ojeda del Arco A (2007) Diversity, floristic composition and endemism in the seasonally altered dry forests of Jaén district, Peru. Appl Ecol 6(1-2):9–22
Martin EM, Fry RC (2018) Environmental influences on the epigenome: exposure-associated DNA methylation inhuman populations. Annu Rev Public Health 39(1):309–333
Matci DK, Kaplan G, Avdan U (2022) Changes in air quality over different land covers associated with COVID-19 in Turkey aided by GEE. Environ Monit Assess 194:762
Meehl GA, Covey C, Delworth T, Latif M, McAvaney B, Mitchell JFB, Taylor KE (2007) The WCRP CMIP3 multimodel dataset: a new era in climate change research. Bull Am Meteorol Soc 88(9):1383–1394
Melesse AM, Weng Q, Thenkabail PS, Senay GB (2007) Review: remote sensing sensors and applications in environmental resources mapping and modelling. Sensors 2007(7):3209–3241
Milesi C, Churkina G (2020) Review: measuring and monitoring urban impacts on climate change from space. Remote Sens 12:3494. https://doi.org/10.3390/rs12213494
Moubarz G, Saad-Hussein A, Shahy EM, Mahdy-Abdallah H, Mohammed AMF, Saleh IA et al (2023) Lung cancer risk in workers occupationally exposed to polycyclic aromatic hydrocarbons with emphasis on the role of DNA repair gene. Int Arch Occup Environ Health 96:313–329
Muazu RI, Rothman R, Maltby L (2021) Integrating life cycle assessment and environmental risk assessment: a critical review. J Clean Prod 293:126120
Mulinari S (2023) Short-circuiting biology: digital phenotypes; digital biomarkers; and shifting gazes in psychiatry. Big Data Soc 10(1). https://doi.org/10.1177/20539517221145680
Mustieles V, Zhang Y, Yland J, Braun JM, Williams PL, Wylie BJ, Messerlian C (2020) Maternal and paternal preconception exposure to phenols and preterm birth. Environ Int 137:105523
Naeem I, Saleem M, Javaid S, Amjad A, Hussain M, Ismail T, Ismail A, Fazal A, Mumtaz Z (2021) Exposure assessment of Aflatoxins in humans. In: Hakeem KR, Oliveira CAF, Ismail A (eds) Aflatoxins in food. Springer, Cham. https://doi.org/10.1007/978-3-030-85762-2_4
Nan HM, Kim H, Lim HS, Choi JK, Kawamoto T, Kang JW et al (2001) Effects of occupation; lifestyle and genetic polymorphisms of CYP1A1; CYP2E1; GSTM1 and GSTT1 on urinary 1-hydroxypyrene and 2-naphthol concentrations. Carcinogenesis 22:787–793. https://doi.org/10.1093/carcin/22.5.78
Neveu V, Nicolas G, Salek RM, Wishart DS, Scalbert A (2020) Exposome-Explorer 2.0: an update incorporating candidate dietary biomarkers and dietary associations with cancer risk. Nucleic Acids Res 48:D908–D912
Nienaber-Rousseau C (2025) Understanding and applying gene–environment interactions: a guide for nutrition professionals with an emphasis on integration in African research settings. Nutr Rev 83(2):e443–e463. https://doi.org/10.1093/nutrit/nuae015
Noyes PD, McElwee MK, Miller HD et al (2009) The toxicology of climate change: environmental contaminants in a warming world. Environ Int 35(6):971–986. https://doi.org/10.1016/j.envint.2009.02.006
Obermeyer Z, Powers B, Vogeli C, Mullainathan S (2019) Dissecting Racial Bias in an algorithm used to manage the health of populations. Science 366(6464):447–453
Orru H, Ebi K, Forsberg B (2017) The interplay of climate change and air pollution on health. Curr Environ Health Rep 4:504–513

Petchpoung K, Kaojarern S, Yoovathaworn K, Sura T, Sirivarasai J (2011) The influence of metabolic gene polymorphisms on urinary 1-hydroxypyrene concentration in Thai bus drivers. Environ Toxicol Pharmacol 31:160–164. https://doi.org/10.1016/j.etap.2010.10.006

Pigini D, Caporossi L, Paci E, Capanna S, Viganò P, Alteri A, Rabellotti E, Buonaurio F et al (2022) Phthalate exposure and biomarkers of oxidation of nucleic acids: results on couples attending a fertility center. Toxics 10(2):61. https://doi.org/10.3390/toxics10020061

Pleil JD, Sheldon LS (2011) Adapting concepts from systems biology to develop systems exposure event networks for exposure science research. Biomarkers 16(2):99–105. https://doi.org/10.3109/1354750X.2010.541565

Poblete-Naredo I, Albores A (2016) Molecular biomarkers to assess health risks due to environmental contaminants exposure. Biomedica 3;36(2):309–335. https://doi.org/10.7705/biomedica.v36i3.2998

Poirier MC (2004) Chemical-induced DNA damage and human cancer risk. Nat Rev Cancer 4:630–637. https://doi.org/10.1038/nrc1410

Prasad SB, Vidyullatha P, Vani GT, Devi RP, Rani UP, Reddy PP et al (2013) Association of gene polymorphism in detoxification enzymes and urinary 8-OHdG levels in traffic policemen exposed to vehicular exhaust. Inhal Toxicol 25:1–8. https://doi.org/10.3109/08958378.2012.745634

Przydatek G, Wota AK (2019) Analysis of the comprehensive management of sewage sludge in Poland. J Mater Cycles Waste Manage:1–9

Ramírez F, Afán I, Davis LS, Chiaradia A (2017) Climate impacts on global hot spots of marine biodiversity. Sci Adv 3(2):e1601198

Rashidi HH, Tran N, Albahra S, Dang LT (2021) Machine learning in health care and laboratory medicine: general overview of supervised learning and Auto-ML. Int J Lab Hematol 43(S1):15–22

Rayner PJ, O'Brien DM (2001) The utility of remotely sensed CO_2 concentration data in surface source inversions. Geophys Res Lett 28(1):175–178

Ribera G, Clarens F, Martínez-Lladó X, Jubany I, Martí V, Rovira M (2014) Life cycle and human health risk assessments as tools for decision making in the design and implementation of nanofiltration in drinking water treatment plants. Sci Total Environ 466–467:377–386. https://doi.org/10.1016/j.scitotenv.2013.06.085

Rocque RJ, Beaudoin C, Ndjaboue R, Cameron L, Poirier-Bergeron L, Poulin-Rheault RA, Fallon C, Tricco AC, Witteman HO (2021) Health effects of climate change: an overview of systematic reviews. BMJ Open 11(6):e046333. https://doi.org/10.1136/bmjopen-2020-046333

Rodríguez-Carrillo A, Rosenmai AK, Mustieles V, Couderq S, Fini JB, Vela-Soria F et al (2021) Assessment of chemical mixtures using biomarkers of combined biological activity: a screening study in human placentas. Reprod Toxicol 100:143–154. https://doi.org/10.1016/j.reprotox.2021.01.002

Rodríguez-Carrillo A; Mustieles V; Salamanca-Fernández E; Olivas-Martínez A; Suárez B; et al (2023). Implementation of effect biomarkers in human biomonitoring studies: a systematic approach synergizing toxicological and epidemiological knowledge. Int J Hyg Environ Health; 249; Article 114140.

Roszak A, Lianeri M, Sowinska A, Jagodzinski PP (2014) CYP1A1 Ile462Val polymorphism as a risk factor in cervical cancer development in the Polish population. Mol Diagn Ther 18:445–450. https://doi.org/10.1007/s40291-014-0095-2

Saba VS, Griffies SM, Anderson WG, Winton M, Alexander MA, Delworth TL, Zhang R (2016) Enhanced warming of the Northwest Atlantic Ocean under climate change. J Geophys Res Oceans 121(1):118–132

Salvia AL, Leal FW, Brandli LL, Griebeler JS (2019) Assessing research trends related to sustainable development goals: local and global issues. J Clean Prod 208:841–849

Sánchez-Guerra M, Pelallo-Martínez N, Díaz-Barriga F, Rothenberg SJ, Hernández-Cadena L et al (2012) Environmental polycyclic aromatic hydrocarbon (PAH) exposure and DNA damage in Mexican children. Mutat Res 742:66–71. https://doi.org/10.1016/j.mrgentox.2011.12.006

Schwarz G (2016) Molybdenum cofactor and human disease. Curr Opin Chem Biol. https://doi.org/10.1016/j.cbpa.2016.03.016

Schwedler G, Murawski A, Schmied-Tobies MIH, Rucic E, Scherer M, Pluym N, Scherer G et al (2021) Benzene metabolite SPMA and acrylamide metabolites AAMA and GAMA in urine of children and adolescents in Germany - human biomonitoring results of the German Environmental Survey 2014–2017 (GerES V). Environ Res 192:110295. https://doi.org/10.1016/j.envres.2020.110295

Shah A, Grimberg DC, Inman BA (2020) Classification of molecular biomarkers. Société Internationale d'Urologie Journal 1(1):8–15. Retrieved from https://siuj.org/index.php/siuj/article/view/44

Sim D, Brothers MC, Slocik JM, Islam AE, Maruyama B, Grigsby CC, Naik RR, Kim SS (2022) Biomarkers and detection platforms for human health and performance monitoring: a review. Adv Sci (Weinh) 9(7):e2104426. https://doi.org/10.1002/advs.202104426

Simpson NP et al (2021) A framework for complex climate change risk assessment. One Earth 4(4):489–501

SLGIS; Sinergise Laboratory for Geographical Information Systems (2023) Sentinel Hub. Sinergise Ltd. http://apps.sentinel-hub.com. Accessed 23 Feb 2023

Sobus JR et al (2018) Integrating tools for non-targeted analysis research and chemical safety evaluations at the US EPA. J Expo Sci Environ Epidemiol 28:411–426

Soleimanifar N, Assadiasl S, Kalateh E et al (2023) Circulating exosomes and ambient air pollution exposure in COPD. Chronic Obstr Pulm Dis 10(4):412–421. https://doi.org/10.15326/jcopdf.2023.0400

Soleimany A, Grubliauskas R, Šerevičienė V (2020) Application of satellite data and GIS services for studying air pollutants in Lithuania (case study: Kaunas city). Air Qual Atmos Health 14:411–429

Song W, Bian L, Xiong M, Duan Y, Wang Y, Zhang X, Li B, Dai Y, Lu J, Li M, Liu Z et al (2023) Association of genetic polymorphisms with mercapturic acids in the urine of young healthy subjects before and after exposure to outdoor air pollution. Int J Environ Health Res 33(9):936–948. https://doi.org/10.1080/09603123.2022.2066068

Srivastava A, Sharma A, Yadav S, Flora SJ, Dwivedi UN, Parmar D (2014) Gene expression profiling of candidate genes in peripheral blood mononuclear cells for predicting toxicity of diesel exhaust particles. Free Radic Biol Med 67:188–194. https://doi.org/10.1016/j.freeradbiomed.2013.10.820

Stoney W (2005) A guide to the global explosion of land-imaging satellites; markets and opportunities. http://www.eijournal.com/, http://www.geog.ucsb.edu/~jeff/115a/remotesensinghistory.html

Suh YJ, Ha EH, Park H, Kim YJ, Kim H, Hong YC (2008) GSTM1 polymorphism along with PM10 exposure contributes to the risk of preterm delivery. Mutat Res 656:62–67. https://doi.org/10.1016/j.mrgentox.2008.07.006

Tang S, Li T, Fang J, Chen R, Cha Y, Wang Y, Zhu M, Zhang Y, Chen Y, Du Y, Yu T et al (2021) The exposome in practice: an exploratory panel study of biomarkers of air pollutant exposure in Chinese people aged 60–69 years (China BAPE Study). Environ Int 157:106866. https://doi.org/10.1016/j.envint.2021.106866

Touzé-Peiffer L, Barberousse A, Le Treut H (2020) The coupled model intercomparison project: history; uses; and structural effects on climate research. WIREs Clim Chang 11(4):e648

Turesky RJ, Lu K (2020) Biomarkers of environmental toxicants: exposure and biological effects. Toxics 8(2):37. https://doi.org/10.3390/toxics8020037

UN (2019) The Sustainable Development Goals Report 2019. https://sdgs.un.org/goals. Last accessed 8 Jan 2023

van den Brink W, Bloem R, Ananth A, Kanagasabapathi T, Amelink A, Bouwman J et al (2021) Digital resilience biomarkers for personalized health maintenance and disease prevention. Front Digit Health 2:614670. https://doi.org/10.3389/fdgth.2020.614670

Vasudevan S, Saha A, Tarver ME et al (2022) Digital biomarkers: Convergence of digital health technologies and biomarkers. NPJ Digit Med 5:36. https://doi.org/10.1038/s41746-022-00583-z

Veale M, Binns R (2017) Fairer machine learning in the real world: mitigating discrimination without collecting sensitive data. Big Data Soc 4(2):2053951717743530

Ventura C, Gomes BC, Oberemm A, Louro H, Huuskonen P, Mustieles V, Fernández MF et al (2021) Biomarkers of effect as determined in human biomonitoring studies on hexavalent chromium and cadmium in the period 2008–2020. Environ Res 197:110998. https://doi.org/10.1016/j.envres.2021.110998

Viana M, de Leeuw F, Bartonova A, Castell N, Ozturk E, González OA (2020) Air quality mitigation in european cities: status and challenges ahead. Environ Int 143:105907

Virolainen SJ, VonHandorf A, Viel KCMF et al (2023) Gene–environment interactions and their impact on human health. Genes Immun 24:1–11. https://doi.org/10.1038/s41435-022-00192-6

Vorkamp K, Castaño A, Antignac JP, Boada LD, Cequier E, Covaci A, Esteban López M et al (2021) Biomarkers; matrices and analytical methods targeting human exposure to chemicals selected for a European human biomonitoring initiative. Environ Int 146:106082. https://doi.org/10.1016/j.envint.2020.106082

Walker B, Barrett S, Polasky S, Galaz V, Folke C, Engström G, De Zeeuw A (2009) Looming global-scale failures and missing institutions. Science 325(5946):1345–1346

Wallace L, Pellizzari E, Hartwell TD, Perritt R, Ziegenfus R (1987) Exposures to benzene and other volatile compounds from active and passive smoking. Arch Environ Health 42(5):272–279. https://doi.org/10.1080/00039896.1987.9935820

Wallace MA, Kormos TM, Pleil JD (2016) Blood-borne biomarkers and bioindicators for linking exposure to health effects in environmental health science. J Toxicol Environ Health B Crit Rev 19(8):380–409. https://doi.org/10.1080/10937404.2016.1215772

WHO (2011) Health statistics and health information systems: global burden of disease. Retrieved August 2011. 29 p

WHO (2022) Household air pollution. Fact Sheets. 2023; 2022. https://www.who.int/news-room/fact-sheets/detail/household-air-pollution-and-health?gad_source=1&gclid=CjwKCAjwrIixBhBbEiwACEqDJU6Fy9plaq1TRy6j1GLeUarwavF2cO2L4E6bvkdiuklYZj8wtpiZzBoC040QAvD_BwE

WHO (2023) W. H. O. Ambient (outdoor) air pollution. https://www.who.int/news-room/fact-sheets/detail/ambient-(outdoor)-air-quali ty-and-health (8 March)

Wu H, Eckhardt CM, Baccarelli AA (2023) Molecular mechanisms of environmental exposures and human disease. Nat Rev Genet 24:332–344. https://doi.org/10.1038/s41576-022-00569-3

Xiao J, Chevallier F, Gomez C, Guanter L, Hicke JA, Huete AR, Zhang X (2019) Remote sensing of the terrestrial carbon cycle: a review of advances over 50 years. Remote Sens Environ 233:111383

Yang Y, Tian Y, Jin X, Yan C, Jiang F, Zhang Y et al (2009) A case-only study of interactions between metabolic enzyme polymorphisms and industrial pollution in childhood acute leukemia. Environ Toxicol Pharmacol 28:161–166. https://doi.org/10.1016/j.etap.2009.03.004

Yang D, Liu Y, Cai Z, Chen X, Yao L, Lu D (2018) First global carbon dioxide maps produced from TanSat measurements. Adv Atmos Sci 35:621–623

Yang M, Wang H, Cheng J (2023) Continuous monitoring of multiple biomarkers with an ultrasensitive 3D-structured wearable biosensor. Cell Rep Method 3:100579

Yi B, Yang JY, Yang M (2007) Past and future applications of CYP450-genetic polymorphisms for biomonitoring of environmental toxicants. J Environ Sci Health C Environ Carcinog Ecotoxicol Rev 25:353–377. https://doi.org/10.1080/10590500701704037

Zare-Jeddi M, Hopf NB, Viegas S, Price AB, Paini A, van Thriel C, Benfenati E, Ndaw S et al (2021) Towards a systematic use of effect biomarkers in population and occupational biomonitoring. Environ Int 146:106257. https://doi.org/10.1016/j.envint.2020.106257

Zhang P, Carlsten C, Chaleckis R et al (2021) Defining the scope of exposome studies and research needs from a multidisciplinary perspective. Environ Sci Technol Lett 8(10):839–852. https://doi.org/10.1021/acs.estlett.1c00648

Zhang S, Yan X, Tang B, Luo W, Chen S, Luo X et al (2023) Human hair as a noninvasive matrix to assess exposure to micro-organic contaminants: state of the art review. Sci Total Environ 892:164341. Epub 2023 May 24. pmid:37236479

Zhao S, Liu M, Tao M, Zhou W, Lu X, Xiong Y, Li F, Wang Q (2023) Review: the role of satellite remote sensing in mitigating and adapting to global climate change. Sci Total Environ 904(2023):166820. https://doi.org/10.1016/j.scitotenv.2023.166820

Zhavoronkov A (2018) Artificial Intelligence for drug discovery; biomarker development; and generation of novel chemistry. Mol Pharm 15(10):4311–4313. https://doi.org/10.1021/acs.molpharmaceut.8b00930. PMID: 30269508

Zin MMT, Lim S-R (2023) A methodology to screen priority toxins in pollutant release inventories. J Environ Manag 341:118068

The manufacturer's authorised representative in the EU is Springer Nature Customer Service Centre GmbH, Europaplatz 3, 69115 Heidelberg, Germany. If you have any concerns regarding our products, please contact ProductSafety@springernature.com

Printed and bound by CPI Group (UK) Ltd, Croydon, CR0 4YY
01/10/2026
02225942-0004